Quality Aspects in Spatial Data Mining

Edited by
Alfred Stein
Wenzhong Shi
Wietske Bijker

CRC Press
Taylor & Francis Group
Boca Raton London New York

CRC Press is an imprint of the
Taylor & Francis Group, an **informa** business

CRC Press
Taylor & Francis Group
6000 Broken Sound Parkway NW, Suite 300
Boca Raton, FL 33487-2742

First issued in paperback 2019

© 2009 by Taylor & Francis Group, LLC
CRC Press is an imprint of Taylor & Francis Group, an Informa business

No claim to original U.S. Government works

ISBN-13: 978-1-4200-6926-6 (hbk)
ISBN-13: 978-0-367-38620-0 (pbk)

Library of Congress Cataloging-in-Publication Data

Quality aspects in spatial data mining / editors, Alfred Stein, Wenzhong Shi, Wietske Bijker.
    p. cm.
Includes bibliographical references and index.
ISBN 978-1-4200-6926-6 (alk. paper)
    1. Geographic information systems. 2. Spatial analysis (Statistics) I. Stein, Alfred. II. Shi, Wenzhong. III. Bijker, Wietske, 1965- IV. Title.

G70.212.Q35 2008
910.285--dc22   2008009919

Visit the Taylor & Francis Web site at
http://www.taylorandfrancis.com

and the CRC Press Web site at
http://www.crcpress.com

Qualitas est nobilior quantitate. Qualitas, non quantitas.

Sêneca, *Epistulae Morales* **17.4**

Quality in a product or service is not what the supplier puts in. It is what the customer gets out and is willing to pay for. A product is not quality because it is hard to make and costs a lot of money, as manufacturers typically believe. This is incompetence. Customers pay only for what is of use to them and gives them value. Nothing else constitutes quality.

Peter Drucker

It is not a question of how well each process works, the question is how well they all work together.

Lloyd Dobyns and Clare Crawford-Mason, *Thinking About Quality*

Contents

SECTION I Systems Approaches to Spatial Data Quality

SECTION II Geostatistics and Spatial Data Quality for DEMs

SECTION III   Error Propagation

SECTION IV   Applications

SECTION V   Communication

Foreword

Quality Aspects in Spatial Data Mining, edited by Alfred Stein, Wenzhong Shi, and Wietske Bijker, and published by CRC Press is a highly impressive collection of chapters that address many of the problems that lie on the frontiers of spatial data mining, classification, and signal processing. The sections are authoritative and up to date. The coverage is broad, with subjects ranging from systems approaches to spatial data quality; quality of descriptions of socially constructed facts, especially legal data, in a GIS; and a multicriteria fusion approach for geographical data matching, to quality-aware and metadata-based decision-making support for environmental health, geostatistical texture classification of tropical rainforests in Indonesia, and formal languages for expressing data consistency rules and implications for reporting of quality metadata.

The wealth of concrete information in *Quality Aspects of Spatial Data Mining* makes it clear that in recent years substantial progress has been made toward the development of effective techniques for spatial information processing. However, there is an important point that has to be made.

Science deals not with reality but with models of reality. As we move further into the age of machine intelligence and automated reasoning, models of information systems, including spatial information systems, become more complex and harder to analyze. An issue that moves from the periphery to the center is that of dealing with information that is imprecise, uncertain, incomplete, and/or partially true. What is not widely recognized is that existing techniques, based as they are on classical, bivalent logic, are incapable of meeting the challenge. The problem is that bivalent logic is intrinsically unsuited for meeting the challenge because it is intolerant of imprecision and partiality of truth.

So what approach can be used to come to grips with information, including spatial information, that is contaminated with imprecision, uncertainty, incompleteness, and/or partiality of truth? A suggestion that I should like to offer is to explore the use of granular computing. Since granular computing is not a well-known mode of computation, I will take the liberty of sketching in the following its underlying ideas.

In conventional modes of computation, the objects of computation are values of variables. In granular computing, the objects of computation are not values of variables but the information about values of variables, with the information about values of variables referred to as granular values. When the information is described in a natural language (NL), granular computing reduces to NL computation. An example of granular values of age is young, middle-aged, and old. An example of a granular value of imprecisely known probability is not very low and not very high.

How can a granular probability described as "not very low and not very high" be computed? This is what granular computing is designed to do. In granular computing, the key to computation with granular values is the concept of a generalized constraint. The concept of a generalized constraint is the centerpiece of granular computing.

The concept of a constraint is a familiar one in science. But, in science, models of constraints tend to be oversimplified in relation to the complexity of real-world constraints. In particular, constraints are generally assumed to be hard, with no elasticity allowed. A case in point is the familiar sign "Checkout time is 1 p.m." This constraint appears to be hard and simple but in reality it has elasticity that is hard to define.

A fundamental thesis of granular computing is that information is, in effect, a generalized constraint. In this nontraditional view of information, the traditional statistical view of information is a special case.

The concept of a generalized constraint serves two basic functions: (a) representation of information and, in particular, representation of information that is imprecise, uncertain, incomplete, and/or partially true; and (b) computation/deduction with information represented as a system of generalized constraints. In granular computing, computation/deduction involves propagation and counterpropagation of generalized constraints. The principal rule of deduction is the so-called extension principle. A particularly important application area for granular computing is computation with imprecise probabilities. Standard probability theory does not offer effective techniques for this purpose.

What I said above about granular computing in no way detracts from the importance of the contributions in *Quality Aspects of Spatial Data Mining*. I have taken the liberty of digressing into a brief discussion of granular computing because of my perception that granular computing is a nascent methodology that has high potential relevance to spatial information processing — and especially to processing of information that is imprecise, uncertain, incomplete, and/or partially true — a kind of information that the spatial information systems community has to wrestle with much of the time.

In conclusion, *Quality Aspects of Spatial Data Mining* is an important work that advances the frontiers of spatial information systems. The contributors, the editors, and the publisher deserve our thanks and loud applause.

Lotfi Zadeh
Berkeley, California

Contributing Authors

Tamiru H. Alemseged
Department of Water Resources
ITC
Enschede, The Netherlands

Karl-Heinrich Anders
Institute of Cartography and
  Geoinformatics
Leibniz Universität Hannover
Hannover, Germany

Mohamed Bakillah
Département des Sciences Géomatiques
Centre de Recherche en Géomatique
Université Laval
Québec City, Québec, Canada

Yvan Bédard
Département des Sciences Géomatiques
Centre de Recherche en Géomatique
Université Laval
Québec City, Québec, Canada

Sarah A. Bekessy
School of Global Studies, Social
  Science and Planning
RMIT University
Melbourne, Australia

Wietske Bijker
Department of Earth Observation
  Science
ITC
Enschede, The Netherlands

Anna T. Boin
Department of Geomatics
Cooperative Research Centre for Spatial
  Information
University of Melbourne
Coburg, Australia

Jean Brodeur
Centre d'Information Topographique
  de Sherbrooke
Sherbrooke, Québec, Canada

Alan Brown
Countryside Council for Wales
Bangor, United Kingdom

James Brown
National Weather Service
NOAA
Silver Spring, Maryland, U.S.A.

Gilberto Câmara
Image Processing Division
National Institute for Spatial Research
São José dos Campos, Brazil

René R. Colditz
German Aerospace Center
German Remote Sensing Data Center
Wessling, Germany
and
Department of Geography
Remote Sensing Unit
University of Wuerzburg
Wuerzburg, Germany

Alexis J. Comber
Department of Geography
University of Leicester
Leicester, United Kingdom

Christopher Conrad
Department of Geography
Remote Sensing Unit
University of Wuerzburg
Wuerzburg, Germany

Robert Corner
Department of Spatial Sciences
Curtin University
Bentley, Western Australia

Sytze de Bruin
Centre for Geo-Information
Wageningen University
Wageningen, The Netherlands

Stefan Dech
German Aerospace Center
German Remote Sensing Data Center
Wessling, Germany
and
Department of Geography
Remote Sensing Unit
University of Wuerzburg
Wuerzburg, Germany

Eduardo S. Dias
SPINlab
Vrije Universiteit
Amsterdam, The Netherlands

Alistair J. Edwardes
Department of Geography
University of Zurich
Zurich, Switzerland

Pete F. Fisher
Department of Information Science
City giCentre
City University
London, United Kingdom

Andrew Frank
Institute of Geoinformation and
  Cartography
Technical University of Vienna
Vienna, Austria

Richard Gloaguen
Remote Sensing Group
Institute for Geology
TU-Bergakademie
Freiberg, Germany

Michael F. Goodchild
Department of Geography
National Center for Geographic
  Information and Analysis
University of California, Santa Barbara
Santa Barbara, California, U.S.A.

Abdelbasset Guemeida
Laboratoire Sciences et Ingénierie
  de l'Information et de l'Intelligence
  Stratégique
Université de Marne-la-Vallée
Marne-la-Vallée, France

Jan-Henrik Haunert
Institute of Cartography and
  Geoinformatics
Leibniz Universität Hannover
Hannover, Germany

Felix Hebeler
Department of Geography
University of Zurich
Zurich, Switzerland

Gerard Heuvelink
Wageningen University and Research
  Centre
Wageningen, The Netherlands
and
Alterra – Soil Science Centre
Wageningen, The Netherlands

Gary J. Hunter
Department of Geomatics
Cooperative Research Centre for Spatial
  Information
University of Melbourne
Parkville, Australia

Kerstin Huth
Faculty of Geomatics
Karlsruhe University of Applied
  Sciences
Karlsruhe, Germany

Robert Jeansoulin
Laboratoire d'Informatique de l'Institut
  Gaspard Monge
Université Paris-EST Marne-la-Vallée
Champs-sur-Marne, France

Simon D. Jones
School of Mathematical and Geospatial
  Sciences
RMIT University
Melbourne, Australia

Martin Knotters
Alterra – Soil Science Centre
Wageningen, The Netherlands

Karin Kollo
Department of Geodesy
Estonian Land Board
Tallinn, Estonia

Alex M. Lechner
School of Mathematical and Geospatial
  Sciences
RMIT University
Melbourne, Australia

Rodrigo Lilla Manzione
National Institute for Spatial Research
Image Processing Division
São José dos Campos, Brazil

Haixia Mao
Department of Land Surveying and
  Geo-Informatics
Advanced Research Centre for Spatial
  Information Technology
The Hong Kong Polytechnic University
Hong Kong SAR, China

Marco Marinelli
Department of Spatial Sciences
Curtin University
Bentley, Western Australia

Prashanth R. Marpu
Remote Sensing Group
Institute for Geology
Freiberg, Germany

Nick Mitchell
Faculty of Geomatics
Karlsruhe University of Applied
  Sciences
Karlsruhe, Germany

Mir Abolfazl Mostafavi
Département des Sciences Géomatiques
Centre de Recherche en Géomatique
Université Laval
Québec City, Québec, Canada

Gerhard Navratil
Institute for Geoinformation and
  Cartography
Vienna University of Technology
Vienna, Austria

Ana-Maria Olteanu
COGIT Laboratory
IGN/France
Paris, France

Alejandro Pauly
Sage Software
Alachua, Florida, U.S.A.

Ekaterina S. Podolskaya
Cartographic Faculty
Moscow State University of Geodesy
  and Cartography
Moscow, Russia

Ross S. Purves
Department of Geography
University of Zurich
Zurich, Switzerland

Tom H.M. Rientjes
Department of Water Resources
ITC
Enschede, The Netherlands

Gabriella Salzano
Laboratoire Sciences et Ingénierie
  de l'Information et de l'Intelligence
  Stratégique (S3IS)
Université de Marne-la-Vallée
Paris, France

Gertrud Schaab
Faculty of Geomatics
Karlsruhe University of Applied
  Sciences
Karlsruhe, Germany

Sven Schade
Institute for Geoinformatics
University of Münster
Münster, Germany

Michael Schmidt
German Aerospace Center
German Remote Sensing Data Center
Wessling, Germany
and
Remote Sensing Unit
Department of Geography
University of Wuerzburg
Wuerzburg, Germany

Markus Schneider
Department of Computer & Information
  Science & Engineering
University of Florida
Gainesville, Florida, U.S.A.

Monika Sester
Institute of Cartography and
  Geoinformatics
Leibniz Universität Hannover
Hannover, Germany

Wenzhong Shi
Department of Land Surveying and
  Geo-Informatics
Advanced Research Centre for Spatial
  Information Technology
The Hong Kong Polytechnic University
Hong Kong SAR, China

Alfred Stein
Department of Earth Observation
  Science
ITC
Enschede, The Netherlands

Rangsima Sunila
Department of Surveying
Laboratory of Geoinformation and
  Positioning Technology
Helsinki University of Technology
Espoo, Finland

Yan Tian
Department of Land Surveying and
  Geo-Informatics
Advanced Research Centre for Spatial
  Information Technology
The Hong Kong Polytechnic University
Hong Kong SAR, China
and
Department of Electronic and
  Information Engineering
Huazhong University of Science and
  Technology
Wuhan, China

Martin Vermeer
Department of Surveying
Helsinki University of Technology
Helsinki, Finland

Jos von Asmuth
Kiwa Water Research
Nieuwegein, The Netherlands

Paul Watson
1Spatial
Cambridge, United Kingdom

Thilo Wehrmann
German Aerospace Center
German Remote Sensing Data Center
Wessling, Germany

Arief Wijaya
Remote Sensing Group
Institute for Geology
TU-Bergakademie
Freiberg, Germany
and
Faculty of Agricultural Technology
Gadjah Mada University
Yogyakarta, Indonesia

Graeme Wright
Department of Spatial Sciences
Curtin University
Bentley, Western Australia

Lotfi A. Zadeh
University of California
Berkeley, California, U.S.A.

Introduction

ABOUT THIS BOOK

Spatial data mining, sometimes called image mining, is a rapidly emerging field in Earth observation studies. It aims at identification, modeling, tracking, prediction, and communication of objects on a single image, or on a series of images. All these steps have to deal with aspects of quality. For example, identification may concern uncertain (vague) objects, and modeling of objects relies, among other issues, on the quality of the identification. In turn, tracking and prediction depend on the quality of the model. Finally, communication of uncertain objects to stakeholders requires a careful selection of tools.

Quality of spatial data is both a source of concern for the users of spatial data and a source of inspiration for scientists. In fact, spatial data quality and uncertainty are two of the fundamental theoretical issues in geographic information science. In both groups, there is a keen interest to quantify, model, and visualize the accuracy of spatial data in more and more sophisticated ways. This interest was at the origin of the 1st International Symposium on Spatial Data Quality, which was held in Hong Kong in 1999, and still is the very reason for the 5th symposium, ISSDQ 2007, in Enschede, The Netherlands. The organizers of this symposium selected the best papers presented at the conference to be published in this book after peer-review and adaptation.

DATA QUALITY—A PERSPECTIVE

The quality of spatial data depends on "internal" quality, the producer's perception, and "external quality," or the perspective of the user. From the producer's point of view, quality of spatial data is determined by currency, geometric and semantic accuracy, genealogy, logical consistency, and the completeness of the data. The user's concern, on the other hand, is "fitness for use," or the level of fitness between the data and the needs of the users, defined in terms of accessibility, relevancy, completeness, timeliness, interpretability, ease of understanding, and costs (Mostafavi, Edwards, and Jeansoulin, 2004).

The field of spatial data quality has come a long way. Five hundred years ago, early mapmakers like Mercator worried already about adequate representation of sizes and shapes of seas and continents to allow vessel routing. Mercator's projection allowed representing vessel routes as straight lines, which made plotting of routes easier and with greater positional accuracy. Ever since, surveyors, cartographers, users, and producers of topographic data have struggled to quantify, model, and increase the quality of data, where accuracy went hand in hand with fitness for use. Next to navigation, description of property, from demarcation of countries to cadastre of individual property, became an important driving force behind the quality of spatial data in general, with emphasis on positional accuracy and correct labeling of objects (e.g., ownership).

In the environmental sciences, the focus on aspects of quality of spatial data differed from the topographic sciences. Of course soils, forests, savannahs, ecosystems, and climate zones needed to be delineated accurately, but acceptable error margins were larger than in the topographic field. Attention was focused on the adequate and accurate description of the content. Well-structured, well-described legends became important, and statistical clustering techniques such as canonical analysis were used to group observations into classes. With a trend toward larger scales (higher spatial resolution), the positional accuracy became more important for the environmental sciences for adequate linking and analyzing of data of different sources, while the need for thematic accuracy and thematic detail increased in the topographic sciences. Thematic and positional accuracy became increasingly correlated.

For a long time scientists have realized that, in reality, objects weren't always defined by sharp boundaries and one class of soils or vegetation will change gradually into another in space as well as in time. Nevertheless, because of a lack of appropriate theory and appropriate tools, everything had to be made crisp for analysis and visualization. In the last decade or so, theories for dealing with vague objects and their relations have been developed (Dilo et al., 2005), such as fuzzy sets, the egg-yolk model (Cohn and Gotts, 1996), the cloud model (Cheng et al., 2005 citing Li et al., 1998) and uncertainty based on fuzzy topology (Shi and Liu, 2004).

The way we look at our world, and the way we define objects from observations, depend on the person, background, and purpose. One remotely sensed image, one set of spatial data, can be a source for many different interpretations. Of course there are a number of common perceptions in society that enable us to communicate spatial information. These common perceptions change with time as the challenges society faces change. A look at a series of land cover maps from the same area but from different decades clearly shows how thinking went from "exploration" and "conservation" to "multiple-use" and the legend and the spatial units changed accordingly, even where no changes happened on the ground. This is where ontology plays a role.

During times when spatial data were scarce, a limited number of producers produced data for a limited well-known market of knowledgeable users with whom they had contact. Now there are many producers of spatial data; some are experts, others are not. Users have easy access to spatial data. Maps and remote sensing images are available in hard copy and via the Internet in ever-growing quantities. Producers have no contact with all users of their data. Spatial data are also easily available to users for whom the data were not intended (fitness for use!) and to nonexpert users, who do not know all the ins and outs of the type of data. Not all producers of spatial data are experts either; yet, their products are freely available. A good example is Google Earth and Google Maps, where everyone with access to the Internet can add information to a specific location and share this with others. The increasing distance between producer and user of spatial data calls for adequate metadata, including adequate descriptions of data accuracy in terms that are relevant to both the user and producer of the data.

This book addresses quality aspects in spatial data mining for the whole flow from data acquisition to the user. A systematic approach for handling uncertainty and data quality issues in spatial data and spatial analyses covers understanding the sources of uncertainty, and modeling positional, attribute, and temporal uncertainties and their integration in spatial data as well as modeling uncertainty relations and completeness errors in spatial data, in both object-based and field-based data sets. Such types of approaches can be found as Section I, "Systems Approaches to Spatial Data Quality." Besides modeling uncertainty for spatial data, modeling uncertainty for spatial models is another essential issue, such as accuracy in DEM. Section II, "Geostatistics and Spatial Data Quality for DEMs," deals specifically with this aspect of data quality. Uncertainties may be propagated or even amplified in spatial analysis processes, and, therefore, uncertainty propagation modeling in spatial analyses is another essential issue, which is treated in more detail in Section III, "Error Propagation."

Quality control for spatial data and spatial analyses should ensure the information can fulfill the needs of the end users. For inspiration to users and producers alike, practical applications of quality aspects of spatial data can be found in Section IV, "Applications." New concepts and approaches should prove their worth in practice. Questions from users trigger new scientific developments. Just like the need to represent routes by straight lines on maps inspired Mercator to develop a map projection, present-day users inspire scientists to answer their questions with innovative solutions, which in turn give rise to more advanced questions, which could not be asked previously.

From a known user, one can get specifications of the data quality that are needed. But what to do with the (yet) unknown users, who may use the data for unforeseen purposes, or the "non-users" or "not-yet users" (Pontikakis and Frank, 2004), from whom we would like to know why they are not using spatial information? Section V, "Communication," focuses on ways to communicate with users about their needs and the quality of spatial data.

ACKNOWLEDGMENTS

This book emerged from a symposium, consisting of presentations, proceedings, and a social program. Prior to the conference we organized a very careful review process for all the papers. At this stage, we thank the reviewers, who were indispensable to having this book reach the standard that it has at the moment: Rolf de By, Rodolphe Devillers, Pete Fisher, Andrew Frank, Michael Goodchild, Nick Hamm, Geoff Hennebry, Gerard Heuvelink, Gary Hunter, Robert Jeansoulin, Wu Lun, Martien Molenaar, Mir Abolfazl Mostavafi, and David Rossiter.

We realize very well that any symposium has its support. At this stage, we would like to thank the ITC International Institute for Geo-Information Science and Earth Observation for hosting this meeting and for all its support. In particular, we thank Saskia Tempelman, Rens Brinkman, Janneke Kalf, Harald Borkent, Frans Gollenbeek, and many others. Without their input the meeting would not have been possible. The International Society for Photogrammetry and Remote Sensing (ISPRS) actively participated in getting the symposium organized, and we thank them for the support given. Finally, we thank the sponsors of the meeting:

CRC Press/Taylor & Francis, the CTIT Research School at Twente University, the PE&RC Research School based at Wageningen University, and the Dutch Kadaster and Geoinformatics Netherlands.

Alfred Stein, Wenzhong Shi, and Wietske Bijker

REFERENCES

Cheng, T., Z. Li, M. Deng, and Z. Xu. 2005. Representing indeterminate spatial object by cloud theory. In: L. Wu, W. Shi, Y. Fang, and Q. Tong (Eds.), *Proceedings of the 4th International Symposium on Spatial Data Quality*, 25th to 26th August 2005, Beijing, China, The Hong Kong Polytechnic University.

Cohn, A. G. and N. M. Gotts. 1996. Geographic objects with indeterminate boundaries, chapter The "egg-yolk" representation of regions with indeterminate boundaries. In: Burrough and Frank (eds.), *GISDATA*, 171–187.

Dilo, A., R. A. de By, and A. Stein. 2005. A proposal for spatial relations between vague objects. In: L. Wu, W. Shi, Y. Fang, and Q. Tong (Eds.), *Proceedings of the 4th International Symposium on Spatial Data Quality*, 25th to 26th August 2005, Beijing, China. The Hong Kong Polytechnic University.

Li, D., D. Cheung, X. Shi, and D. Ng. 1998. Uncertainty reasoning based on cloud model in controllers. *Computers Math. Application*, 35, pp. 99–123.

Mostafavi, M. A., G. Edwards, and R. Jeansoulin. 2004. An ontology-based method for quality assessment of spatial databases. In: A. U. Frank and E. Grum (compilers), *Proceedings of the ISSDQ '04*, Vol. 1. Geo-Info 28a, pp. 49–66. Dept. for Geoinformation and Carthography, Vienna University of Technology.

Pontikakis, E. and A. Frank, 2004. Basic spatial data according to users' needs: Aspects of data quality. In: A. U. Frank and E. Grum (compilers), *Proceedings of the ISSDQ '04*, Vol 1. Geo-Info 28a, pp. 13–29. Dept. for Geoinformation and Carthography, Vienna University of Technology.

Shi, W. Z. and K. F. Liu, 2004. Modeling fuzzy topological relations between uncertain objects in GIS, *Photogrammetric Engineering and Remote Sensing*, 70(8), pp. 921–929.

Section I

Systems Approaches to Spatial Data Quality

INTRODUCTION

Spatial data quality is a concept that is partly data- and object-driven and partly based on fitness for use. In order to integrate, the systems approach is likely to be useful. A systems approach is well known in geo-information science (one may think of the GEOSS initiative) as well as in several other fields of science, like agriculture, economy, and management sciences. Its approach thus serves as a guiding principle for spatial data quality aspects. For spatial data, geographical information systems found their place in the 1980s, and these systems are still potentially useful to serve the required purposes. But here the word "system" largely expresses the possibilities of storing, displaying, handling, and processing spatial data layers. This is not sufficient for the emerging field of spatial data quality, requiring in its current development a full systems approach. In fact, data can be different as compared to previously collected and analyzed data, and the objects will be inherently uncertain. As compared to the traditional GIS, a systems approach to spatial data quality should be able to deal with uncertainties. These uncertainties are usually expressed either by statistical measures, by membership functions of fuzzy sets, or they are captured by metadata.

A first and foremost challenge is thus to be able to extract, i.e., to query, vague spatial objects from databases. Common GIS, still seen as a spatial database with some specific functionalities, do not allow one to do so. This field is, at the moment, therefore, still very much an area of research rather than an issue of production. As concerns the data aspect, socially constructed facts are recognized as being important. This refers in part to social objects, but also to legal facts.

More recently, semantic issues have found their place in spatial research, thus acknowledging that the traditional fuzzy and statistical measures may fall short. Modern and prospective approaches toward spatial data quality are thus governed by semantic aspects of data and maps. In the frame of this section, semantic issues are approached along two lines. First, a conceptual framework for quality assessment is presented. Such a framework may be different from the ordinary conceptual frameworks, which did not include data quality aspects explicitly. In this sense, one chapter considers semantic mapping between ontologies. Next it is recognized that a semantic reference system should account for uncertainty. A requirement analysis is thus appropriate in that sense.

Section I of the book thus considers modern aspects of a systems approach to spatial data quality.

1 Querying Vague Spatial Objects in Databases with VASA

Alejandro Pauly and Markus Schneider

CONTENTS

1.1 INTRODUCTION

Many man-made spatial objects such as buildings, roads, pipelines, and political divisions have a clear boundary and extension. In contrast to these crisp spatial objects, most naturally occurring spatial objects have an inherent property of vagueness or indeterminacy of their extension or even of their existence. Point locations may not be exactly known; paths or trails might fade and become uncertain at intervals. The boundary of regions might not be certainly known or simply not be as sharp as that of a building or a highway. Examples are lakes (or rivers) whose extensions (or paths) depend on pluvial activity, or the locations of oil fields that in many cases can only be guessed. This inherent uncertainty brings to light the necessity of more adequate models that are able to cope with what we will refer to as *vague spatial objects*.

Existing implementations of geographic information systems (GIS) and spatial databases assume that all objects are crisply bounded. With the exception of a few domain-specific solutions, the problem of dealing with spatial vagueness has no widely accepted practical solution. Instead, different conceptual approaches exist for

3

which researchers have defined formal models that can deal with a closer approximation of reality where not all objects are crisp. For the treatment of vague spatial objects, our *vague spatial algebra* (*VASA*), which can be embedded into databases, encompasses data types for *vague points*, *vague lines*, and *vague regions* as well as for all operations and predicates required to appropriately handle objects of these data types. The central goal of the definition of VASA is to leverage existing models for crisp spatial objects, resulting in robust definitions of vague concepts derived from proven crisp concepts.

In order to fully exploit the power of VASA in a database context, users must be able to pose significant queries that will allow retrieval of data that are useful for analysis. In this chapter, we provide an overview of VASA and the capabilities it provides for handling vague spatial objects. Based on these capabilities, we describe how users can take full advantage of an implementation of VASA by proposing meaningful queries on vague spatial objects. We use sample scenarios to explain how the queries can be posed with a moderate extension of SQL.

This chapter starts in Section 1.2 by summarizing related work that covers relevant concepts from crisp spatial models as well as other concepts for handling spatial vagueness. In Section 1.3 we introduce the VASA concepts for data types, operations, and predicates. Section 1.4 shows how a simple extension to SQL will be of great benefit when querying vague spatial data. Finally, in Section 1.5 we derive conclusions and expose future work.

1.2 RELATED WORK

Existing concepts relevant to this work can be divided into two categories: (1) concepts that provide the foundation for the work presented in this chapter and (2) concepts that are defined with goals similar to those of the work in this chapter.

Related to the former, we are interested in crisp spatial concepts that define the crisp spatial data types for *points, lines,* and *regions* [25]. We are also interested in the relationships that can be identified between instances of these types. Topological relationships between spatial objects have been the focus of much research, and we concentrate on the concepts defined by the *9-intersection model* originally defined in [10] for simple regions, and later extended for simple regions with holes in [11]. The complete set of topological relationships for all type combinations of complex spatial objects is defined in [25] on the basis of the 9-intersection model.

We categorize available concepts for handling spatial vagueness by their mathematical foundation. Approaches that utilize existing exact (crisp) models for spatial objects include the *broad boundaries* approach [6, 7], the *egg-yolk* approach [8], and the *vague regions* concept [12]. These models extend the common assumption that boundaries of regions divide the plane into two sets (the set that belongs to the region, and the set that does not) with the notion of an intermediate set that is not known to certainly belong or not to the region. Thus we say that these models extend crisp models that operate on the Boolean logic (*true, false*) into models that handle uncertainty with a three-valued logic (*true, false, maybe*). VASA, our concept for handling spatial vagueness (Section 1.3), is based on exact models for crisp spatial objects. Although fundamentally different from the exact-based approaches, rough set theory [22] provides tools for deriving concepts with a close relation to what can be achieved with exact models. Rough set theory–based approaches include early work by Worboys in [26], the concepts for deriving *quality measures* presented in [4], and the concept of *rough classification* in [1].

One of the advantages of fuzzy set theory is the ability to handle *blend-in* type boundaries (such as that between a mountain and a valley). Approaches in this category include earlier *fuzzy regions* [3]; the formal definition of *fuzzy points*, *fuzzy lines*, and *fuzzy regions* in [23]; and an extension of the rough classification from [1] to account for fuzzy regions [2]. A recent effort for the definition of a *spatial algebra* based on fuzzy sets is presented in [9]. Finally, probabilistic approaches [13] focus on an *expected* membership to an object that can be contrasted to the membership values of fuzzy sets that are objective in the sense that they can be computed formally or determined empirically.

Concepts even closer to that dealt with in this chapter, namely, querying with vagueness, are discussed in [17] where it is proposed that vagueness does not necessarily appear only in the data being queried but can also be part of the query itself. The work in [24] proposes classifications of membership values in order to group sets of values together (near fuzzy concepts). For example, a classification could assign the term "mostly" to high membership values (e.g., 0.95–0.98). In the context of databases in general, the approaches in [15, 16, 18, 19] all propose extensions to query languages on the basis of an operator that enables vague results under different circumstances. For example, in [15] the operator *similar-to* for QBE (Query-by-Example) is proposed alongside relational extensions so that related results can be provided in the event that no exact results match a query. In [18] the operator ~ is used in a similar way to the *similar-to* operator. All these approaches require additional information to be stored as extra relations and functions about distance that allow the query processor to compute close enough results. Although some of these approaches are extended to deal with fuzzy data, the general idea promotes the execution of vague queries over crisp data.

1.3 VASA

In this section we describe the concepts that compose our vague spatial algebra. The foundation of VASA is its data types, which we specify in Section 1.3.1. Spatial set operations and metric operations are introduced in Section 1.3.2. Finally, the concept of vague topological predicates is briefly introduced in Section 1.3.3.

1.3.1 VAGUE SPATIAL DATA TYPES

An important goal of VASA (and of all approaches to handling spatial uncertainty that are based on exact models) is to leverage existing definitions of crisp spatial concepts. In VASA, we enable a generic vague spatial type constructor v that, when applied to any crisp spatial data type (i.e., *point*, *line*, *region*), renders a formal syntactic definition of its corresponding vague spatial data type. For any crisp spatial object x, we define its composition from three disjoint point sets, namely the interior $(x°)$, the boundary (∂x) that surrounds the interior, and the exterior (x^-) [25]. We

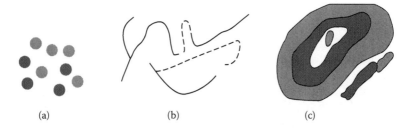

(a)   (b)   (c)

FIGURE 1.1   A vague point object (a), a vague line (b), and a vague region (c). Kernel parts
are symbolized by dark gray points, straight lines, and dark gray areas. Conjecture parts are
symbolized by light gray point, dashed lines, and light gray areas.

also assume a definition of the geometric set operations union ($\oplus$), intersection ($\otimes$),
difference ($\ominus$), and complement ($\boxminus$) between crisp spatial objects such as that from
[14].

> **Definition 1** Let $\alpha \in \{point, line, region\}$. A *vague spatial data type* is given
> by a type constructor v as a pair of equal crisp spatial data types α, i.e.,
>
> $$v(\alpha) = \alpha \times \alpha$$
>
> such that, for $w = (w_k, w_c) \in v(\alpha)$,
>
> $$w_k^{\circ} \cap w_c^{\circ} = \varnothing$$
>
> holds.

We call $w \in v(\alpha)$ a (two-dimensional) *vague spatial object* with *kernel part* w_k and
conjecture part w_c. Further, we call $w_o := (w_k, w_c)$ the *outside part* of w. For $\alpha =$
point, $v(point)$ is called a *vague point* object and denoted as *vpoint*. Correspond-
ingly, for *line* and *region* we define $v(line)$ resulting in *vline* and $v(region)$ resulting
in *vregion*.

Syntactically, a vague spatial object is represented by a pair of crisp spatial
objects of the same type. Semantically, the first object denotes the kernel part that
represents what certainly belongs to the object. The second object denotes the conjecture part that represents what is not certain to belong to the object. We require
both underlying crisp objects to be disjoint from each other. More specifically, the
constraint described above requires the interiors of the kernel part and the conjecture
part to not intersect each other. Figure 1.1 illustrates instances of a vague point, a
vague line, and a vague region as objects of the data types defined above.

1.3.2   VAGUE SPATIAL OPERATIONS

For the definition of the vague spatial set operations that compute the *union, intersection*, and *difference* between two vague spatial objects, we leverage crisp spatial
set operations to reach a generic definition of vague spatial set operations.

TABLE 1.1

Components Resulting from Intersecting Kernel Parts, Conjecture Parts, and Outside Parts of Two Vague Spatial Objects with Each Other

union	*k*	*c*	*o*	*intersection*	*k*	*c*	*o*	*difference*	*k*	*c*	*o*	*complement*	*k*	*c*	*o*
k	*k*	*k*	*k*	*k*	*k*	*c*	*o*	*k*	*o*	*c*	*k*	*k*	*o*	*c*	*k*
c	*k*	*c*	*c*	*c*	*c*	*c*	*o*	*c*	*o*	*c*	*c*				
o	*k*	*c*	*o*	*o*	*o*	*o*	*o*	*o*	*o*	*o*	*o*				

We define the syntax of function $h \in$ [*intersection, union, difference*] as $h: v(\alpha) \times v(\alpha) \rightarrow v(\alpha)$. The complement operation is defined as *complement*: $v(\alpha) \rightarrow v(\alpha)$. Semantically, their generic (type-independent) definition is reached by considering the individual relationships between kernel parts, conjecture parts, and the outside part (i.e., everything that is not a kernel part or conjecture part) of the vague spatial objects involved in the operations. The result of each operation is computed using one of the tables in Table 1.1. For each operation, the rows denote the parts of one object and the columns the parts of another, and we label them k, c, and o to denote the kernel part, conjecture part, and outside part, respectively. Each entry of the table denotes the intersection of kernel parts, conjecture parts, and outside parts of both objects, and the label in each entry specifies whether the corresponding intersection belongs to the kernel part, conjecture part, or outside part of the operation's result object.

Each table from Table 1.1 can be used to generate an executable specification of the given crisp spatial operations. For each table, the specification operates on the kernel parts and conjecture parts to result in a definition of its corresponding vague spatial operation. Following are such definitions as executable specifications of geometric set operations over crisp spatial objects:

Definition 2 Let $u, w \in v(\alpha)$, and let u_k and w_k denote their kernel parts and u_c and w_c their conjecture parts. We define:

u **union** w $:= (u_k \oplus w_k, (u_c \oplus w_c) \ominus (u_k \oplus w_k))$
u **intersection** $w := (u_k \otimes w_k, (u_c \otimes w_c) \oplus (u_k \otimes w_c) \oplus (u_c \otimes w_k))$
u **difference** w $:= (u_k \otimes (\boxminus(w_k \oplus w_c)), (u_c \otimes w_c) \oplus (u_k \otimes w_c) \oplus u_c \otimes (\boxminus(w_k \oplus w_c)))$
complement u $:= (\boxminus(u_k \oplus u_c), u_c)$

1.3.3 VAGUE TOPOLOGICAL PREDICATES

For the definition of topological predicates between vague spatial objects (*vague topological predicates*), it is our goal to continue leveraging existing definitions of crisp spatial concepts, in this case topological predicates between crisp spatial objects. Topological predicates are used to describe purely qualitative relationships such as *overlap* and *disjoint* that describe the relative position between two objects and are preserved under continuous transformations.

For two vague spatial objects $A \in v(\alpha)$ and $B \in v(\beta)$, and the set $T_{\alpha\beta}$ of all crisp topological predicates between objects of types α and β [25], the topological relationship between A and B is determined by the 4-tuple of crisp topological relationships (p,q,r,s) such that $p,q,r,s \in T_{\alpha\beta}$ and

$$p(A_k, B_k) \wedge q(A_k \oplus A_c, B_k) \wedge r(A_k, B_k \oplus B_c) \wedge s(A_k \oplus A_c, B_k \oplus B_c)$$

We define the set $V_{\alpha\beta}$ of all vague topological predicates between objects of types $v(\alpha)$ and $v(\beta)$. Due to inconsistencies that can exist between elements within each tuple, not all possible combinations result in 4-tuples that represent valid vague topological predicates in the set $V_{\alpha\beta}$. An example is the 4-tuple

$$(overlap(A_k, B_k), disjoint(A_k, B_k \oplus B_c), disjoint(A_k \oplus A_c, B_k),$$
$$disjoint(A_k \oplus A_c, B_k \oplus B_c))$$

In this example, the implications of $overlap(A_k, B_k) \Rightarrow A_k^\circ \cap B_k^\circ \neq \varnothing$ and $disjoint(A_k, B_k \oplus B_c) \Rightarrow A_k^\circ \cap (B_k \oplus B_c)^\circ = \varnothing$ clearly show a contradiction.

In [21], we present a method for identifying the complete set of vague topological predicates. At the heart of the method, each 4-tuple is modeled as a *binary spatial constraint network* (BSCN). Each BSCN is tested for *path-consistency*, which is used to check, via constraint propagation, that all original constraints are consistent; otherwise, the inconsistency indicates an invalid 4-tuple.

For each type combination of *vpoint*, *vline*, and *vregion*, possibly thousands of predicates are recognized. Sets of 4-tuples are created into clustered vague topological predicates. Clusters can be defined by the user who specifies three rules for each cluster: One rule is used to determine whether the clustered predicate certainly holds between the objects, the second to determine whether the cluster certainly does not hold, and the third to determine when the cluster maybe holds, but it is not possible to give a definite answer. This effectively symbolizes the three-valued logic that is central to our definition of vague spatial data types.

1.4   QUERYING WITH VASA

We propose two ways of enabling VASA within a database query language: The first, as presented in Section 1.4.1, works by adapting VASA to partially work with SQL, currently the most popular database query language. The second, presented in Section 1.4.2, extends SQL to enable handling of vague queries.

1.4.1   CRISP QUERIES OF VAGUE SPATIAL DATA

One of the advantages of being able to use VASA in conjunction with popular DBMSs is the availability of a database query language such as SQL. We focus on querying with SQL as it represents the most popular and widely available database query language. SQL queries can be used to retrieve data based on the evaluation of Boolean expressions. This obviously represents a problem when dealing with vague spatial objects because their vague topological predicates are based on a three-valued logic. On the other hand, the current definitions of numeric vague spatial operations do

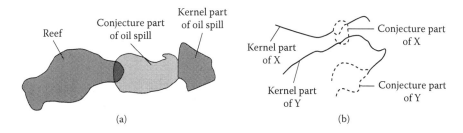

FIGURE 1.2 (a) A representation of an ecological scenario using vague regions. (b) Scenario illustrating the use of vague lines to represent routes of suspected terrorists X and Y.

not suffer from this issue because the operations return crisp values that are later interpreted by the user (e.g., the user posing a query must know that *min-length* returns the length associated with the kernel part of a vague line object). Thus, these concepts are already adapted to provide crisp results of vague data.

In the case of vague topological predicates, the first step in order to allow querying of vague spatial objects through SQL is to adapt the results of the predicates to a form understandable by the query language. The adaptation of the three-valued vague topological predicates to Boolean predicates can be done with the following six transformation predicates that are defined for each vague topological predicate P that can operate over vague spatial objects A and B (see Figure 1.2):

$$True\_P(A,B) = true \quad \Rightarrow \quad P(A,B) = true$$
$$True\_P(A,B) = false \quad \Rightarrow \quad P(A,B) = maybe \vee P(A,B) = false$$
$$Maybe\_P(A,B) = true \quad \Rightarrow \quad P(A,B) = maybe$$
$$Maybe\_P(A,B) = false \quad \Rightarrow \quad P(A,B) = true \vee P(A,B) = false$$
$$False\_P(A,B) = true \quad \Rightarrow \quad P(A,B) = false$$
$$False\_P(A,B) = false \quad \Rightarrow \quad P(A,B) = true \vee P(A,B) = maybe$$

With this transformation in place, queries operating on vague spatial objects can include references to vague topological predicates and vague spatial operations. For example, for the purpose of storing scenarios such as that in Figure 1.2a, assume that we have a table *spills*(*id* : *INT*, *name* : *STRING, area* : *VREGION*) where the column representing oil spills is denoted by a vague region where the conjecture part represents the area where the spill may extend to. We also have a table *reefs*(*id* : *INT, name* : *STRING, area* : *VREGION*) with a column representing coral reefs as vague regions. We can pose an SQL query to retrieve all coral reefs that are in any danger of contamination from an oil spill. We must find all reefs that are not certainly *Disjoint* from the *Exxon-Valdez* oil spill:

```
SELECT r.name FROM reefs r, spills s
WHERE s.name = "Exxon-Valdez" and NOT True_Disjoint
(r.area,s.area);
```

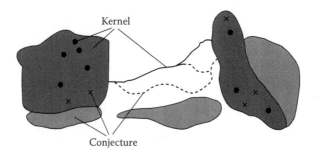

FIGURE 1.3 The vague spatial object representation of an animal's roaming areas, migration routes, and drinking spots.

Vague topological predicates can also be used to optimize query performance. Assume that, as illustrated in Figure 1.2b, we have data of terrorists' routes represented by vague lines in the table *terrorists(id : INT , name : STRING, route : VLINE)*. We want to retrieve the minimum length of the intersections of all pairs of intersecting routes of terrorists. To do so, we choose to compute the intersection of only those pairs that are certainly not *Disjoint* and neglect the computation of the intersection of those pairs that have been determined to not certainly intersect:

```
SELECT a.name, b.name, min-length(intersection(a.route,
b.route))
FROM terrorists a, terrorists b WHERE False_Disjoint
(a.route,b.route);
```

Other queries can include retrieval based not only on spatial data but also based on common type data (i.e., numbers, characters) stored alongside the spatial objects. Being able to relate both data domains (spatial and nonspatial) in queries is one of the main advantages of providing VASA as an algebra that can extend current DBMSs that are well-proven to provide the necessary services for dealing with data of common types. We can provide such queries based on Figure 1.3, where the data can be stored in the table *animals(id : INT , name : STRING, roam area : VREGION , mig route : VLINE , drink spot : VPOINT)*.

For example, we wish to retrieve all species of animals whose average weight is under 40 lbs. Their last count was under 100 and may have roaming areas completely contained within the roaming areas of carnivore animals whose average weight is above 80 lbs. This information might recognize animal species with low counts that could be extinct due to larger predators. The extinction of the smaller species can be catastrophic even for the larger species that depend on the smaller for nutrition. This retrieval uses data elements that are both spatial and nonspatial:

```
SELECT s.name FROM animals s, animals l
WHERE s.avgsize<40 AND l.avgsize>80 AND s.count<100;
```

Queries can also be posed to test elements from within single tuples in the database. For example, we would like to retrieve all animal species that do not have drinking spots that are certainly lying inside their roaming areas. For any of these species, environmentalists must create artificial drinking spots where the animals can hydrate:

```
SELECT s.name,
FROM animals s
WHERE NOT False_Disjoint(s.drink_spot,s.roam_area);
```

1.4.2 A VAGUE QUERY LANGUAGE EXTENSION FOR VAGUE QUERIES ON VAGUE SPATIAL DATA

We analyze the approaches introduced in Section 1.2 and notice that, in the context of VASA, we are not trying to solve the problem of dealing with vague queries, but we need to query vague data. Thus, we propose to extend a common query language such as SQL with the operator ~. However, our data themselves are vague and thus we do not need the extra relations and functions of distance required by previous approaches. As a result, the semantics of the operator ~ is not the same as in the existing literature, where it allows for vague queries to be executed on crisp data. Instead, we will allow for the execution of vague queries over vague data.

Boolean predicates in SQL and in fact in many programming languages implicitly assume truth values unless otherwise noted, which is commonly done with the negation operators NOT or !. We propose ~ to operate syntactically similar to !, but semantically, instead of negating the result, it opens the possibility for uncertain results. For example, let us assume we have the table *tempzones*(*id* : *INT* , *name* : *STRING, shape* : *VREGION*) that contains information about different temperature zones, including their representation as vague regions in the column named *shape*. We pose the following query:

```
SELECT a.name, b.name
FROM tempzones a, tempzones b
WHERE Overlap(a.shape,b.shape);
```

This query will return only those regions that certainly overlap. But instead we want to include in the result all those regions that might overlap as well, so we pose the query again as

```
SELECT a.name, b.name
FROM tempzones a, tempzones b
WHERE ~Overlap(a.shape,b.shape);
```

In this case, the interpretation of ~ should allow the retrieval of all temperatures that may or may not overlap in addition to those that definitely overlap. For the use of numeric values in queries, the query processor should be able to handle number ranges as an atomic data type such that we can combine the minimum and maximum area operations on vague regions into one operator and pose the following query:

```
SELECT a.name
FROM tempzones a
WHERE a.shape.area()~300;
```

That is, the result of this query will include all temperature zones whose area range includes 300. The inclusion of this operator and the management of number ranges does not preclude the use of exact operators that would allow dealing with crisp spatial regions. Because crisp spatial objects represent simply a specific instance of vague spatial data types, a query such as the following can still be executed with the result set including all those temperature zones that were modeled as vague regions with no conjecture, thus representing crisp regions:

```
SELECT a.name
FROM tempzones a
WHERE a.shape.area()=300;
```

This extension, of course, would require actual re-implementation of the query language within the DBMS in order to enable the handling of numeric ranges and three-valued logic operations.

1.5 CONCLUSIONS AND FUTURE WORK

The conceptual design of VASA that we have presented in this chapter shows the clear goal of leveraging existing crisp concepts. There is more than one reason behind this goal. The first reason is to take advantage of existing robust concepts for handling crisp spatial objects. Second, at the conceptual level, the correctness of the definitions for vague concepts largely rests on the correctness of the defined crisp concepts; thus, we reduce the chance of errors in our definitions. As an example, see Definition 2, where vague spatial operations are defined as an executable specification on the basis of crisp spatial operations. Third, the executable specification translates easily to the implementation level. Having an existing correct implementation of crisp spatial data types, their operations, and predicates, we can implement VASA by instantiating existing crisp spatial data types and executing operations on them.

In Section 1.2, we mentioned current approaches to handling spatial vagueness and imprecision. VASA's concepts feed from all these and thrive in providing a complete type system that includes a systematic approach to vague spatial operations, and most importantly to vague topological predicates. The main advantages of VASA include conceptual simplicity, robustness derived from existing robust crisp concepts, and viability of implementation. In contrast, VASA's main disadvantage is its inability to effectively deal with situations that would seem appropriate for fuzzy set–based systems. Nonetheless, we believe that future work can be directed toward more general definitions based on exact models that would be more near to the capabilities of fuzzy set–based systems but that can take advantage of existing crisp concepts.

Based on these concepts, we have proposed ideas for database querying of objects from VASA. While these ideas are simple, they are able to fully exploit the capabilities of VASA and allow the user to pose significant queries that can handle spatial vagueness. The proposed language extension and transformation mechanisms further reassure the advantages of defining VASA on the basis of existing exact spatial models. These advantages include robustness of formal concepts that can directly transfer into an implementation that also benefits from simplicity.

Other future work related to VASA stems in at least two directions that are worth following: The first involves enabling similar querying ideas to systems that attempt to handle vagueness with a higher precision, such as fuzzy set theory–based systems or even systems with finite multivalued logics (i.e., more than three values). The other direction involves the performance aspect of implementing indexes that can operate on vague spatial objects and whether it is possible to extend current indexing concepts for crisp spatial objects, thus following the design of VASA.

ACKNOWLEDGMENT

This work was partially supported by the National Science Foundation under grant number NSF-CAREER-IIS-0347574.

REFERENCES

[1] O. Ahlqvist, J. Keukelaar, and K. Oukbir. Rough Classifcation and Accuracy Assessment. *Int. Journal of Geographical Information Science*, 14:475–496, 2000.

[2] O. Ahlqvist, J. Keukelaar, and K. Oukbir. Rough and Fuzzy Geographical Data Integration. *Int. Journal of Geographical Information Science*, 17:223–234, 2003.

[3] D. Altman. Fuzzy Set Theoretic Approaches for Handling Imprecision in Spatial Analysis. *Int. Journal of Geographical Information Systems*, 8(3):271–289, 1994.

[4] T. Beaubouef and F. Petry. A Rough Set Foundation for Spatial Data Mining Involving Vague Regions. In *IEEE Int. Conf. on Fuzzy Systems*, pp. 767–772, 2002.

[5] P. A. Burrough and A. U. Frank, editors. *Geographic Objects with Indeterminate Boundaries*. Taylor & Francis, Boca Raton, FL, 1996.

[6] E. Clementini and P. Di Felice. A Spatial Model for Complex Objects with a Broad Boundary Supporting Queries on Uncertain Data. *Data & Knowledge Engineering*, Vol. 37, Issue 3, pp. 285–305, 2001.

[7] E. Clementini and P. Di Felice. An Algebraic Model for Spatial Objects with Indeterminate Boundaries. In Burrough and Frank, editors, *Geographic Objects with Indeterminate Boundaries*. Taylor & Francis, Boca Raton, FL, 1996, pp. 153–169.

[8] A. G. Cohn and N. M. Gotts. The "Egg-Yolk" Representation of Regions with Indeterminate Boundaries. In Burrough and Frank, editors, *Geographic Objects with Indeterminate Boundaries*. Taylor & Francis, Boca Raton, FL, 1996, pp. 171–187.

[9] A. Dilo, R., A. de By, and A. Stein. A System of Types and Operators for Handling Vague Spatial Objects. *Int. Journal of Geographical Information Science*, 21(4):397–426.

[10] M. J. Egenhofer. A Formal Definition of Binary Topological Relationships. In *Int. Conf. on Foundations of Data Organization and Algorithms*. Springer-Verlag, New York, 1989, pp. 457–472.

[11] M. J. Egenhofer, E. Clementini, and P. Di Felice. Topological Relations between Regions with Holes. *Int. Journal of Geographical Information Systems*, 8:128–142, 1994.

[12] M. Erwig and M. Schneider. Vague Regions. In *5th Int. Symp. on Advances in Spatial Databases*. Springer-Verlag, New York, 1997, pp. 298–320.

[13] J. T. Finn. Use of the Average Mutual Information Index in Evaluating Classification Error and Consistency. *Int. Journal of Geographical Information Systems*, 7(4):349–366, 1993.

[14] R. H. Guting and M. Schneider. Realm-Based Spatial Data Types: The ROSE Algebra. *VLDB Journal*, 4:100–143, 1995.

[15] T. Ichikawa and M. Hirakawa. ARES: A Relational Database with the Capability of Performing Flexible Interpretation of Queries. *IEEE Trans. on Software Engineering*, 12:624–634, 1986.

[16] J. Kung and J. Palkoska. Vague Joins — An Extension of the Vague Query System VQS. In *9th Int. Workshop on Database and Expert Systems Applications*, IEEE Computer Society, Los Alamitos, Ca, USA, 1998, pp. 997–1001.

[17] D. H. Lee and M. H. Kim. Accommodating Subjective Vagueness through a Fuzzy Extension to the Relational Data Model. *Information Systems*, 18:363–374, 1993.

[18] A. Motro. VAGUE: A User Interface to Relational Databases That Permits Vague Queries. *ACM Trans. on Information Systems*, 6:187–214, 1988.

[19] J. Palkoska and J. Kung. VQS — A Vague Query System Prototype. In *Int. Workshop on Database and Expert Systems Applications*, IEEE Computer Society, Los Alamitos, CA, USA, 1997, pp. 614–618.

[20] A. Pauly and M. Schneider. Vague Spatial Data Types, Set Operations, and Predicates. In *East-European Conf. on Advances in Databases and Information Systems*, Springer-Verlag, Berlin/Heidelberg, 2004, pp. 379–392.

[21] A. Pauly and M. Schneider. Topological Reasoning for Identifying a Complete Set of Topological Predicates between Vague Spatial Objects. In *FLAIRS Conference*, AAAI Press, Menlo Park, CA, USA, 2006, pp. 731–736.

[22] Z. Pawlak. Rough Sets. *Int. Journal of Computer and Information Sciences*, pp. 341–356, 1982.

[23] M. Schneider. Uncertainty Management for Spatial Data in Databases: Fuzzy Spatial Data Types. In *Int. Symp. on Advances in Spatial Databases*. Springer-Verlag, New York, 1999, pp. 330–351.

[24] M. Schneider. Fuzzy Topological Predicates, Their Properties, and Their Integration into Query Languages. In *ACM Symp. on Geographic Information Systems*. ACM Press, New York, 2001, pp. 9–14.

[25] M. Schneider and T. Behr. Topological Relationships between Complex Spatial Objects. *ACM Trans. on Database Systems (TODS)*, 31:39–81, 2006.

[26] M. Worboys. Imprecision in Finite Resolution Spatial Data. *GeoInformatica*, 2(3):257–279, 1998.

2 Assessing the Quality of Data with a Decision Model

Andrew Frank

CONTENTS

2.1 INTRODUCTION

Research in data quality is hindered by a lack of understanding of what quality for data means. The slogan "data quality is 'fitness for use'" is not giving an answer because it leaves open the question to what use the data should be fit. Data, especially GIS data, can be used in many ways; remember that a precursor of GIS was called a "multi-purpose cadastre" (Arentze et al., 1992; Harvey, 1997)! Data are used to improve decisions; decisions can be made without pertinent information (case of "null" information, e.g., none, inappropriate), and decisions are not necessarily

15

changed after data are needed—only confidence is increased (Frank, to appear 2007). GIS data can be used to improve many decisions, from ordinary, everyday decisions in wayfinding (left or right here?) to complex decisions about the location of a new nuclear power plant or a new factory or the violation of an international treaty (Abushady and Frank, 2005).

The quality of the information influences the decision—it must be assessed with respect to the decision-making process: Can it be used to make this decision? Does the lack of quality influence the outcome?

The diversity of the decisions GIS data are used for makes it difficult to understand how the quality of the data affects the decision. This is further complicated by the psychological complexity of how people actually make decisions. A number of studies have shown how data quality propagates from the data stored to data derived from a GIS to help make decisions (Karssenberg and De Jong, 2005). De Bruin et al. (2001, 2003) investigated whether acquiring better data for a particular decision is worthwhile.

Schneider (1999) and Frank (2007) have been able to reduce decisions as they are made by engineers when designing technical artifacts to a statistical test. Once the engineer has selected the model and parameters to include, the decision itself can be reduced to a comparison of two desired quantities. This approach is generalized here to as broad a range of decisions as possible.

This approach to data quality from the perspective of a user is different from describing data quality from the perspective of the data producer working with a specification, which typically emphasizes precision of location (Timpf et al., 1996). Unfortunately, such quality descriptions from the producer perspective are seldom relevant for users of the data (Shyllon and Hunter, 2004).

In this chapter I briefly review in Section 2.2 the model for engineering decisions as proposed before (Frank, 2007). In Section 2.3 different types of decisions are analyzed. Twaroch and Achatschitz (2005) investigate how the user's situation can be captured separately in an interactive process; the models their work produces can be used to assess the propagation of data quality to decision quality as described here. Ignoring the psychological complexity of decision, especially if made in a group, a similar reduction to a comparison of values devised from the data stored can be achieved. Section 2.4 then generalizes the model for random errors in the data, and Section 2.5 discusses the propagation of different data quality aspects from stored data to desired quantities.

As a result, the chapter shows a reduced model of decision making, which separates the psychological complexities of taking a decision into a first phase in which the "problem" is conceptualized into a decision test and a model selected. This process is in most decisions not consciously performed or verbalized. In the second phase, the decision is computed according to the model selected. It is possible to construct the model used "after the facts," when the decision is made and one can reconstruct the process. This reconstructed model can then be used to assess how data quality has influenced the decision, which makes the method described not only of theoretical interest but also practically applicable.

With this division of a complex decision into two steps the propagation of data quality can be computed, because error propagation affects only the second one and can be formalized. This chapter identifies the processing steps for which the propagation of imperfections is necessary and points to the research needed to give general rules for the ones not currently well understood.

A note on terminology: I prefer to speak of imperfections of the data (Frank, 2007) and to characterize these. This is focusing on the effects such imperfections have on the (imperfect) result, and I avoid statements like "low data quality" or "lack of data quality." All data contain imperfections, and it seems conceptually simpler to address these imperfections, rather than talk about data quality, which describes the degree of absence of imperfections.

2.2   ENGINEERING DESIGN DECISIONS

Engineering design, for example, for buildings, bridges, sewage systems, etc., is based on physical observations that are combined in formulas. The results are used to decide if a design satisfies the requirements and is acceptable or not. Error propagation is applicable here, and one can ask how much every value computed is influenced by the error in the data. Schneider has analyzed the influence of assumptions about load, strength of materials, or required safety levels (Schneider, 1999).

In engineering design, decisions can be abstracted to a comparison between the load on a system S compared with the resistance of the system R as designed. A design is acceptable if the resistance is larger than the load: $R > S$ resp. $R - S > 0$.

For a bridge, this means that the resistance R of the structure (i.e., maximum capacity) must be higher than the maximally expected load S (e.g., assumed maximum high water event). For a more environmental example, the opening under a bridge is sufficient and inundation upstream is avoided when the maximally possible flow R under the bridge is more than the maximal amount of water S expected from rainfall on the watershed above the bridge. To assess the influence of data quality on the decision, one computes the error on $(R - S)$ using the law of error propagation and applies test statistics to conclude whether the value is lager than zero with probability p (e.g., 95%).

The law of error propagation for a formula

$$r = f(a, b, \ldots)$$

for random uncorrelated errors e_a, e_b, e_c on values a, b, c, … was given by C. F. Gauss as

$$e_r^2 = \left(\frac{\partial f}{\partial a}\right)^2 e_a^2 + \left(\frac{\partial f}{\partial b}\right)^2 e_b^2 + \ldots \tag{2.1}$$

where e_i is the standard deviation of value i. If the observations are correlated, the correlation must be included (Ghilani and Wolf, 2006). The test on $R - S > 0$ is then

$$\frac{R - S}{\sqrt{\sigma_R^2 + \sigma_S^2}} > C \tag{2.2}$$

where C is determined by the desired significance, e.g., for 95%, $C = 1.65$.

In such engineering design decisions, a number of poorly known values must be used, e.g., the expected maximum rainfall in the next 50, 100, or 500 years; the maximum load on the bridge; the expected derivation from the plan in the building process; etc., and these may be correlated. The law or standards of engineering practice fix values for them. The accuracy of such general, fixed values to describe a concrete case is low and the effect of these uncertainties in a design decision high. This explains why more precision in observations is rarely warranted, because gains in a reduced construction are minimal. The uncertainties in the assumption about the load dominate the design decision. A rule of thumb for the law of error propagation engineers use is as follows: Error terms that are one order of magnitude less than others have no influence on the result; this is the effect of squaring the standard deviations before adding them! For the formulas used to design an opening under a bridge to avoid inundations upstream, the comparison of the maximally possible flow with the largest flow expected in a period of 50 years gives, for example, $R = 200$ m³/sec and $S = 80$ m³/sec, which satisfies $R > S$. Assuming error in the values used in the computation and propagating then to compute the standard derivation for R and S, we obtain, e.g., $\sigma_R = 60$ m³/sec (30%) and $\sigma_S = 16$ m³/sec (20%), and a test at the 95% level gives

$$\frac{200 - 80}{\sqrt{60^2 + 16^2}} = \frac{120}{62} = 1.93 > 1.65$$

This design is therefore satisfactory.

Schneider (1999) discusses the selection of security levels, which are traditionally mandated as security factors, increasing the load and reducing the bearing capacity of a design. He shows that current values lead to designs that satisfy expectations, but a statistical viewpoint would result in similar levels of security for different subsystems and therefore a higher overall security level with less overall effort and for a better price.

2.3   OTHER DECISION SITUATIONS

Navratil has applied error propagation to simple derivations from observed values (Navratil and Achatschitz, 2004). For example, the computation of the surface area of a parcel given the coordinates of the corners can be computed, if the standard derivations for the observations and their correlations are given. This uncertainty in the area is then sometimes multiplied by the going price per square meter and leads to critical comments by landowners about the quality of a land surveyor's work. The argument is false, because it does not consider a decision. In this section, some often occurring decisions are reformulated in the model proposed above and error propagation applied.

2.3.1   DECISION TO ACQUIRE A PLOT OF LAND

The error in the computed area of a parcel (Figure 2.1) seems high, e.g., some square meters, when one considers the price per square meter one has to pay (i.e., €550). Would more precise measurements be warranted?If one rephrases the question as a

9'150.05 ± 5.3 m²

Price € 950'000

Price per m² € 103.81

FIGURE 2.1   An example parcel.

decision, e.g., whether one should buy the land, this can be seen as a test: Are the benefits derived from the parcel larger than the cost? For simplicity assume that we intend to develop the land and build an office building, where we earn 200 €/m2 when we sell it (cost of the construction deduced). The test of whether this business opportunity is worthwhile is therefore if benefit is larger than price $(B > F)$ or $B > F > 0$ (i.e., is anything left after the transaction?).

Assuming the standard deviation on the benefit to be $\sigma_B = 0.3 \cdot 1'830'010 = 549'000$ we obtain

$$t = \frac{880'010}{\sqrt{550^2 + 549'00^2}} = \frac{880'010}{549'000} = 1.62$$

which will occur with a probability of ~ 94%. Note that for reason of constructing useful tables it is usual to fix the level of probability and then test, but it is also possible to ask what the probability of a given t value is. In this case, for an acceptable risk of 10%, the decision to buy is acceptable (significance 90%).

2.3.2   FIND OPTIMAL CHOICE

Many decision situations—especially personal decisions—consist of selecting the best choice from several variants. This can be seen as finding the variant with the highest benefit, computed with a formula, including weights to indicate the importance of various aspects (Twaroch and Achatschitz, 2005). For this formula the propagation of error for both data values and for the weights can be computed, using the methods described before. One can determine, with the method shown above, the probability that variant 1 with benefit v_1 and standard deviation σ_1 is indeed better than variant 2 (with v_2 and σ_2, respectively). Achatschitz has proposed applying sensitivity analysis and informing the user about how much his preferences (weights) had to change to make variant 2 be the best.

2.3   LEGAL DECISIONS

In a recent court case in Austria, the question was whether a building was constructed too close to the parcel boundary or not. Abstracting from a number of technical issues of surveying engineering, the distance between the boundary and building is established as 3.98 m with a standard deviation of 0.015 m. The law stipulates the required distance has to be 4.00 m. Is the building too close? A test for 4.00 – 3.98 > 0 at 95% significance gives

$$\frac{4.00 - 3.98}{1.5} = \frac{2}{1.5} < 1.65$$

The probability that the distance is shorter is ~ 91%. Whether this is considered sufficient evidence or not depends on the particulars of the case and the judge. I hope that if such cases are approached statistically, the courts will over time develop some standards.

2.4 OTHER DECISION SITUATIONS

A complex decision process can be split into a phase to select a model to use to make the decision and a phase of using the selected model to arrive at the decision. The discussion of examples in the previous section suggested that the influence of random errors can be computed with the regular error propagation formula if the decision is modeled formally. This section gives a generalized description.

2.4.1 Model of a Decision

By model of a decision we mean the formal model of a particular decision; Section 2.3 gave several examples. In general, a decision can be reduced to a test of a value being positive ($v > 0$). The acceptance of an engineering design immediately has the form $R - S > 0$, and other "yes/no," "go/no go" decisions can be brought to this form. The selection of an optimal solution from a series of variants can be seen as the selection of the variant i with the highest value v_i. It seems easier to describe the two situations separately, but they can be merged into a single approach.

2.4.2 Binary Decisions

A decision to do something or not has a decision model with the test $v > 0$ (or it can be rewritten to conform to this form; see Figure 2.2). v is computed as a function

$$v = f(a_1, a_2, \ldots a_n, s_1, s_2, \ldots s_n) \qquad (2.3)$$

of input values $a_1, a_2, \ldots a_n$ describing the situation, which comes, for example, from the GIS, and values describing other factors $s_1, s_2, \ldots s_n$, for material constants, security factors, etc. If $v > 0$, the action is carried out, the design is built, etc. The influence of random errors in the data ($a_1, a_2, \ldots a_n$ and $s_1, s_2, \ldots s_n$) on the decision is computed by the law of error propagation (Equation 2.1) and a statistical test. From the standard deviations of the data ($\sigma_{a1}, \sigma_{a2}, \ldots \sigma_{an}$ and $\sigma_{s1}, \sigma_{s2}, \ldots \sigma_{sn}$) and the partial derivatives

$$\frac{\partial f}{\partial a_1}, \frac{\partial f}{\partial a_2} \ldots$$

of Equation 2.3, the standard derivation σ_v of v is computed. From v/σ_v results a probability p that $v > 0$ is the integral of the normal distribution curve with σ_v up to v

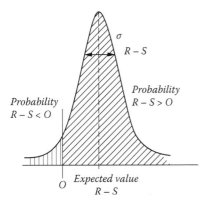

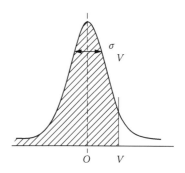

FIGURE 2.2 Statistical test for $R - S$.   **FIGURE 2.3** Probability of $v < V$.

(Figure 2.3). Usually a significance level is set first and only checks that v/σ_v is larger than the corresponding value $\Phi(p)$.

2.4.3 SELECTION

Selecting the optimal v_0 from a set of variants v_i described with data values $a_i = (a_1, a_2, \dots a_n)$ uses a valuation function f:

$$u_i = f(a_{i1}, a_{i2}, \dots a_{in}, s_1, s_2, \dots s_n, w_1, w_2, \dots w_n) \tag{2.4}$$

where the data describing the variant, constants, and weights describing the importance of the criterion for the decision $w_1, w_2, \dots w_n$ are combined. For each variant v_i we obtain a value u_i and select the variant v_0 for which $u_0 > u_i$ for any $i \neq 0$.

The effect of random errors in the data on the values u_i is computed as the standard derivations of u_i using error propagation as above (Equation 2.1). The uncertainty of the selection of v_0 compared to v_p (assuming it is the second best) can be tested as $v_0 - v_p > 0$ and the probability computed (Equation 2.2). From this, one must separate the test $v_0 > 0$, which gives the uncertainty that the best variant is an acceptable solution.

2.4.4 ASSUMPTION

Two assumptions must be stressed:

1. The data values are assumed to be influenced by random error, which follows a normal distribution and the errors are not correlated. If correlation is present, it can be taken into account by extensions of the formula for error propagation.
2. The decision model is fixed and not influenced by the imperfections in the data. The decision model includes all aspects of the decision, including subjective elements, which are assumed to be fixed for one decision. Note that for the computation of the effects of random errors in the data only the uncertainty of weights is required, not the exact weights, representing personal preferences.

2.5   GENERALIZATION: ERROR PROPAGATION IN DECISION FOR RANDOM ERRORS

The law of error propagation applies only to random errors in data values; in this section the approach is generalized to other types of imperfections.

2.5.1   OMISSIONS AND COMMISSIONS

Omissions and commissions influence the computation of aggregates differently than the "best solution."

2.5.1.1   Aggregate Values

If a number of values are summed (generally aggregated) to a single value, the effect of omission is a sum too small and the effect of commission is a sum too high. With given probabilities p_0 and p_C for omissions and commissions, the effect on the sum is $s' = (\Sigma v_i) \cdot (p_0 - p_C)$. This assumes that omissions and commissions are random, not systematic.

2.5.1.2   Selections

If v_0 is the best variant, then commissions could invalidate the choice. If p_{0i} are the probabilities for the data a_i to be the result of commission errors, then the probability is the sum of the p_{0i}, because any single committed value invalidates the selection. For omission, the effects are less devastating. A statistical test against a possible better variant not showing due to omissions can be constructed; intuitively, it seems only warranted if the probability of omissions is high.

2.5.2   PROBABILITY OF NORMAL AND ORDINAL DISCRETE VALUES

Values not measured on a continuous scale do not have an error distribution, but only a probability to be in error, for example, as a confusion matrix (Colwell, 1983). For such values in a formula $v = f(a, ...)$ (Equation 2.3) the computation must bifurcate and compute with each possible value a_i (e.g., land use values "forest," "agricultural") the corresponding v_i and then compute the sum of the products of values times probability $v = \Sigma p_i \cdot f(... v_i)$.

2.5.3   FUZZY MEMBERSHIP

For linguistic variables, e.g., "large," sharp boundaries of applicability are impossible, and modeling with a fuzzy membership function is appropriate. Formulas for propagating imperfection modeled with fuzzy membership functions are known (Zadeh, 1965). Schneider (1999) has shown that different distinctions for error do not influence the results. Further studies on effects of better models for fuzzy values seem warranted (Viertl, 2006).

2.6   CONCLUSION

Separating the complex decision process in a step to establish a model for this decision and then to apply this case-specific decision model to compute the outcome allows to one assess the influence of imperfections in the data on the decision, and this achieves a realistic assessment of the quality of data as fitness for a particular use. For the assessment of data "fitness for use," the decision model need not be known with precision; it is only necessary to know the formulas used and plausible values for the weights. Detailed knowledge is not required, and users may make decisions "intuitively" and guided by subjective considerations—the assessment of the quality of the decision as a result of the imperfections in the dataset remains valid if the used decision model approximates the personal decision process sufficiently. It is possible, and probably more realistic, to deduce the decision model after the decision has been taken and use it only for the assessment of the influence the data quality has on the decision. A detailed study on how decisions in other application areas can be modeled with decision models of this type will show how easily the concept generalizes beyond what was discussed here. The examples studied in Frank (2007) and Schneider (1999) give results that correspond to intuition, which is promising. Schneider (Petschacher, 1996) has constructed software to treat engineering decisions, and extensions to other cases seem possible and are planned as future work. It seems possible to include such tools in general GIS packages in the future.

ACKNOWLEDGMENTS

Prof. J. Schneider at ETH Zürich taught me as a student about engineering decision; his lifelong interest in risks of technical systems inspired the present approach to data quality. Thoughtful reviewers helped to improve the presentation.

REFERENCES

Abushady, A., and Frank, A. U., 2005. How Can Remote Sensing and GIS Help in the Verification of International Treaties? In: *Proceedings of RAST Conference*, Turkey.

Arentze, T., Borgers, A., and Timmermans, H., 1992. Geographical Information Systems, Accessibility, and Multi-Purpose, Multi-Stop Travel: A New Measurement Approach. In: *Proceedings of EGIS '92*, Munich, EGIS Foundation, pp. 438–450.

Colwell, R. N., 1983. *Manual of Remote Sensing*, ASPRS.

de Bruin, S., Bregt, A., and van de Ven, A., 2001. Assessing Fitness for Use: The Expected Value of Spatial Data Sets. *Int. Journal of Geographical Information Science* 15 (5): 457–471.

de Bruin, S., and Hunter, G. J., 2003. Making the Trade-Off between Decision Quality and Information Cost. *Photogrammetric Engineering & Remote Sensing* 69 (1): 91–98.

Frank, A., 2007. Incompleteness, Error, Approximation, and Uncertainty: An Ontological Approach to Data Quality. NATO Advanced Research Workshop, Kiev, Ukraine.

Ghilani, C. D., and Wolf, P. R., 2006. *Adjustment Computations Spatial Data Analysis*. Hoboken, NJ, John Wiley & Sons, Inc.

Harvey, F., 1997. Improving Multi-Purpose GIS Design: Participative Design. In: *Proceedings of Spatial Information Theory—A Theoretical Basis for GIS*. Berlin and Heidelberg, Springer-Verlag, pp. 313–328.

Karssenberg, D., and De Jong, K., 2005. Dynamic Environmental Modelling in GIS: 2. Modelling Error Propagation. *Int. Journal Geographical Information Systems* 19 (6): 623–637.

Navratil, G., and Achatschitz, C., 2004. Influence of Correlation on the Quality of Area Computation. ISSDQ, Bruck a.d. Leitha, Austria, Department of Geoinformation and Cartography.

Petschacher, M., 1996. *Programm VaP for Windows*. IBK ETH Zürich.

Schneider, J., 1999. Zur Dominanz der Lastannahmen im Sicherheitsnachweis. Festschrift zum 60. Geburtstag von Eduardo Anderheggen, Institut für Baustatik und Konstruktion der ETH Zürich.

Shyllon, E. A., and Hunter, G. J., 2004. Handling Spatial Data Quality Semantics. ISSDQ'04, Bruck a. d. Leitha, Austria, Department of Geoinformation and Cartography.

Timpf, S., Raubal, M., and Kuhn, W., 1996. Experiences with Metadata. In: *7th Int. Symposium on Spatial Data Handling, SDH'96*, Delft, The Netherlands (August 12–16, 1996).

Twaroch, F., and Achatschitz, C., 2005. Conceptual Model for a Hotel Seeking Agent. IWWPST '05 International Workshop on Web Portalbased Solutions for Tourism, Tampere, Finnland, Deaprtment of Geoinformation and Cartography, Faculty of Geodetic Engineering, Delft University of Technology, Delft, The Netherlands.

Viertl, R., 2006. Univariate Statistical Analysis with Fuzzy Data. *Computational Statistics & Data Analysis* 51: 133–147.

Zadeh, L. A., 1965. Fuzzy Sets. *Information and Control* 8: 338–353.

3 Semantic Reference Systems Accounting for Uncertainty
A Requirements Analysis

Sven Schade

CONTENTS

3.1   INTRODUCTION

Attribute reference systems and, more generally, semantic reference systems have been introduced as a means for reaching formal models of semantics (Chrisman, 2002; Kuhn, 2003). Such systems are proposed in order to overcome semantic heterogeneities between information communities. An information community is defined as a group of people who agree on a shared set of concepts (OGC, 2003). For the geospatial domain, these concepts are used to define data models for geospatial information (Kuhn and Raubal, 2003). Considering geospatial information, at least two dimensions of the information space (Fauconnier, 1994) refer to geographical space. Another dimension refers to temporal space and others to thematic spaces (Chrisman, 2002).

The expressions, which represent the concepts used by a distinct community, are defined as their specialized vocabulary. Semantic reference systems contribute to semantic interoperability by providing methods to explain meanings of vocabularies and to translate expressions between communities (Kuhn, 2003, 2005; Kuhn and Raubal, 2003). Current efforts on designing semantic reference systems aim at dissolving ambiguities in geospatial data models (Kuhn, 2005). This work contributes to the design by discussing requirements relating to a more general notion of imperfection. Semantic referencing, i.e., referencing user vocabularies to semantic reference frames, entails *ambiguity*, *generality*, and *open-texture*. Ontologies, which constitute these reference frames, have to account for *data acquisition errors* and to provide descriptions of *vague concepts*. Both are required to be grounded in a semantic datum. Semantic projection and semantic transformation, i.e., methods translating expressions between reference frames, have to include the *propagation of imperfect information*. The kinds of imperfections apply to all dimensions of the information space. The thematic attribute temperature and its possible values are used as an example. Challenges evolving from this dimension can be generalized to any other dimension, also geographic or temporal once. This chapter investigates the relation between the notions of imperfection and semantic reference systems. The definition of imperfection comprises unintended interpretation of expressions, characteristics of conceptual modeling, and the nature of observation. Considering data modeling, such a comprehensive analysis was never provided before. Section 3.2 provides background to the work. Imperfections causing uncertainty are identified and categorized in Section 3.3. Section 3.4 discusses each category in relation to the core elements and processes of semantic reference systems. Conclusions and a discussion of future work follow.

3.2   BACKGROUND

The distinction of expressions, concepts, and the real world is essential for the presented work. The main challenges addressed arise in the use of expressions to define conceptual data models (ISO, 2001). Imperfection in the description of data models causes uncertainty. Semantic reference systems aim to account for uncertainties by specifying data model semantics and supporting the definition of translation rules.

3.2.1 CONCEPTS AND EXPRESSIONS

As all expressions, the symbols and symbol combinations that are used to define feature names, attribute values, relations, and operators of conceptual data models *represent* concepts that are present in human minds. They are meant to symbolize the concepts that the developer had in mind when designing the model. Inversely, concepts *provide meaning to* these symbols and symbol combinations. The concepts evolved from human experiences in the real world. They *classify* (real-world) entities, also termed referents. Sets of entities become *conceptualized*. In general, the membership of each entity to the set of entities that is conceptualized as a certain concept can always be evaluated *true* or *false*. The expression, which represents a concept that classifies a certain set of entities, is said to *denote* the set.

The outlined relations between the notions of expression, concept, and referent emerged in the domain of semiotics (Ogden and Richards, 1946). They are commonly known as the semiotic triangle (Figure 3.1). The expression "temperature" (composed from the symbols "a," "e," "m," "p," "r," "t," and "u"), which is used as an attribute in a data model, serves as an example. Perceiving the symbol invokes a certain concept of a temperature in the reader's mind. The concept of temperature may classify many distinct sets of entities, reaching from the set of current air temperature, over monthly averages of air temperature, to human body temperature. Similarly, expressions like "cold," "warm," and "hot," which are used as attribute values in a data model, invoke concepts in the mind of the person perceiving the expressions.

3.2.2 UNCERTAINTY AND IMPERFECTION

Although *uncertainty* is often considered present in data or data models (Heuvelink, 1998; Fisher, 1999; Devillers and Jeansoulin, 2006), this work considers uncertainty to appear exclusively to living beings. Here especially humans are uncertain because of limited knowledge. As uncertainty, *knowledge* is strictly bound to living beings (Schutz, 1967; Wilson, 2002). Knowledge evolves by experience, which is based on the perception of information (Schutz, 1967). *Information* is interpreted data, while

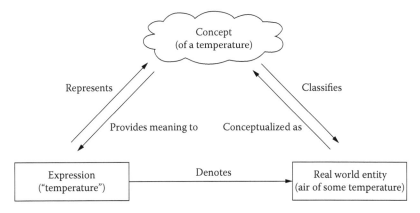

FIGURE 3.1 The semiotic triangle, also known as a semantic triangle and as a meaning triangle.

data is a pure carrier of information (Gerber and Weibel, 1995). Following this view, uncertainty is present if the perceived information is imperfect. In other words, *imperfection* of information causes uncertainty (see also Chapter 1).

Uncertainty may be caused by an unintended interpretation of expressions, characteristics of conceptual modeling, and the nature of observation, i.e., value assignments based on measurement. Unintended interpretation arises as soon as an expression represents a different concept in the data receiver's mind than in the mind of the developer(s) of the data model. For example, the expression "temperature" may be intended to invoke the concept of air temperature, but for a person from the medical domain, the expression may represent body temperature.

Conceptual models are simplifications of the real world (Cook and Daniels, 1994; ISO, 2001). A model always abstracts, and thus entities are conceptualized in respect to a specific purpose. When sensing a phenomenon, it can never be ensured to exactly represent the phenomenon's "true value" (Finkelstein, 1982). Considering air temperature measurement, for example, a normal distribution of measurement errors and a precisely measuring sensor with a standard error deviation of 0.5°C is assumed. A measurement result of 14°C approximates some unknown "true temperature value." With roughly 68% probability the "true temperature value" lies between 13.5°C and 14.5°C. In philosophy, this "true value" is also known as *quale* (Goodman, 1951; Masolo et al., 2003). A more detailed categorization of imperfections is presented in Section 3.3.

3.2.3   SEMANTIC REFERENCE SYSTEMS

Intended interpretations of vocabularies, i.e., collections of expressions, can be specified using the structural elements (semantic reference frame and semantic datum) and the process of semantic referencing. Semantic projection and semantic transformation are means to support the translation of expressions. The following explanations are based on Kuhn (2003).

Structural elements provide the backbone of semantic reference systems. They constitute the conceptual structure that is required to reduce doubt in possible interpretations of expressions. Upper-level, also called top-level, ontologies are used as *semantic reference frames*. The vocabulary used in an upper-level ontology serves as a formally defined, application-independent conceptual structure. In this context, an ontology is defined as "an engineering artefact, constituted by a specific vocabulary used to describe a certain reality, plus a set of explicit assumptions regarding the intended meaning of the vocabulary words" (Guarino, 1998).

As each ontology, the reference frame is constituted of building blocks. The basic building blocks are called primitives. The meaning of primitives applied in the semantic reference frame is grounded in the *semantic datum*. The semantic datum is part of the semantic reference system, but not of the ontologies constituting the frame.

Interpretation rules for vocabularies used by information communities are defined by establishing a reference between the communities' vocabulary and the semantic reference frame. The process of establishing these links is called *semantic referencing*.

Once a vocabulary is semantically referenced, *semantic projection* may be used to simplify the underlying formal model. If a second vocabulary is semantically referenced, a *semantic transformation* may be used to switch from one vocabulary to the other.

So far, the translation rules required to implement semantic transformations have to be built manually. In an overall mathematical structure based on category theory (Barr and Wells 1990), these rules can be precisely defined (Kuhn and Raubal, 2003).

3.3 FACETS OF IMPERFECTION

Before the relation between imperfection in geospatial information and semantic reference systems is analyzed, possible facets of imperfection need to be defined. For this purpose imperfections in the use of concepts and of expressions are separated. The use of expressions within vocabularies is exploited.

The notion of imperfection (also mostly labeled uncertainty) is used in distinct ways. According to Fisher and others, imperfection can be understood as a general concept that can be separated into varying kinds (Fisher et al., 2006). In this section, Fisher's categorization of imperfect data objects (Figure 3.2) is projected to imperfection regarding expressions. The expressions are used to define the data models that the objects are built upon. At the highest level of abstraction, Fisher distinguishes poorly defined and well-defined objects. In the case of well-defined expressions, the distinct meaning (i.e., the intended interpretation) of the expressions is the only one possible. This criterion is clarified by changing the label for the two upper-level categories into "unique interpretation" and "manifold interpretation." In the following, the lower-level categories are defined. Further adaptations of Fisher's proposal are considered if appropriate.

3.3.1 MANIFOLD INTERPRETATION

Given an expression, it may be unclear which concept it represents. Such imprecision can be further separated, depending on the nature of the concept or characteristics of the used expression.

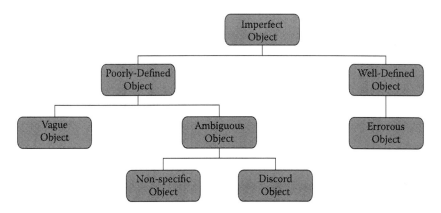

FIGURE 3.2 Fisher's categories of imperfection of objects. (Adapted from Fisher et al., 2006; Fisher, 1999; Klir and Yuan, 1995.)

3.3.1.1   Ambiguity

Ambiguity arises if two expressions have the same outward appearance, but one denotes something and the other does not (Scheffler, 1979). Accordingly, expressions used in information communities might be ambiguous. For example, the attribute "temperature" might be used to express the measurement of air temperature at a point in space and time, with the unit degree Celsius and resulting from a one-time measurement with a quicksilver thermometer. But, it might as well express the monthly average temperature.

Other authors suggest adding discord as a subconcept of ambiguity (Figure 3.2). Discord arises if the possible interpretations of an expression are clearly specified, but the denotations refer to different sets of entities (according to varying interpretations used). In this work, discord is conceptualized as a situation that may be caused by ambiguity instead of being a kind of it.

3.3.1.2   Generality

General concepts conceptualize multiple entities, no matter how dissimilar these may be or what similarity criteria are used (Scheffler, 1979). The concept of being *cold*, for example, can be used to conceptualize air temperature at a certain location in space-time, but it can also be used to characterize the personality of a person. In the same way, the concept of *temperature* can be used in one context to conceptualize the temperature of air and in another the one of water or of a human body.

General expressions are symbols that denote many things, no matter how dissimilar these may be and what similarity criteria are used (Scheffler, 1979). They represent generality at the concept level. The attribute value "cold," for example, is a general expression. This kind of imperfection is frequently confounded with ambiguity. For example, generality appears as "nonspecificity" and is considered a specific kind of ambiguity (Fisher, 1999; Fisher et al., 2006; Klir and Yuan, 1995). In case of ambiguity, the membership of entities to the set of referents, which are denoted by an ambiguous expression, may shift from *true* to *false* if changing interpretation. This does not hold for general expressions. Being "cold," for example, is always evaluated "true" or "false," after the membership was assigned once.

3.3.1.3   Open-Texture

Open-texture targets membership assignments to sets of entities that are classified by a certain concept (Scheffler, 1979). It is present, if the criteria, which are required for assigning membership to a specific set of referents, cannot be defined distinctively. Thus, it can be seen as an underspecification on purpose. In case of open-textured concepts, a human being is required to assign the membership value (*true* or *false*). Similarly, open-texture might be defined on expressions, which denote a specific set of entities.

Open-textured concepts (respectively, expressions) frequently arise in the geospatial domain. It is common practice that domain experts, like landscape ecologists, assign attribute values based on personal experiences. Consider, for example, *land use* as a dimension of the concept *earth's surface* (Comber, 2002).

3.3.2 Unique Interpretation

Assuming that the expressions used in the vocabulary are defined clearly, each facet of imperfection discussed here is orthogonal to any other category introduced in Section 3.3.

3.3.2.1 Incompleteness

Error is the most prominent reason for incompleteness in the geospatial domain. It does not apply to expressions and concepts, but to the assignments of values to attributes. Following Heuvelink and others (Heuvelink, 1998; Goodchild, 2004), error is conceptualized as the difference between the "true value" and the represented one. Error is present in each observation (Finkelstein, 1982). Measurement errors occur in defining the spatio-temporal location of an observation, as well as during the measurement of temperature values.

Anyhow, incompleteness covers more than just errors. It acknowledges the fact, that the complete truth of a situation cannot be accessed in many cases. This view corresponds to what is called uncertainty in the artificial intelligence (AI) community (Russell and Norvig, 2003). In order to reason under incompleteness, data models may explicitly represent dependencies between the values of two or more different attributes (Russell and Norvig, 2003).

3.3.2.2 Vagueness

A concept is vague if membership assignments may cause "borderline cases" (Tomai and Kavouras, 2004). A borderline case occurs if the membership of a certain entity to the set of represents, which are classified by a concept, cannot be assigned by *true* or *false*. For example, consider the concepts *cold*, *warm*, and *hot* used as property values for temperature. An air temperature of +20°C might be conceptualized as *warm* and one of –5°C as *cold*. How do we conceptualize 14°C?

Vague expressions denote vague concepts in human minds. Scheffler defines the vagueness of expressions as indeterminacy and concomitant interpretations and as ambivalence in deciding the applicability of an expression to an entity (Scheffler, 1979). Take "cold," "warm," and "hot" as used in everyday language as an example. Unlike Fisher's suggestion, this kind of vagueness is not categorized as "poorly defined," i.e., as having many possible interpretations. Whereas Fisher targets objects as subject to poor definitions, here expressions used to define object models are in focus. Vague expressions represent the vague nature of a concept in an object model, not the possibility of misinterpretation. Consequently, vagueness is a facet of imperfection that relates to unique interpretation.

The projected categorization of imperfections is summarized in Figure 3.3. The connections indicate subcategory relations. In contrast to Fisher's proposal for categorizing object imperfection, vague expressions are not categorized as poorly defined. The notion of ambiguous expressions is more specific and separated from other facets of manifold interpretation. Effects caused by ambiguous expressions have been discarded.

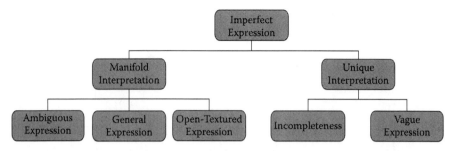

FIGURE 3.3 Fisher's categories of uncertainty projected from objects to expressions, in order to accentuate semantic aspects.

3.4 IMPERFECTION MEETS SEMANTIC REFERENCE SYSTEMS

Requirements for the system design and implementation are revealed by relating the categories of imperfection to the elements of semantic reference systems.

3.4.1 REQUIREMENTS TO CLARIFY INTENDED INTERPRETATIONS

The reduction of possible interpretations to the intended ones is topic to the reference frame, the datum, and to the process of semantic referencing. Intended interpretations need to face ambiguity, generality, and open texture.

3.4.1.1 Requirements Posed by Ambiguity

Natural language is subject to ambiguity, as are semi-formal concept representation approaches. Requirements are

A1. Usage of formal approaches, in order to reduce possible interpretations of the vocabulary to the one intended by a specific community.
A2. Formal specifications, i.e., ontologies, need to be free of cyclic references.
A3. Primitives used in the ontologies need to be grounded.

3.4.1.2 Requirements Posed by Generality

General expressions are loosely defined. Their specific meaning is only given by a particular context. The same expression may appear in varying contexts, but with distinct specialized meanings. Accounting for general expressions requires

G1. The possibility of using one expression in varying contexts with different meanings.
G2. The exact definition of the context of use related to a specific vocabulary.
G3. The distinction between contexts.
G4. The assurance that general expressions are put into the intended context.

3.4.1.3 Requirements Posed by Open Texture

The use of open-textured expressions is highly subjective, because it depends on personal interpretation of facts given in a distinct situation. One requirement is as follows: 01: Identifying the agent that assigned a membership, i.e., that interpreted the available expressions and assigned the membership value for a specific entity as *true* or *false*.

3.4.2 REQUIREMENTS IF INTENDED INTERPRETATIONS ARE CLARIFIED

Incompleteness and vagueness constrain the formalisms and languages, which suit implementing semantic reference systems. Error models need to be included in the semantic frame and the propagation of errors related to semantic projection and transformation.

3.4.2.1 Requirements Posed by Incompleteness

Errors affect spatial, temporal, and thematic information. When considering measurement errors, **all** of the following requirements need to be concerned:

E1. Extending the models of attribute values in the ontologies, which constitute reference frames, with error models. This includes grounding in a semantic datum.
E2. Defining methods to describe the propagation mechanisms that are used in the data models.
E3. Enabling the propagation of error values for semantic projection and transformation.

In complex application scenarios, usually only partial information is available (Russell and Norvig, 2003). The state of the model cannot be generated completely. Since data models may explicitly represent dependencies between data values, semantic reference systems are required to include: U1: A possibility to express data models that capture information dependencies.

3.4.2.2 Requirements Posed by Vagueness

Vague concepts need to be represented appropriately in the semantic reference system. Appropriately means explicitly ignoring the vague characteristics or using a formalization and an according representation language accounting for vagueness. If vagueness is ignored, concept definitions remain crisp. Membership to the set of referents is restricted to *true* and *false*. Only crisp links from community vocabularies, i.e., expressions used in data models, to the reference frame are allowed. Experiences in AI showed that ignoring vagueness leads to systems without extensive practical use (Russell and Norvig, 2003). Accounting for vagueness requires **at least one** of the following:

V1. The semantic reference frame offers the possibility to represent vague concepts. The required primitives are grounded in a semantic datum.
V2. A method for representing vague membership is included in the reference frame. The required primitives are grounded in a semantic datum.
V3. The semantic reference frame stays crisp, but possibilities for vague semantic referencing are provided.

3.5   CONCLUSIONS AND FUTURE WORK

Imperfection in data models leads to imperfect information and results in uncertainty. The notion of semantic reference systems has been related to categories of imperfection. The developed categorization projects an earlier attempt by Fisher and others (Fisher, 1999; Fisher et al., 2006; Klir and Yuan 1995) accentuating semantic issues. Ambiguity, generality, or open-texture may cause an expression to require further specification inside semantic reference frames. Once the intended interpretations are defined, incompleteness and vagueness require additional consideration.

All five categories of imperfection pose distinct requirements toward the design of semantic reference systems for geospatial information. Surveying the possibilities to account for generality, incompleteness, and vagueness in semantic reference systems is part of ongoing work. Practically, experimentation regarding models of air temperature will be focused on. On the theoretical level, the suggested categorization of uncertainties will be completed by adding *metaphors* to the branch of *manifold interpretation*, and by separating *incompleteness* from *error*. It is planned to add one additional level to the categorization, where especially *ambiguity* will be examined in more detail. This work will be based on achievements by Scheffler (1979).

ACKNOWLEDGMENTS

The work was carried out while the author was funded by the European Commission under the SWING project (FP6-26514, http://www.swing-project.org). I thank various members of the MUSIL group (http://musil.uni-muenster.de) for their valuable input.

REFERENCES

Barr, M. and C. Wells. 1990. *Category Theory for Computer Science*. London: Prentice Hall.
Chrisman, N. 2002. *Exploring Geographic Information Systems (2nd edition)*. London: Wiley.
Comber, A. 2002. *Automated Land Cover Change*. PhD diss., University of Aberdeen, Scotland.
Cook, S. and J. Daniels. 1994. *Designing Object Systems: Designing Object Oriented Modelling with Syntropy*. New York: Prentice Hall.
Devillers, R. and R. Jeansoulin. 2006. *Fundamentals of Spatial Data Quality*. London: ISTE.
Fauconnier, G. 1994. *Mental Spaces*. Cambridge: Cambridge University Press.
Finkelstein, L. 1982. Theory and Philosophy of Measurement. In: *Handbook of Measurement Science, Volume 1*, P. Sydenham and R. Thorn, Eds., Chapter 1, pp. 1–30. Chichester: John Wiley & Sons.
Fisher, P. 1999. Models of Uncertainty in Spatial Data. In: *Geographical Information Systems—Principles and Technical Issues. Volume 1*, P. Longley, M. Goodchild, D. Maguire, and D. Rhind, Eds., pp. 191–205. New York: John Wiley & Sons.
Fisher, P., A. Comber, and R. Wadsworth. 2006. Approaches to Uncertainty in Spatial Data. In: *Fundamentals of Spatial Data Quality*, R. Devillers and R. Jeansoulin, Eds., pp. 43–59. London: ISTE.
Gerber, K. and Weibel, P. 1995. *Mythos Information. Ars Electronica 95*. Vienna: Springer.
Goodchild, M. 2004. Special Issue on Measurement-Based GIS. *Journal of Geographical Systems* 6(4): 323–428.
Goodman, N. 1951. *The Structure of Appearance*. Cambridge, MA: Harvard University Press.

Guarino, N. 1998. Formal Ontology in Information Systems. In: *Proc. of Formal Ontology and Information Systems (FOIS'98)*, N. Guarino, Ed., pp. 3–15. Amsterdam: IOS Press.

Heuvelink, G. 1998. *Error Propagation in Environmental Modelling with GIS*. London: Taylor & Francis.

ISO: 2001. 19101 *Geographic Information—Reference Model*. Final Draft, International Standard.

Klir, G. J. and B. Yuan, 1995. Fuzzy Sets and Fuzzy Logic: Theory and Applications. Englewood Cliffs, NJ: Prentice Hall.

Kuhn, W. 2003. Semantic Reference Systems. *International Journal of Geographic Information Science* 17(5): 405–409.

Kuhn, W. 2005. Geospatial Semantics: Why, of What, and How? *Journal on Data Semantics III*. Springer Lecture Notes in Computer Science 3534, pp. 1–24.

Kuhn, W. and M. Raubal. 2003. Implementing Semantic Reference Systems. In: *Proc. of the 6th AGILE Conference on Geographic Information Science (AGILE 2003)*, M. Gould, R. Laurini, and S. Coulondre, Eds., pp. 63–72. Lyon: Presses Polytechniques et Universitaires Romandes.

Masolo, C., S. Borgo, A. Gangemi, N. Guarino, and A. Oltramari. 2003. *WonderWeb Deliverable D18, Ontology Library (final version)*. http://wonderweb.semanticweb.org/deliverables/documents/D18.pdf.

OGC. 2003. *The OpenGIS Guide—Introduction to Interoperable Geoprocessing and the OpenGIS Specification (3rd Edition)*. Wayland, MA: OpenGIS Consortium.

Ogden, C. and I. Richards. 1988. *The Meaning of Meaning*. 8th ed. Orlando, FL: Harcourt, Brace Jovanovick.

Russell, S. and P. Norvig. 2003. *Artificial Intelligence—A Modern Approach (2nd edition)*. Englewood Cliffs, NJ: Prentice Hall.

Scheffler, I. 1979. *Beyond the Letter—A Philosophical Inquiry into Ambiguity, Vagueness and Metaphor in Language*. London: Routledge & Kegan Paul.

Schutz, A. 1967. *The Phenomenology of the Social World*. Evanston, IL: Northwestern University Press.

Tomai, E. and M. Kavouras. 2004. From "Onto-GeoNoesis" to "Onto-Genesis": The Design of Geographic Ontologies. *Geoinformatica* 8(3): 285–302.

Wilson, T. 2002. The nonsense of knowledge management. In: *Information Research 8*, no. 1 (October). http://informationr.net/ir/8-1/paper144.html.

4 Elements of Semantic Mapping Quality
A Theoretical Framework

Mohamed Bakillah, Mir Abolfazl Mostafavi,
Yvan Bédard, and Jean Brodeur

CONTENTS

4.1 INTRODUCTION

The increasing volume of spatial data available to users emphasizes the importance of spatial data integration. This becomes even more important in the context of spatial decision making, where a fast and effective processing of data is necessary. The quality of integrated data depends on approaches that are used to resolve semantic, geometric, and structural heterogeneities between different sources. For semantic integration, many semantic mapping models were proposed to relate ontologies describing these sources (Noy and Musen, 2001; Maedche and Staab, 2002; Doan et al., 2004; Mostafavi, 2006) or schemas (Do and Rahm, 2001; Madhavan et al., 2001). The motivation of this chapter is the problem of semantic mapping quality, which is a new concept that we will define in this chapter. Semantic mapping quality affects the quality of the integrated data since semantic mappings between ontologies are used to rewrite queries on a first source for another source (Bouquet et al., 2005). For example, a mapping may involve a loss of precision (i.e., the semantic mapping model does not exploit the finest degree of definition of concepts). The

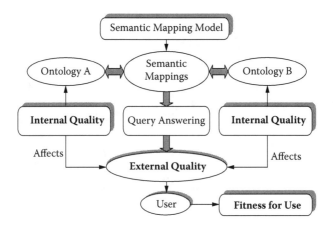

FIGURE 4.1   Impact of semantic mapping quality on external quality.

computed mappings may also be inconsistent with each other. Consequently, the external quality of the integrated data (perceived by users) is affected not only by internal data quality, which depends on each source (Wand and Wang 1996), but also by semantic mapping quality (Figure 4.1).

A framework for semantic mapping quality can help to indicate the quality of data resulting from semantic integration and to interpret resulting mappings. However, current methods for the assessment of semantic mapping quality focus toward a global performance evaluation, generally using precision and recall metrics (Do et al., 2003). Precision is the proportion of correct mappings while recall is the proportion of reference mappings (i.e., manually identified by experts) that were identified. However, these metrics cannot evaluate the quality of an individual mapping and they cannot be used when no reference is available. We don't know of any approach that defines or describes the concept of semantic mapping quality. In this chapter, we present a new conceptual framework where we define and represent semantic mapping quality for the first time. It includes a metamodel for the assessment of semantic mapping quality, a semantic mapping model that explicitly includes semantic mapping quality, and it defines the elements of semantic mapping quality. This chapter is structured as follows: Section 4.2 is a review of ontology mapping and data quality; Section 4.3 presents the metamodel; Section 4.4 the semantic mapping model; Section 4.5 presents the new elements of semantic mapping quality; and Section 4.6 concludes this chapter.

4.2   ONTOLOGY MAPPING AND DATA QUALITY

Semantic mapping consists of identifying with a semantic mapping model the semantic relationship between elements of different ontologies or schemas. Semantic mapping models can differ according to the input data they use, the characteristics of the reconciliation process, and the results produced (Bouquet et al., 2005). Current semantic mapping approaches use one or more semantic similarity models to identify relationships between concepts of different ontologies (Do and Rahm, 2001;

Madhavan et al., 2001; Brodeur and Bédard, 2001; Maedche and Staab, 2002; Doan et al., 2004; Mostafavi, 2006; Bakillah et al., 2006). Semantic similarity expresses commonality between concepts. It supports the identification of concepts that have a similar meaning and refer to a similar entity of the reality. Examples of elements that contribute to semantic similarity are common properties between concepts (Rodriguez and Egenhofer, 2003), or similar positions of concepts in their respective hierarchies (Madhavan et al., 2001). Each semantic mapping model will provide different results with varying quality depending on its characteristics. A framework for semantic mapping quality needs to consider these characteristics and to be anchored to research on ontology quality in order to be coherent. There are several classification frameworks for quality of ontology (Cross and Pal, 2005; Mostowfi and Fotouhi, 2006). Methods were also proposed to assess the internal quality and consistency of ontologies (Mostafavi et al., 2004) and communicate the elements of data quality to users in order to avoid misuse of data, for example, by comparing metadata and users' needs at various levels of detail (Devillers et al., 2005). However, to our knowledge, no complete work has been done on how to define and effectively assess the quality of semantic mapping between different ontologies, which will be addressed in this chapter.

4.3 A METAMODEL FOR ASSESSMENT OF SEMANTIC MAPPING QUALITY

The metamodel for the assessment of semantic mapping quality (Figure 4.2) defines the relationships between the different entities that influence the semantic mapping quality. It is designed in UML (Universal Modeling Language), where composition, generalization, and association relationships between those entities are defined. The proposed semantic mapping model is composed of: (1) a *semantic similarity model* that gives *semantic similarity* between concepts; (2) *semantic mapping rules* that

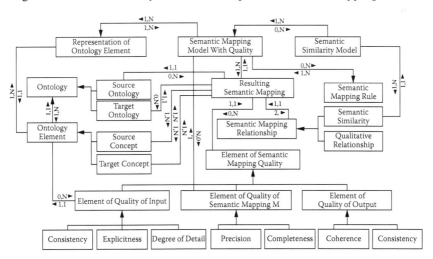

FIGURE 4.2   The proposed metamodel for the assessment of semantic mapping quality.

allow finding the *qualitative relationship* between concepts; and (3) *representation of the ontology element*, where *ontology elements* are the abstract constructs that constitute an ontology, such as concepts, attribute, relationships, etc. These components of the *semantic mapping model* will be defined in Section 4.4. The *resulting semantic mappings* depend on the *elements of semantic mapping quality*, which are divided in three categories: *quality of input of the semantic mapping model*, *quality of the semantic mapping model* itself, and *quality of output of semantic mapping model*. For example, quality of input refers to the quality of the definitions of the *ontology elements* that are used by the *semantic mapping model*. It is a general class for some elements of quality such as *consistency* of input ontology element. In Section 4.5 we will explain these categories of elements of semantic mapping quality and give definitions for some of them.

4.4   SEMANTIC MAPPING MODEL ENHANCED WITH ELEMENTS OF QUALITY

This section introduces the semantic mapping model for ontologies. Ontology is defined by $O = (C, A, \Gamma, R)$, where C is a set of concepts, A is a set of attributes of concepts, Γ is a set of instances for each concept, and R is a set of hierarchical relationships between concepts, such as generalization or inclusion relationships. Concepts are defined by three sets of features: internal features P_i, which are the descriptive attributes and name of concepts; relational features P_r, which denote relationships to other concepts and external features; and P_e, which are features of concepts related to the concept in the ontology. The task of computing a semantic mapping is defined as follows: Given two ontologies, find semantically related concepts between them and indicate the quality of the relationship established between the concepts. The output of the semantic mapping model is a tuple, $m = (c_1, c_2, s, r, Q)$ with $Q = (q_1, q_2, \ldots, q_n)$, where concepts c_1 and c_2 are linked by a qualitative relationship r, a semantic similarity value s, and a set of elements of semantic mapping quality Q that will be defined in Section 4.5. Semantic similarity s is a function of three parameters: s_{jj} terms, which compute the overlap of the same sets of features j (ex: between sets of internal features of both concepts); s_{jk} terms, which compute the overlap of different sets of features j and k (ex: between internal features of first concept and external features of second concept); and, finally a weight ω for each term, which is a function of the frequency of the features in their respective ontologies. Semantic similarity is given by

$$s(c_1, c_2) = \sum_j \omega_{jj} s_{jj}(c_1, c_2) + \sum_j \sum_{k \neq j} \omega_{jk} s_{jk}(c_1, c_2) \tag{4.1}$$

with $j, k = i$ (for internal), r (for relational), or e (for external). The overlaps computed in Equation 4.1 are distributed in the matrix of relation (Equation 4.2), whose state gives the qualitative relationship r that holds between concepts (Table 4.1) and thus the semantic mapping rules of the proposed model:

TABLE 4.1

Semantic Mapping Rules in the Semantic Mapping Model

State of Matrix of Relation	Relationship r	Symbol
$M(c_1, c_2) = M(c_2, c_1) = identity\ matrix$	equivalence	$\equiv$
$M(c_1, c_2) = identity\ matrix; M(c_2, c_1) = diagonal\ matrix$	c_1 is included in c_2	$\subseteq$
$M(c_1, c_2) = diagonal\ matrix; M(c_2, c_1) = identity\ matrix$	c_1 includes c_2	$\supseteq$
$M(c_1, c_2) \neq identity\ matrix; M(c_2, c_1) \neq identity\ matrix$	overlap	$\cap$
$M(c_1, c_2) = M(c_2, c_1) = zero\ matrix$	disjoint	$\perp$

$$M(c_1,c_2) = \begin{pmatrix} s_{ii}(c_1,c_2) & s_{ir}(c_1,c_2) & s_{ie}(c_1,c_2) \\ s_{ri}(c_1,c_2) & s_{rr}(c_1,c_2) & s_{re}(c_1,c_2) \\ s_{ei}(c_1,c_2) & s_{er}(c_1,c_2) & s_{ee}(c_1,c_2) \end{pmatrix} \tag{4.2}$$

4.5 ELEMENTS OF SEMANTIC MAPPING QUALITY

As expressed in the metamodel in Section 4.3, elements of semantic mapping quality are divided into three categories defined in this section: quality of input of the semantic mapping model, quality of the semantic mapping model, and quality of output of the semantic mapping model.

4.5.1 ELEMENTS OF QUALITY OF INPUT OF THE SEMANTIC MAPPING MODEL

This category refers to the quality of the data exploited by the semantic mapping model, that is, concepts, attributes, relationships, etc. More specifically, a mapping input is an ontology element that is used by the semantic mapping model to produce a semantic mapping. Consequently, the quality of input is related to the quality of the source and target ontologies. Starting from elements of quality of ontologies (Cross and Pal, 2005; Mostowfi and Fotouhi, 2006; Mostafavi et al., 2004), we define appropriate elements for the quality of input of the semantic mapping model. The first element of this category is the **consistency of mapping input** (i.e., an ontology element is consistent if its definition creates no conflict with the integrity constraints defined in the ontology). Integrity constraints give some conditions that must be verified in the ontology in order to preserve its consistency. For example, an instance of *Geographic Feature concept* must have a *Spatial Feature* (i.e., it must be located in space). If this condition is not respected by some instance, the mapping that takes this instance for input will be less consistent. A second element of this category is the **explicitness of mapping input**, which refers to the clarity of the definition of ontology elements. More specifically, the explicitness of a mapping input is high if the knowledge we have about this input is explicit, i.e., no inference is necessary in order to get this knowledge. For example, subconcepts of a concept can be discovered with inference reasoning while these generalization relationships are not stated explicitly in the ontology. Finally, the **degree of detail of mapping input** refers to the quantity of knowledge we have about an ontology element. Several cases can be considered: for example, the degree of detail can refer to the presence of annotation (i.e., metadata) of ontology elements, which can be very useful in specifying the meaning of ontology elements that have to be mapped. It can also refer to the number of attributes of a concept, i.e., more attributes can contribute to augment the knowledge we have about this concept, thus augmenting the information carried by the mapping.

4.5.2 ELEMENTS OF QUALITY OF THE SEMANTIC MAPPING MODEL

The quality of the semantic mapping model is related to the adequacy between the characteristics of the semantic mapping model and the characteristics of the ontology element compared by this model. The first element of this category is the **precision of the semantic mapping model**. A semantic mapping model preserves the precision of the mapping inputs (i.e., ontology elements) when it uses their finer level of definition. For example, suppose an attribute of a concept is associated to a domain value; the semantic mapping model is imprecise if it only evaluates the correspondence at the attribute level and not at the domain value level. The second element is the **completeness of the semantic mapping model**. A mapping preserves the completeness of the mapping input when it takes into account all aspects of its definition. For example, a mapping does not preserve completeness if it does consider instances of concepts.

4.5.3 ELEMENTS OF QUALITY OF OUTPUT OF THE SEMANTIC MAPPING MODEL

The quality of the output is related to the quality of the result produced by the semantic mapping model. In this framework, we choose to define the consistency of the semantic mappings. A mapping preserves consistency when it does not create conflict with the integrity constraints defined in ontologies. This is different from the consistency of the mapping input because it is the semantic mapping that creates the conflict rather than the definitions of the ontology elements. We consider a specific case of constraints, i.e., constraints defined by relationships between concepts in an ontology (they can be relations of generalization or part-of relations). For the purpose of this discussion, we call the relationships between concepts of a single ontology *internal relationships*. Suppose that we have concepts $\{a_0, a_1, a_2\}$ and $\{b_0, b_1, b_2\}$, respectively, from ontologies A and B with the following internal relationships: $a_0 \supseteq a_1 \supseteq a_2$ and $b_0 \supseteq b_1 \supseteq b_2$. Moreover, suppose that the mappings $m = (a_1, b_1, r = \supseteq)$ and $m' = (a_1, b_0, r = \supseteq)$ were computed. m' and m are conflicting since together they assert that b_0 is included in b_1, which is contradictory with internal relationships of ontology B. We have established a set of conditions, called Mapping Conflict Predicates (MCPs), for expressing this kind of conflict between mappings considering the preceding internal relationships. When defining these predicates we have considered mappings that involve adjacent concepts of ontology, that is, concepts that are directly linked by internal relationships. Table 4.2 lists the five categories of Mapping Conflict Predicates (one for each semantic relationship given in Table 4.1). The third column of Table 4.2 contains the mappings that are conflicting with the mapping of the second column, knowing that $a_0 \supseteq a_1 \supseteq a_2$ and $b_0 \supseteq b_1 \supseteq b_2$. For

Table 4.2
Categories of Mapping Conflict Predicates

Category of MCP	Mapping $m(a_1,b_1)$	Mappings in Conflict with $m(a_1,b_1)$
MCP1	$a_1 \equiv b_1$	$m(a_0, b_0)$ is conflicting if $a_0 \perp b_0$.
		$m(a_0, b_1)$ is conflicting if $a_0 \equiv b_1$; $a_0 \subseteq b_1$; $a_0 \perp b_1$; $a_0 \cap b_1$.
		$m(a_0, b_2)$ is conflicting if $a_0 \equiv b_2$; $a_0 \subseteq b_2$; $a_0 \perp b_2$; $a_0 \cap b_2$.
		$m(a_1, b_0)$ is conflicting if $a_1 \equiv b_0$; $a_1 \subseteq b_0$; $a_1 \perp b_0$; $a_1 \cap b_0$.
		$m(a_1, b_2)$ is conflicting if $a_1 \equiv b_2$; $a_1 \subseteq b_2$; $a_1 \perp b_2$; $a_1 \cap b_2$.
		$m(a_2, b_0)$ is conflicting if $a_2 \equiv b_0$; $a_2 \subseteq b_0$; $a_2 \perp b_0$; $a_2 \cap b_0$.
		$m(a_2, b_1)$ is conflicting if $a_2 \equiv b_1$; $a_2 \subseteq b_1$; $a_2 \perp b_1$; $a_2 \cap b_1$.
MCP2	$a_1 \subseteq b_1$	$m(a_0, b_0)$ is conflicting if $a_0 \perp b_0$.
		$m(a_0, b_1)$ is conflicting if $a_0 \perp b_1$.
		$m(a_0, b_2)$ is conflicting if $a_0 \perp b_2$.
		$m(a_1, b_0)$ is conflicting if $a_1 \equiv b_0$; $a_1 \supseteq b_0$; $a_1 \perp b_0$; $a_1 \cap b_0$.
		$m(a_2, b_0)$ is conflicting if $a_2 \equiv b_0$; $a_2 \supseteq b_0$; $a_2 \perp b_0$; $a_2 \cap b_0$.
		$m(a_2, b_1)$ is conflicting if $a_2 \equiv b_1$; $a_2 \supseteq b_1$; $a_2 \perp b_1$; $a_2 \cap b_1$.
MCP3	$a_1 \supseteq b_1$	$m(a_0, b_0)$ is conflicting if $a_0 \perp b_0$.
		$m(a_0, b_1)$ is conflicting if $a_0 \equiv b_1$; $a_0 \subseteq b_1$; $a_0 \perp b_1$; $a_0 \cap b_1$.
		$m(a_0, b_2)$ is conflicting if $a_0 \equiv b_2$; $a_0 \subseteq b_2$; $a_0 \perp b_2$; $a_0 \cap b_2$.
		$m(a_1, b_0)$ is conflicting if $a_1 \perp b_0$.
		$m(a_1, b_2)$ is conflicting if $a_1 \equiv b_2$; $a_1 \subseteq b_2$; $a_1 \perp b_2$; $a_1 \cap b_2$.
MCP4	$a_1 \cap b_1$	$m(a_0, b_0)$ is conflicting if $a_0 \perp b_0$.
		$m(a_0, b_1)$ is conflicting if $a_0 \equiv b_1$; $a_0 \subseteq b_1$; $a_0 \perp b_1$.
		$m(a_0, b_2)$ is conflicting if $a_0 \equiv b_2$; $a_0 \subseteq b_2$.
		$m(a_1, b_0)$ is conflicting if $a_1 \equiv b_0$; $a_1 \supseteq b_0$; $a_1 \perp b_0$.
		$m(a_1, b_2)$ is conflicting if $a_1 \equiv b_2$; $a_1 \subseteq b_2$.
		$m(a_2, b_0)$ is conflicting if $a_2 \equiv b_0$; $a_2 \supseteq b_0$.
		$m(a_2, b_1)$ is conflicting if $a_2 \equiv b_1$; $a_2 \supseteq b_1$.
MCP5	$a_1 \perp b_1$	$m(a_0, b_1)$ is conflicting if $a_0 \equiv b_1$; $a_0 \subseteq b_1$.
		$m(a_0, b_2)$ is conflicting if $a_0 \equiv b_2$; $a_0 \subseteq b_2$.
		$m(a_1, b_0)$ is conflicting if $a_1 \equiv b_0$; $a_1 \supseteq b_0$.
		$m(a_1, b_2)$ is conflicting if $a_1 \equiv b_2$; $a_1 \subseteq b_2$; $a_1 \supseteq b_2$; $a_1 \cap b_2$.
		$m(a_2, b_0)$ is conflicting if $a_2 \equiv b_0$; $a_2 \supseteq b_0$.
		$m(a_2, b_1)$ is conflicting if $a_2 \equiv b_1$; $a_2 \subseteq b_1$; $a_2 \supseteq b_1$; $a_2 \cap b_1$.
		$m(a_2, b_2)$ is conflicting if $a_2 \equiv b_2$; $a_2 \subseteq b_2$; $a_2 \supseteq b_2$; $a_2 \cap b_2$.

example, consider two spatial ontologies of environmental health, the first with concepts *region* $\supseteq$ *sanitary region* $\supseteq$ *medical territory* and the second with concepts *region* $\supseteq$ *sanitary zone* $\supseteq$ *local medical territory*. If the mapping m = (*sanitary region, sanitary zone, r* = $\supseteq$) is computed, the mapping m' = (*sanitary region, local medical territory, r* = $\supseteq$) is inconsistent with m since it verifies a condition of category 3 of the Mapping Conflict Predicates.

We have proposed a conceptual framework for semantic mapping quality that can help to obtain better information about the meaning of semantic mappings. The quality of input of the semantic mapping model can represent the quality of the data coming from multiple sources. The quality of the semantic mapping model can indicate if the model is precise and complete enough for the ontologies being compared. Finally, the quality of the output of the semantic mapping model can be used to verify the consistency of the automatically computed mappings.

4.6   CONCLUSION AND FUTURE WORK

This is a first attempt to define semantic mapping quality, which is important because it has an impact on the quality of querying between multiple geospatial data sources. We have proposed a conceptual framework for semantic mapping quality and given a semantic mapping model including quality elements that we have defined. This approach can help to indicate to users the quality of the data resulting from the semantic integration of multiple sources. In future work, we will further precise the elements for semantic mapping quality and provide quantitative measurements for them and explore how semantic mapping quality affects the semantic interoperability measure.

ACKNOWLEDGMENT

This research was made possible by an operating grant from the Natural Sciences and Engineering Research Council of Canada (NSERC).

REFERENCES

Bakillah, M., Mostafavi, M. A., and Bédard, Y., 2006. A Semantic Similarity Model for Mapping between Evolving Geospatial Data Cubes. In *On the Move to Meaningful Internet Systems Workshops*, LNCS 4278, R. Meersman, Z. Tari, P. Herrero, et al., Eds., pp. 1658–1669. Berlin, Heidelberg: Springer-Verlag.

Bouquet, P., Mikalai, Y., and Zanobini, S., 2005. Critical Analysis of Mapping Languages and Mapping Techniques. Technical Report DIT-05-052, University of Trento, Italy.

Brodeur, J., and Bédard, Y., 2001. Geosemantic Proximity, a Component of Spatial Data Interoperability. In *Int. Workshop on Semantic of Enterprise Integration*, pp. 14–18. Florida, USA.

Cross, V., and Pal, A., 2005. Metrics for Ontologies. In *Annual Meeting of the NAFIPS*, pp. 448–453, Tokyo, Japan.

Devillers, R., Bédard, Y., and Jeansoulin, R., 2005. Multidimensional Management of Geospatial Data Quality Information for Its Dynamic Use within Geographical Information Systems. *American Society for Photogrammetry and Remote Sensing* 71(2): 205–215.

Do, H. H., and Rahm, E., 2001. COMA—A System for Flexible Combination of Schema Matching Approaches. In *Proc. of the 28th Conf. on VLDB*, pp. 610–621. Hong Kong, China.

Do, H. H., Melnik, S., and Rahm, E., 2003. Comparison of Schema Matching Evaluation. In *Proc. of the 9th Annual ACM International Workshop on Web Information and Data Management LNCS 2593*, A. B. Chaudhri et al., Eds., pp. 221–237. Berlin, Heidelberg: Springer-Verlag.

Doan, A., Madhavan, J., Domingos, P., and Halevy. A. Y., 2004. Ontology Matching: A Machine Learning Approach. In *Handbook of Ontologies*, International Handbooks on Information Systems, S. Staab and R. Studer, Eds., pp. 385–404. Berlin, Heidelberg: Springer-Verlag.

Madhavan, J., Bernstein, P., and Rahm, E., 2001. Generic Schema Matching with Cupid. In *Proc. of the 28th Conf. on VLDB*, pp. 49–58. Hong Kong, China.

Maedche, A., and Staab, S., 2002. Measuring Similarity between Ontologies. In *Proc. of Int. Conf. on Knowledge Engineering and Knowledge Management*, pp. 251–263, Siguenza, Spain.

Mostafavi, M. A., 2006. Semantic Similarity Assessment in Support of Geospatial Data Integration. In *7th Int. Symposium on Spatial Accuracy Assessment in Natural Resources and Environmental Sciences*, pp. 685–693. Lisbon, Portugal.

Mostafavi, M. A., Edwards, G., and Jeansoulin, R., 2004. An Ontology-Based Method for Quality Assessment of Spatial Data Bases. In *ISSDQ'04 Proc.*, pp. 49–66. Bruck am der Leitha, Austria.

Mostowfi, F., and Fotouhi, F., 2006. Improving Quality of Ontology: An Ontology Transformation Approach. In *Proc. of the 22nd ICDEW*, p. 61, Miami, FL.

Noy, N. F., and Musen, M. A., 2001. Anchor-Prompt: Using Non-Local Context for Semantic Matching. In *Proc. of Workshop on Ontologies and Information Sharing at International Joint Conference on Artificial Intelligence*, pp. 63–70. Seattle, WA.

Rodriguez, M. A., and Egenhofer, M. J., 2003. Determining Semantic Similarity among Entity Classes from Different Ontologies. *IEEE Trans. on Knowledge and Data Engineering* 15(2): 442–456.

Wand, Y., and Wang, R., 1996. Anchoring Data Quality Dimensions in Ontological Foundation. *Communications of the ACM* 39(11): 86–95.

5 A Multicriteria Fusion Approach for Geographical Data Matching

Ana-Maria Olteanu

CONTENTS

5.1   INTRODUCTION

The integration of geographical databases (GDB) describing the same real world from different sources is becoming a growing issue. At present, many geographical databases exist to jointly describe the same reality at different spatial and semantic resolutions. Many applications, such as quality evaluations, incoherence detection, propagation of updates, and the study of several adjacent zones, require the integration of geographical databases. The integration process (Devogele et al., 1998; Sheeren et al., 2004) is usually carried out in three stages. The first stage is pre-integration, consisting of enrichment of the schema sources and standardization of the models. The second stage consists of schema matching (Gesbert, 2005) and data matching (Mustière, 2006; Olteanu et al., 2006), which are not completely independent. The last stage is defining an integrated schema and specifications, and populating the integrated GDB.

The aim of this chapter is to structure the data matching process. We use the Dempster-Shafer theory (Shafer, 1976) to fuse different knowledge components provided by different criteria, such as geometry, name, and semantic of the spatial feature in order to make a decision on the matching process.

5.2   RELATED WORK

Data matching is a geoinformation process to find homologous features in different databases (Walter and Fritsch, 1999). It is useful in many fields handling spatial information such as integration of spatial data (Voltz, 2006; Mustière, 2006), automated updating, quality analysis (Bel Hadj Ali, 2001), and inconsistency detection between databases (Sheeren et al., 2004).

Matching methods developed in the literature have revealed good results and efficiency on certain types of data in selected test areas. Matching algorithms depend on the geometry of the features to match, the spatial relations, and, last but not least, on semantic attributes. Bel Hadj Ali (2001) proposed an approach for polygons, which mainly relies on the distances between geometries. Beeri et al. (2004) proposed four methods for matching while comparing locations of points and relying on probabilistic considerations. Feature names can also be compared using a string distance (Levenshtein, 1965; Cohen et al., 2003). With respect to linear networks, Walter and Fritsch (1999) developed a statistical approach based on geometric and topologic criteria to match features in two different databases at similar scales. Matching algorithms for two networks at similar (Voltz, 2006) or different scales (Devogele, 1997; Mustière, 2006) were proposed. Matching is based on different criteria, including an analysis of topological relations, and it is carried out in several steps. Comparing the semantics of classes is necessary to match schemas and is also useful to match data, even if this approach is rarely used (Gesbert, 2005). Matching methods generally use one or several of the geometrical, semantic, and topological criteria.

All criteria are usually applied successively, and, more importantly, most approaches do not explicitly model data imperfections such as uncertainty, incompleteness, and imprecision. Therefore, we aim to develop a matching approach taking all criteria jointly into account and to model imperfection. We consider the Dempster-Shafer theory that is relevant to achieve these objectives because, on the one hand, it allows us to model imprecision, uncertainty, and incompleteness and to combine criteria and hypotheses, and, on the other hand, a decision is made based on beliefs.

5.3   THE FRAME OF THE DEMPSTER-SHAFER THEORY

The Dempster-Shafer theory was introduced by Dempster in 1967. Based on Dempster's work, Shafer (Shafer, 1976) introduced the Dempster-Shafer model, which is based on belief functions.

5.3.1 THE FRAME OF DISCERNMENT

Let Θ be a finite set of N hypotheses, H_i, $i = 1, ..., N$, called the frame of discernment that corresponds to the potential solution of a decision problem, $\Theta = \{H_1, H_2, ..., H_N\}$. From the frame of discernment, let 2^Θ denote the set of all subsets of Θ defined by

$$2^\Theta = \{\phi, \{H_1\}, \{H_2\}, \{H_1, H_2\} ... \{H_1 ... H_N\}, \Theta\} \tag{5.1}$$

where $\{H_i, H_j\}$ represents the hypothesis that the solution of a problem is one of them, i.e., either H_i or H_j. The Dempster-Shafer theory is based on the basic belief assignment, i.e., a function that assigns to a proposition, $A \in 2^\Theta$, and a value named the basic belief mass, denoted $m(A)$, that represents the extent to which a criterion believes in it. As an example, we consider a matching that is based on the geometry of features. The closer two features are, the stronger the criterion believes that the features are homologous, and thus the value of the basic belief mass is important.

The basic belief assumption satisfies

$$m : 2^\Theta \to [0,1]$$

$$\sum_{A \subseteq \Theta} m(A) = 1 \tag{5.2}$$

Every proposition $A \in 2^\Theta$ with $m(A) > 0$ is called a focal proposition of m. Hence, from now on, we will consider only the focal propositions to combine knowledge and to make a decision.

5.3.2 DEMPSTER'S RULE OF COMBINATION

The Dempster-Shafer theory allows one to fuse several criteria using Dempster's rule (Shafer, 1976). Let us consider two sources, S_1 and S_2. Each source supports a proposition with a certain basic belief mass, called $m_1(A)$ and $m_2(A)$, respectively. We denote by m_{12} the basic belief mass resulting from the combination of two sources by Dempster's rule and that supports the same proposition A. If, for example, we wish to find out if two features belonging to two different databases should be matched, two criteria are used, e.g., geometry and comparison of toponyms. Assuming that the two features are homologous, the first criterion believes that they are homologous because the features are close and assigns to it an important basic belief mass, whereas the second criterion is not sure because the two toponyms are not similar and thus it assigns to this assumption a less important basic belief mass. To take a decision, different criteria have to be combined, potentially leading to a conflict situation. If this is the case, the conflict is assigned to the empty set, and it is used by Dempster to normalize the basic belief mass resulting from the combination of the two sources, m_{12}. Thus, the basic belief mass assigned to the conflict is redistributed according to the focal elements (Shafer, 1976; Smets and Kennes, 1994).

5.4   THE DEMPSTER-SHAFER THEORY IN A DATA MATCHING CONTEXT

Matching generally consists in searching for potential candidates for each feature belonging to the reference database and then analyzing them in order to determine the final results. Spatial data have imperfections: The location could be imprecise, or the toponyms might have different versions due to omission, abbreviation, and substitution. Using several criteria successively during matching, errors could propagate, leading to an erroneous matching. Therefore, imperfection should be taken into account and criteria should be applied at the same time to obtain more relevant information. The Dempster-Shafer theory offers tools for modeling imperfection through the belief functions and to combine different knowledge through Dempster's rule of combination.

Data matching based on the Dempster-Shafer theory consisting of three steps is now described. Firstly, computation of the basic belief mass for each candidate and each criterion is performed. Secondly, criteria are combined per candidate, i.e., for each candidate the basic belief masses are combined to provide a combined basic belief mass synthesizing the knowledge of the different criteria. Finally, the results of the second step are combined, i.e., candidates are combined to provide an overall view.

5.4.1   LOCAL APPROACH: DEFINITION OF THE FRAME OF DISCERNMENT

We compare two datasets made for different purposes and with different scales. We consider the more detailed dataset as a comparison and the less detailed as a reference dataset. Therefore, for each reference feature, we look into the comparison dataset for geometrically close features that are candidates for matching. The distance that determines how far we look for candidates is an empirical one, and it depends on the nature of the reference feature.

In our case, for each candidate C_i, the assumption "C_i is the homologous feature" is added to the local frame of discernment of the reference feature. Because a feature may have no homologue, a new hypothesis, NM, standing for "the feature is not matched at all," is defined. A similar approach was presented in Royère et al. (2001).

For each feature belonging to the reference database, named a reference feature, a global frame of discernment is now defined as follows:

$$\Theta = \left\{ C_1,\ C_2,\ ...,\ C_i,\ ...,\ C_N,\ NM \right\} \tag{5.3}$$

In Equation 5.3, N defines the number of candidates, C_i represents the assumption that the candidate C_i is the homologous of the reference feature, and NM is the assumption that the reference feature is not matched. To compute the basic belief assignments, we use a local approach by analyzing each candidate separately. To do so, we use a particular case of the Dempster-Shafer theory, named the specialized sources (Appriou, 1991). Each source specializes on a candidate and assigns a belief to it. In our case, a source coincides with a criterion of data matching (geometric, toponym, and semantic).

We consider S_i, a subset of 2^Θ, defined as $S_i = \{C_i, \neg C_i, \Theta\}$, where

- C_i is the hypothesis that the homologous of the reference feature is C_i.
- $\neg C_i = \{C_1, C_2, ..., C_{i-1}, ..., C_N, NM\}$ is the hypothesis that C_i is not the homologous of the reference feature. The solution could be another candidate or the hypothesis NM.
- $\Theta = \{C_1, ..., C_N, NM\}$ is the hypothesis that the criterion does not know if C_i is the right candidate or not, i.e., it models ignorance.

5.4.2 COMPUTATION OF BASIC BELIEF MASSES

We propose three criteria to match data, and in our case these criteria represent the sources in the frame of the Dempster-Shafer theory. There are described below.

5.4.2.1 The Geometrical Criterion

The geometrical criterion is based on the Euclidean distance, d_E, between the location of the reference feature and the location of the candidate. We suppose that the closer the candidate is to the reference feature, the more probable it is that this one is the homologous of the reference feature, as defined in Figure 5.1a. In Figure 5.1a, T_2 represents the selection threshold, i.e., the distance determining how far we look for candidates, while T_1 defines a confidence threshold that gives less weight for candidates that are fairly far. Due to the imprecision of data location, some homologous features can be relatively far. To avoid definitively eliminating a candidate that is

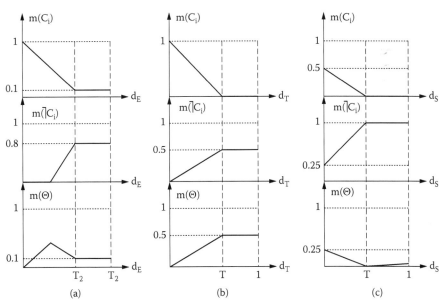

FIGURE 5.1 Modeling of the basic belief assumption for geometrical a), toponym b), and semantic c) criteria.

too far as compared to the reference feature, we consider that the basic belief mass allocated to the hypothesis "the candidate C_i is the homologue feature" is never 0, but it ranges from 0.1 to 1.

5.4.2.2   The Toponym Criterion

The second criterion is based on the comparison of toponyms. A distance, d_T, between two toponyms is computed using the Levenshtein distance (Levenshtein, 1965):

$$d_T = \frac{d_L(toponym_1, toponyme_2)}{\max(L_1, L_2)} \tag{5.4}$$

where d_L = Levenshtein distance, L_1 = toponym$_1$'s length, and L_2 = toponym$_2$'s length.

Figure 5.1b shows the basic belief assignment for the toponym criterion. Curves are different from the geometrical criterion, in order to express that we are less confident in this source. Thus, we manage the case of ambiguity, when two toponyms indicating the same features are compared, with one having the official name, whereas the other has a nonofficial name. It consists in decreasing the mass of belief associated with the assumption "C_i is not the homologue of the feature" and increasing the basic belief mass associated with ignorance. Thus, if the distance d_T is greater than the threshold T (e.g., 30% of letters are different), the basic belief mass assigned to both hypotheses "C_i is not the right candidate" and ignorance are equal to 0.5.

5.4.2.3   The Semantic Criterion

Features may have the same toponyms, i.e., being close to each other, but present different natures and thus are not homologous, like, for example, a summit and a mountain pass. The semantic criterion uses semantic properties to be able to do this. A semantic distance, d_S, based on an ontology that is obtained by automatic extraction from text specifications of the two databases is computed (Abadie et al., 2007). In Figure 5.1c, the basic belief assignment for the semantic criterion is modeled. Thus, if the semantic distance between the reference feature and the candidate C_i is equal to 0 (e.g., the features are similar, semantically speaking), a basic belief mass equal to 0.5 is assigned to the assumption "C_i is the homologous feature," so that this criterion does not allocate a strong belief to this candidate. On the contrary, if the semantic distance is greater than a threshold T, a basic belief mass equal to 1 is assigned to the hypothesis "C_i is not the homologous of the reference feature."

5.4.3   COMBINATION OF THE CRITERIA AND CANDIDATES

Once masses of belief have been initialized, a local approach could be carried out. It merges the criteria for each candidate individually using Dempster's rule, without taking the other candidates into account. In the third step, called the global approach, the results of the local approach are combined, again using Dempster's rule. Thus, the results for the two candidates are combined, these results are combined with the results for the third candidate, and so on and so forth. In our application, a decision has generally to be made in favor of a simple hypothesis. Within the context of the Transferable Belief Model, Smets defines and justifies the use of the "pignistic" probability decision rule, which transforms a belief function into a probability 1 (Smets and Kennes, 1994).

5.5   ANALYSIS OF RELIEF DATA AND RESULTS

We now illustrate and discuss some matching results. Our experiments concern two different spatial databases containing punctual spatial data representing relief. The approach is illustrated with an example of relief points of interest of the relief taken from two databases from IGN, the French mapping agency: BD CARTO®, considered as the reference database, and BD TOPO®, considered as the comparison database. Each database has a specific representation of the real world according to its applications, perception, and purposes. Five datasets representing five *"Départements"* in France (approximately 100 × 100 km) were used for validation. Different types of imperfections are encountered within the data. First, a point location is imprecise and different points may have a different precision. For example, the location of peaks is more precise then the location of valleys. The differences are due to the differences in the sizes of the valleys and peaks. Second, it is useful to take the toponym into account during matching. The toponym, however, as well as its location, has imperfection. For example, used names and official names may differ; the same toponym may be used at several places; or there can be various interpretations of the pronunciation, word, and character omission. The toponym may also be uncertain (e.g., "peyrelue **or** sallent") and incomplete (e.g., "col de louesque" or only "louesque"). Third, the attribute "nature" does not present the same level of details. For example, in BD CARTO®, there is a grouping of the concepts represented with the same value of the attribute nature "summit, ridge, hill," whereas in the BD TOPO®, these concepts are dissociated.

Usage of a single geometrical criterion based on the comparison of the locations does not lead to satisfactory results. The homologous feature is not always the closest one. In the same way, using only a toponym or a semantic criterion gives inconsistencies. Therefore, a combination of this information is necessary to make a sound decision.

Evaluation of the results is an important step for data matching, and the two indicators Precision and Recall are used for this purpose. Precision is the percentage of the matching links discovered by the process that are correct, compared to an interactive matching (Beeri et al., 2004). Recall is the percentage of the correct matching links (defined by an interactive matching) that are actually discovered by the automatic process. Evaluation is done in terms of the number of features.

The first and second lines of Table 5.1 show Precision and Recall values for each dataset when two (geometric and toponym) and three (geometric, toponym, and semantic) criteria are used, respectively. Both Precision and Recall improve when the three criteria are used for all datasets. These results are satisfactory and show the importance of the semantic properties during matching. In both Precision and Recall, conflict (0.7% from all matching results) is not taken into account, leading to less elevated Recall values.

Table 5.1

Qualitative Evaluation of the Matching Process

	Dataset 1		Dataset 2		Dataset 3		Dataset 4		Dataset 5	
	Two Criteria	Three Criteria	Two Criteria	Three Criteria	Two Criteria	Three Criteria	Two Criteria	Three Criteria	Two Criteria	Three Criteria
Precision	100%	100%	96.4%	99.1%	97%	98.3%	100%	100%	97%	99.7%
Recall	99%	99%	95.9%	99.1%	97%	97.5%	97.7%	97.7%	96.7%	98.7%

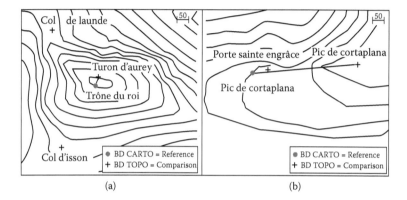

(a)   (b)

FIGURE 5.2   Results of the data matching process.

Figure 5.2 illustrates two matching results obtained using this approach. We observe from Figure 5.2a that two homologous features are matched even if the two toponyms are written in two different languages. Figure 5.2b shows that matching does not necessarily match the closest candidate.

5.6   CONCLUSION

In this chapter, we presented data matching based on the Dempster-Shafer theory. We tested it on five different datasets representing relief data. A high Precision and a Recall larger than 95% are obtained, and they increase when three criteria are combined.

ACKNOWLEDGMENT

The author would like to thank Sébastien Mustière for the reviews of this paper and the quality of his advice.

REFERENCES

Abadie, N., Olteanu, A.-M., and S. Mustière. 2007. Comparaison de la Nature d'Objets Géographiques. Paper presented at the meeting of the Ingénierie des Connaissances, Grenoble.

Appriou, A. 1991. Probabilités et Incertitudes en Fusion de Données Multi-senseurs. *Revue Scientifique et Technique de la Défense* 11:27–40.

Beeri, C., Kanza, Y., Safra, E., and Y. Sagiv. 2004. Object Fusion in Geographic Information Systems. In *Proceedings of the 30th VLDB Conference*, Toronto, Canada, pp. 816–827.

Bel Hadj Ali, A. 2001. Qualité Géométrique des Entités Géographiques Surfacique—Application à l'Appariement et Définition d'Une Typologie des Écarts Géométriques. PhD diss., Marne la Vallée.

Cohen, W. W., Ravikumar, P., and S. E. Fienberg. 2003. A Comparison of String Distance Metrics for Name-Matching Tasks. In *Proceedings of the International Joint Conference on Artificial Intelligence*, Washington, DC, pp. 73–78.

Devogele, T. 1997. Processus d'Intégration et d'Appariement de Bases de Données Géographiques-Applications à une Base de Données Routières Multi-échelles. PhD diss., Université de Versailles.

Devogele, T., Parent, C., and S. Spaccapietra. 1998. On Spatial Database Integration. *International Journal of Geographical Information Science* 12:335–352.

Gesbert, N. 2005, Formalisation des Spécifications de Bases de Données Géographiques en Vue de Leur Intégration, PhD diss., Marne la Vallée.

Levenshtein, V. 1965. Binary Codes Capable of Correcting Deletions, Insertions and Reversals. *Soviet Physics–Doklady* 10(8):707–710.

Mustière, S. 2006. Results of Experiments on Automated Matching of Networks at Different Scales. In *Proceedings of the ISPRS Workshop on Multiply Representation and Interoperability of Spatial Data*, Hanover, Germany, pp. 92–100.

Olteanu, A.-M., Mustière, S., and A. Ruas. 2006. Matching Imperfect Data. In *Proceedings of International Symposium on Spatial Accuracy Assessment in Natural Resources and Environmental Sciences*, Lisbon, Portugal, pp. 694–704.

Royère, C., Gruyer, D., and V. Cherfaoui. 2001. Data Association with Belief Theory. In *Proceedings of International Conference of Information Fusion*, Washington, DC, pp. 23–29.

Shafer, G. 1976. *A Mathematical Theory of Evidence*. Princeton University Press, Princeton, NJ.

Shereen, D., Mustiere, S., and J.-D. Zucker. 2004. How to Integrate Heterogeneous Spatial Databases in a Consistent Way? In *Proceedings of Advanced Databases and Information Systems*, Budapest, Hungary, pp. 364–378.

Smets, P., and R. Kennes. 1994. The Transferable Belief Model. *Artificial Intelligence* 66:191–234.

Voltz, S. 2006. An Iterative Approach for Matching Multiple Representations of Street Data. In *Proceedings of the ISPRS Workshop on Multiply Representation and Interoperability of Spatial Data*, Hanover, Germany, pp. 101–110.

Walter, V., and D. Fritsch. 1999. Matching Spatial Data Sets: Statistical Approach. *International Journal of GIS* 13:445–473.

Section II

Geostatistics and Spatial Data Quality for DEMs

INTRODUCTION

This section includes four chapters, addressing various issues related to geostatistics and spatial data quality of DEMs. The mathematical theories used for this section cover both geostatistics based on probability and fuzzy. The techniques used include spatial sampling schemes, DEM and risk mapping improvement, quantifying data quality by combination of plausible distribution functions, and a comparison study of maps generated by fuzzy set theory and those generated from geostatistics.

In Chapter 6, Mao, Shi, and Tian propose a spatial sampling scheme for topographic data. First, the topographic data are classified into several areas with different values of error based on experience. Second, a multiscale filtering algorithm is explored to peel off error within each area, in order to achieve the reference data for each part of the original topographic data. Third, the fitted function of each subarea of topographic data is integrated together to obtain the estimation of the error-free topographic data. This research actually proposes a spatial sampling scheme, where spatial data are treated as signals in the process. One of the advantages of the method is that the natural error of the dataset in terms of value and spatial distribution of the error can be taken into the considerations in the sampling strategy design.

In order to analyze the hydrological system and areas with potential risks of future water table depths, Manzione et al. conduct a predictive risk mapping of water table depths and quantifying the associated uncertainty, and this is presented in Chapter 7. Several factors that affect the quality are studied, including: the relatively short time series on the quality of the map, the sampling frequency and length of series on the quality of the time series models, and the number and the configuration of well locations on the quality of the model of spatial structure. DEM is used to improve the quality of the final risk maps to some extent.

Chapter 8 introduces a study on modeling data quality by using the possibility distribution by Navratil. The source of the errors is from (a) the technological restrictions, (b) the legal factor, and (c) the usage of the data. The error sources are specified by the corresponding possibility distributions. The overall data quality is quantified by the combination of influences from the three error sources; this involves a combination of the three plausible distribution functions.

From the mathematical point of view, geostatistics is based on probability while fuzzy set is based on possibility. In Chapter 9, Sunila and Kollo present a comparison study on maps generated based on fuzzy set theory and those generated from geostatistics. Based on their experiments, it is concluded that when the number of observations is adequate, the kriging technique in geostatistics provides a better fit in elevation surface modeling than the fuzzy approach.

6 A Preliminary Study on Spatial Sampling for Topographic Data

Haixia Mao, Wenzhong Shi, and Yan Tian

CONTENTS

6.1 INTRODUCTION

What about the quality of geographic maps? Can the quality of maps satisfy users' requirements? Similar questions are raised more and more frequently, because to know the quality of maps is necessary and significant for many applications. Usually, sampling and full inspection are the main methods; sampling, in particular, is a usual and effective way to solve such problems.

The estimation and assessment of the quality of topographic maps have been researched before and several standards have been established and distributed. Early in 1947, the National Map Accuracy Standards (NMAS) were worked out for well-defined points (U.S. Bureau of the Budget, 1947). In 1990 and 1998, respectively, the American Society for Photogrammetry and Remote Sensing (ASPRS) Accuracy

Standards (ASPRS Specifications and Standards Committee, 1990) and the National Standard for Spatial Data Accuracy (NSSDA) (Federal Geographic Data Committee, 1998) from the Federal Geographic Data Committee (FGDC) were established and distributed. The significance of these standards is to lay out a regulation for surveyors and inspectors to understand the quality information of the data.

In practice, the standards for accuracy estimation of maps are applied widely around the world and improved upon based on the practical situation. In China, topographic data producers check the quality of the maps in three steps, namely, interteam checking, cross-team checking, and quality inspection station accepting. Simple random sampling or judge-based sampling methods are widely used to select the sample points. However, simple random sampling alone cannot reflect the error distribution, and only judge-based sampling that results in points located in easy-going places are sampled carefully, while the points located in hard-going places are neglected and misestimated. The direct result is that the sampling scheme cannot reflect the actual accuracy of the data.

Although simple random sampling and judge-based sampling are combined and supplement each other, superfluous sample points will be selected. We can use such methods to reflect the quality of the maps to a certain extent, but we also have to consider some other factors, such as the error distribution, as well as the compromise between sampling accuracy and the cost. The sample cost has to give way to the sample accuracy, whereas the cost should not be sacrificed without limit. Consequently, to balance these factors, especially to assess the general error of the topographic data, an effective and comprehensive sample scheme is in demand.

In this chapter, a novel spatial sampling scheme is described. We considered the error of topographic data as the noise in the signal processing field and try to obtain an error surface. The extreme points on the error surface are selected as the optimal sample points. Research work on error and noise are reviewed in Section 6.2. The proposed spatial sampling method is introduced in detail in Section 6.3. An experimental simulation is provided in Section 6.4. Conclusions are drawn in Section 6.5.

6.2 A BRIEF REVIEW AND ANALYSIS

Error research on GIS raster data has been discussed for a long time. On the basis of literature review, we find that there are already some mature technologies and methodologies in use (Bogaert and Russo, 1999; Muller and Zimmerman, 1999; Banjevic and Switzer, 2002; Royle, 2002; Wiens, 2005; Zhu and Stein, 2005). On the other hand, research on GIS vector data includes the error modeling on the point, linear, and polygonal features (Shi, 1994, 1997, 1998; Shi and Liu, 2000), as well as the error propagation modeling (Shi et al., 2003, 2004). In addition, some software for analyzing, evaluating, and even visualizing the quality of GIS data has been developed (Devillers and Jeansoulin, 2006).

Besides the error model, how to lay out the spatial sampling scheme is another important part of error estimation. For example, spatial simulated annealing (SSA) was introduced to optimize spatial environmental sampling schemes, which are optimized at two levels and take into account sampling constraints and preliminary observations (Van Groenigen and Stein, 1998). A statistical model was used to select an optimal set of sample points for parameter estimation and spatial interpolation (extrapolation) (Angulo and Bues, 2001; Angulo et al., 2005).

To design a spatial sampling scheme is mainly to decide the distribution and the size of the sample points. Traditionally, well-defined or apparent feature points are always the candidates for sample points. Another important part of spatial sampling design is the sampling methods. There are many available methods, such as simple random sampling, stratified sampling, systematic sampling, etc., as well as combinations of these single sampling methods.

In a signal processing field, noise is useless information and should be removed to get a clear signal. Similarly, errors in spatial data are also useless and should be removed to get clear data. There are a lot of mature methodologies to remove the noise from the signal (Godtliebsen et al., 1997; Ishii et al., 2000; Rousseau et al., 2005; Gonzalez and Arce, 2001). Consequently, we can utilize and improve on some mature work, and remove the error like removing the noise.

In this study, we hope to analyze the positional accuracy of the topographic data quantitatively; in a further step, we hope to explore a spatial sampling scheme, using the least number of sample points while depicting the accurate error of the digital topographic data.

6.3 PROPOSED SPATIAL SAMPLING SCHEME

6.3.1 STATISTICAL DISTRIBUTION OF ERROR

Error in GIS data obeys the normal distribution, the L^P norm distribution (Meng et al., 1998; Zhou et al., 2003), and so on. Generally, the error obeying a normal distribution is the widely accepted conclusion. The French mathematician De Moivere and the German mathematician Gauss had produced landmark work on error theory and pointed out the normal distribution of the random error. Gauss especially contributed the basis for the distribution pattern of random error and brought forward the error processing method (Bolstad et al., 1990; Goodchild, 1991). In this study, suppose we have a set of surveyed data $(a_1, a_2, ..., a_n)$ and their corresponding errors $(\varepsilon_1, \varepsilon_2, ..., \varepsilon_n)$. We want to focus on finding an optimal sampling scheme assuming the error obeys the normal distribution.

6.3.2 OVERALL FLOWCHART OF THE PROPOSED SAMPLING METHOD

In this study, the error is regarded as noise in the signal. Our purpose is to bring forth a reasonable spatial sampling scheme that can perfectly reflect the spatial distribution of error and get as few sample points as possible. To solve this problem, we have two steps to take, namely, fitting the estimation of the topographic data and seeking a description of the error surface. The flowchart of this study is shown in Figure 6.1.

First, we perform a rough partition on the sample area. For example, we can divide the sample area into several subareas on the basis of error levels with difference of the data capture procedure. Second, supposing the error obeys a normal distribution, we can utilize multiscale filters to remove the different magnitudes of error for each sample area. The multiscale filtering (MSF) shown in Figure 6.1 is assigned with different parameters. The filter in the signal processing field is always used to

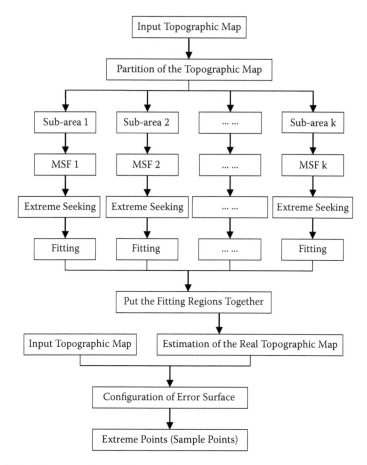

FIGURE 6.1   Flowchart for spatial sampling scheme.

filter the noise; here it is used to filter the error. Third, after removing the error, we anticipate that the reserved data are only a representation of the error-free data and we can obtain the error surface subtracted from the topographic data. Finally, if we can extract the extreme points on the error surface, we can fit a function for the real topographic data. So we put all the fitted functions from each sample area together and obtain an error-free assessment of the entire topographic data.

After filtering the error, we get an approximation of the topographic data. Obviously, the difference between the original measured data and the approximation data represents the error. Consequently, we can configure the error surface and further find out the extreme points. Finally, we can find the appropriate amount of sample points in an optimal location for this spatial sampling scheme. These key components of the overall sampling flowchart including partitioning of the topographic map, the multiscale filtering technique, extreme seeking, the fitting technique, estimation of the real topographic data, configuration of the error surface, and determination of sample points will be discussed in detail in the following subsections.

6.3.3 PARTITION AND MULTISCALE FILTER

6.3.3.1 Prior Partition of the Sample Area

Different data capturing means can bring different errors to the topographic data; it can even have different effects on the same regions in the same topographic map. Usually, the means of field survey, aerial photogrammetry and paper map digitization, are always used to produce the digital topographic map. The errors of a topographic map will differ with the change of data producers. In addition, the different degrees of complexity of the terrain, and the geographic features captured by the same means and even identical producers, result in different errors of the topographic map.

Consequently, there are a lot of situations resulting in different errors of a topographic map; fortunately, we can obtain prior knowledge about the error distribution. Although we cannot know the exact error distribution and the value of the error, it is possible for us to roughly distinguish which part of a map has high errors, which part has medium errors, and which part has low errors. For example, in aerial photogrammetry, the fringes of some features are easy to distinguish and the accuracy of the stabbing points is comparatively higher, and vice versa. As another example, in field surveys, the higher the accuracy of the traverse control survey, the lower the error the features surveyed around the traverse.

On the basis of such prior knowledge, we can plan a rough partition on the topographic data. In this study, we conduct the research on a map obtained from a field survey as an example. As the accuracies of all features vary with the accuracy of a traverse and the degree of difficulty of the feature points to be surveyed, we divide the map extent into two regions with different level of errors (Figure 6.2).

Figure 6.2 is the topographic map around the Hong Kong Polytechnic University, divided by a bold solid line. PolyU buildings inside the red line are deemed to have a similar error, and so do roads outside the bold solid line. The reason for such a prior partition is that we have laid out two traverses in each partition.

FIGURE 6.2 Prior partition for a sampling extent.

6.3.3.2 Multiscale Filter and Extreme Point Seeking on the Survey Data

Based on the accuracy of the traverse, we have divided the sample area into several subareas. Although we have no idea about the true value of the error, we can roughly distinguish the distribution of the error and then take further multiscale filters on such measuring data. As we think of the measuring data with error as the signal with noise, in this study we will use the method of error removing based on the Gauss filter.

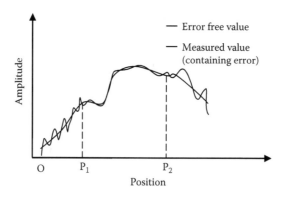

FIGURE 6.3 Error simulation on measuring data.

We need to seek the locations of the extreme points on the error surface, which are also contained in the original topographic data. The purpose of this is to fit more accurate topographic data. However, in practice, the real extreme point is covered by the error. So we want to use a particular filtering method to reveal the extreme point. Unfortunately, the real extreme point may be deduced by the filter to a certain extent, if the scale of the filter is too large. If the scale of the filter is too small, then the noise may not be removed completely. In fact, this is a classical problem of noise removing. In this section, we adopt a multiscale filtering method to resolve this problem. We describe the measured data as a continuous curve in red and the error free data in blue.

In Figure 6.3, we find the error between the interval [P1, P2] is less than that outside this interval. Actually, such a condition is accordant with the case described in Section 6.3.3.1. As the rough error distribution is known, we can process the error dividually. Here the Gaussian filter is used to remove the noise. In the spatial domain, the isotropic Gaussian can be expressed as follows:

$$G(x,y) = \frac{1}{\sqrt{2\pi\sigma^2}} e^{-\frac{x^2+y^2}{2\sigma^2}} \tag{6.1}$$

Here, σ is used to control the filter degree based on the rough error partition.

Consideri that in different subregions the error is also different. Therefore, it is not reasonable to use the Gaussian filter with a fixed scale (that is, the value of σ is given and fixed) to filter the whole region. A natural idea is to use different scale filters for different subregions, which is the multiscale filtering technique and is designed as follows:

Step 1: The original topographic data are divided into K different subregions and denoted by r_i^0 ($i = 1, ..., K$). Their corresponding rough errors are denoted by E_i ($i = 1, ..., K$), and we suppose $E_1 > E_2 > ... > E_K$ without loss of generality.

Step 2: Given the filter scales in Equation 6.1 for the above subregions σ_i^0 ($i = 1, ..., K$), we let $\sigma_1^0 > \sigma_2^0 > ... > \sigma_K^0$.

Step 3: The subregions are filtered by using the filters given in Equation 6.1 with the scales given in step 2 and the filtered result is denoted by r_i^1 ($i = 1, ..., K$).

Step 4: Set another groups with scales as $\sigma_i^2 = \sigma_i^0/2$ ($i = 1, 2, ..., K$), the same procedure as in step 3 is performed, and the corresponding results are denoted by r_i^2 ($i = 1, ..., K$).

Step 5: Set and repeat another group with the scales as

$$\sigma_i^n = \frac{\sigma_i^0}{n+1} \ (i = 1, 2, ..., K);$$

the corresponding filtered result is then represented as r_i^{n+1} ($i = 1, ..., K$).

Step 6: Extract the extreme points of $r_i^1, r_i^2, ..., r_i^{n+1}$ ($i = 1, ..., K$) and check the extreme points that have disappeared. Those across many layers are selected as robust extreme points. And then the corresponding locations are recorded.

Remark 1: The reason for step 6 is based on the understanding that the robust extreme point has a strong correlations with the neighboring point and will disappear slowly. Those points that disappeared quickly (that is, remained in a few layers) are not the real extreme points. The geometric illustration in the one-dimensional case is given in Figure 6.4. From Figure 6.4, we see that the points labelld by star symbols ("*") have weakened slowly, while the points labeled by cross symbols ("+") have dis-

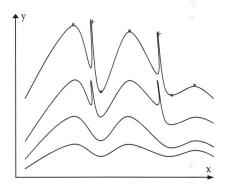

FIGURE 6.4 Extreme point across layers.

appeared quickly, so we have reason to conclude that the star points are really the extreme points while the cross points are false extreme points or noise points.

Remark 2: In practice, the number of the filtered layers should be considered. That is, the filtered time used must be determined. There are two simple strategies for this problem. One is to give a filtered time in advance, and another is that the filtering time can be given by comparing two adjacent filtered results; if the difference of their extreme points is smaller than a given threshold, then the filtered procedure can be phased out. Certainly, the filtered times in each subregion may be different.

6.3.4 FITTING TECHNIQUE AND ESTIMATION OF REAL TOPOGRAPHIC DATA

6.3.4.1 Fitting Technique

We proposed different filters for different subareas and obtained the function curve of the measuring data after filtering the error. After the Gauss filtering, we were able to get several peak points on the curve filtered, which represent the truer situation of the real world contrasted against the measuring data with error.

Although some extreme points on the measured topographic data will be weakened, and may even be eliminated after the filtering, we can say the remaining extreme points are the real extremes of the topographic data with high confidence.

After we extract several extreme points from the measured topographic data in each subregion, the next task is to fit these extreme points into a dense point set by a particular method. In fact, there are many methods that can be employed, such as interpolation techniques and other fitting methods. For the sake of simplicity, the bicubic interpolation method that is used frequently in many applications is adopted. Its formula is expressed as

$$
\begin{aligned}
z = a_1 + a_2 x + a_3 y + a_4 x^2 + a_5 xy + a_6 y^2 \\
+ a_7 x^3 + a_8 x^2 y + a_9 xy^2 + a_{10} y^3
\end{aligned}
\tag{6.2}
$$

6.3.4.2   Estimation of the Real Topographic Data

By the steps introduced in Section 6.3.3.2, approximated subtopographic data are obtained. As mentioned before, on the basis of prior knowledge about the rough error distribution on the topographic data, we have divided the map into several subareas. In order to assess the integrated accuracy of the topographic data, we have to put all the fitted functions together to represent the integrated representative expression of the whole topographic data.

Remember that the above interpolation method is implemented on each of the subregions; then, for each location on the boundary of each subregion, there are two interpolated results. Which interpolated result should be accepted? A simple method is to take the average value of these two results. An illustration for this point is given in Figure 6.5.

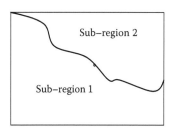

FIGURE 6.5   Illustration of boundary point.

6.3.5   Construction of Error Surface

After fitting the true function for the real world, denoted as $F_1(x)$, and since we already have the function of the measured data, denoted as $F_0(x)$, we can then obviously obtain the error surface for the research data, which can be denoted as $E = F_0(x) - F_1(x)$.

During the spatial sample on the topographic data, we would like to select the points that are representative of the population to the most extent. To satisfy this, such points in the sample areas will be characteristic of the large change in the error value. We would not like to use the maximum error or the minimum error to represent the whole error of the topographic data, as it will magnify or shorten the actual error and cannot reflect the true situation.

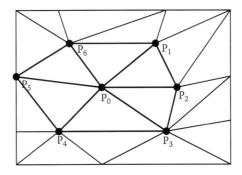

FIGURE 6.6 Extraction of peaks on error surface.

We have thought the error is similar to some extent and we have treated the error in this way; in another words, the error in one area has the equivalent floatability. If we can find the error with the obvious change corresponding to the adjacent one, we can describe the amount of the error more accurately for the topographic data.

The function description of the error is obtained by the method introduced above. The problem of finding an error with a sharp change is equal to the problem of finding the peak points on the error surface. Consequently, we can select such points as our sample points. So the key problem is how to select the peak points on the error surface.

A method can be designed as follows:

Step 1: For the points located on the error surface, construct a Delauney triangle network.
Step 2: For a point p, compare it with the points linked to this point; if it is the smallest or the biggest, then it is a peak, otherwise it is not a peak.

An example is given in Figure 6.6. If we want to determine whether the point P_0 is the peak point, the points P_i $(i = 1, ..., 6)$ should be compared with the point P_0; if the point P_0 is the smallest or the biggest among these six points, then the point P_0 is a peak point. This procedure is implemented over each of the points, and then all the peak points can be extracted.

The peak points extracted are considered to be representative of the accuracy of the topographic data. As the error contributions from such points can be used to describe the integrated accuracy, they can also be used to decide the selection of the sample points in this study. In other words, if we can find out these peak points, the spatial sampling scheme can be decided.

6.4 SIMULATION AND DISCUSSIONS

In this section, a simulation will be conformed to illustrate the effectiveness of the proposed method. A function used for testing is given as follows:

$$f(x) = \begin{cases} \sin 4x, & 0 \le x < 3\pi \\ \sin 8x, & 3\pi \le x \le 5\pi \end{cases} \tag{6.3}$$

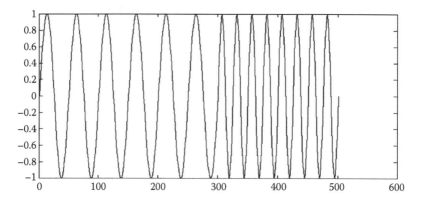

FIGURE 6.7   The function of $f(x)$.

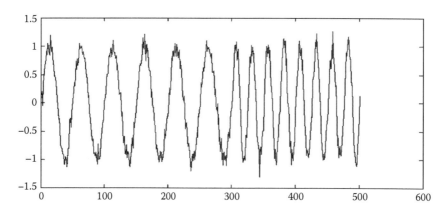

FIGURE 6.8   The function of $F(x)$.

Suppose it is polluted by a Gaussian error and changed as

$$F(x) = f(x) + \text{Gaussian} \tag{6.4}$$

Here, $F(x)$ and Gaussian are the observed data and Gaussian error, respectively. Equation 6.3 and Equation 6.4 are shown in Figure 6.7 and Figure 6.8, respectively.

In this case, the whole region is divided into two parts: the first part is $(0,3\pi)$ and the second part is $[3\pi,5\pi]$. The initial scales of multiscale Gaussian filters used for the first part are set $\sigma_1^0 = 1.2$ and $\sigma_2^0 = 2.75$, respectively. And the times for multiscale filtering are 3 and 5, respectively. After the multiscale filtering procedure, the extreme points are determined and shown in Table 6.1.

By the extreme points, the fitting technique is implemented and the result is shown in Figure 6.9.

After calculating, the error surface and its extreme points are as shown in Table 6.2.

TABLE 6.1
Position of the Extreme Points of the Filtered Data

	1	2	3	4	5	6	7	8	9	10	11	12	13	14
Maximum	18	68	117	169	216	268	309	334	359	384	409	433	459	484
Minimum	42	92	142	193	243	293	321	346	371	397	422	447	471	497

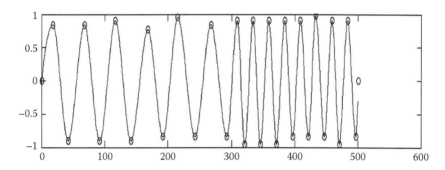

FIGURE 6.9   Fitting results of the two parts.

TABLE 6.2
The Extreme Points of the Error Surface

	1	2	3	4	5	6	7	8	9	10	11	12	13	14
Maximum	13	63	113	163	213	264	307	332	357	382	407	432	457	482
Minimum	38	88	138	188	238	289	320	345	370	395	420	445	470	495

The points listed in Table 6.2 are the sampling points. To illustrate the validity of selecting the sampling points, we can obtain a surface by fitting the sampling points and comparing them with the original surface expressed by Equation 6.3 (Figure 6.10). It can be found that these two surfaces are very close (the original surface is in red, the fitting surface is in blue).

6.5   CONCLUSIONS

In this chapter, we propose a novel spatial sampling scheme. The proposed approach aims at sampling as few points as possible, as well as reflecting the true error of the topographic data. Based on the proposed method, not only the appropriate number but also the location of the sample points have been decided. Compared with existing spatial sampling methods for spatial data in GIS, the approach presented in this chapter provides a possible method to give a generic solution for error removing.

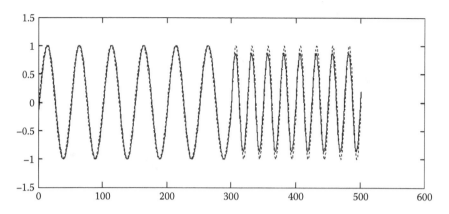

FIGURE 6.10   Fitting surface by the sampling points and the original surface.

The novel idea of this method is that the error of the topographic data is regarded as noise in the signal viewpoint. On the basis that we suppose that the error obeys the Gaussian distribution, we try to remove the error as if removing the noise using different Gaussian filters in signal processing. The error-free data can help us to obtain the error surface. The optimal sample points would be selected based on the search of the extreme points on the error surface. Furthermore, the error surface is the right representation of the error distribution of topographic data.

An important application of this method is to evaluate the quality of the topographic data. When we only have the measured topographic data at hand and prior knowledge of the rough error distribution of the topographic data, we can design a good solution for spatial sampling to select the right sample points to reflect the accuracy and the error distribution of the topographic data.

ACKNOWLEDGMENTS

This work was supported by funds from The Hong Kong Polytechnic University (Project No. G-YF24 and RGMG) and the National Natural Science Foundation (No.: 40629001).

REFERENCES

American Society for Photogrammetry and Remote Sensing (ASPRS) Specifications and Standards Committee, 1990. ASPRS Accuracy Standards for Large-Scale Maps. *Photogrammeric Engineering and Remote Sensing*, 56(7), pp. 1068–1070.

Angulo, J. M. and Bueso, M. C., 2001. Random perturbation methods applied to multivariate spatial sampling design. *Environmetrics*, 12, pp. 631–646.

Angulo, J. M., Ruiz-Medina, M. D., Alonso, F. J., and Bueso, M. C., 2005. Generalized approaches to spatial sampling design. *Environmetrics*, 16, pp. 523–534.

Banjevic, M. and Switzer, P., 2002. Bayesian network designs for fields with variance as a function of the location. *Proceedings of the 2002 JSM Conference*. Section on Statistics and the Environment, New York, USA.

Bogaert, P. and Russo, D., 1999. Optimal spatial sampling design for the estimation of the variogram based on a least squares approach. *Water Resources Research*, 35, pp. 1275–1289.

Bolstad, P. V., Gessler, P., and Lillestand, T. M., 1990. Positional uncertainty in manually digitized map data. *International Journal of Geographical Information System*, 5, pp. 159–168.

Devillers, R. and Jeansoulin, R., 2006. Introduction, In *Fundamentals of Spatial Data Quality*. R. Devillers and R. Jeansonlin, eds, ISTE, London, UK, pp. 17–20.

Federal Geographic Data Committee, 1998. Part 3, National Standard for Spatial Data Accuracy, Geospatial Positioning Accuracy Standards. FGDC-STD-007.3-1998: Federal Geographic Data Committee, Washington, DC, 24 p.

Godtliebsen, F., Spjtvoll, E., and Marron, J. S., 1997. A nonlinear gaussian filter applied to images with discontinuities, *Journal of Nonparametric Statistics*, 8(1), pp. 21–43.

Gonzalez, J. G. and Arce, G. R., 2001. Optimality of the myriad filter in practical impulsive-noise environments, *IEEE Transactions on Signal Processing*, 49(2), pp. 438–441.

Goodchild, M. F., 1991. Keynote address: Symposium on Spatial Database Accuracy. In *Proceedings of Symposium on Spatial Database Accuracy*, edited by G. J. Hunter. Department of Geometrics, University of Melbourne, Australia, pp. 1–16.

Ishii, H., Kimura, T., Sone, M., and Taguchi, A., 2000. An edge detection from images corrupted by mixed noise using fuzzy inference. *Electronics and Communications in Japan Part III—Fundamental Electronic Science*, 83(8), pp. 39–50.

Meng, X. L., Shi, W. Z., and Liu, D. J., 1998. Statistical tests of the distribution of errors in manually digitized cartographic lines. *Geographic Information Sciences*, 4, pp. 52–58.

Muller, W. G. and Zimmerman, D. L., 1999. Optimal designs for variogram estimation. *Environmetrics*, 10, pp. 23–37.

Rousseau, D., Varela, J. R., Duan, F., and Chapeau-Blondeau, F., 2005, Evaluation of a nonlinear bistable filter for binary signal detection, *International Journal of Bifurcation and Chaos*, 15(2), pp. 667–679.

Royle, J. A., 2002. Exchange algorithms for construction large spatial designs. *Journal of Statistical Planning and Inference*, 100(2), pp. 121–134.

Shi, W. Z., 1998. A generic statistical approach for modeling error of geometric features in GIS. *International Journal of Geographical Information Science*, 12, pp. 131–143.

Shi, W. Z., 1997. Statistical modeling uncertainties of three-dimensional GIS features. *Cartography and Geographic Information Systems*, 24, pp. 21–26.

Shi, W. Z., 1994. Modeling positional and thematic error in integration of GIS and remote sensing. *ITC Publications*, Enschede, 22, p. 147.

Shi, W. Z. and Liu W. B., 2000. A stochastic process-based model for the positional error of line segments in GIS. *International Journal of Geographical Information Science*, 14, pp. 51–66.

Shi, W. Z., Cheung, C. K., and Tong, X. H., 2004. Modelling error propagation in vector-based overlay analysis. *ISPRS Journal of Photogrammetry & Remote Sensing*, 59, pp. 47–59.

Shi, W. Z., Cheung, C. K., and Zhu, C. Q., 2003. Modelling error propagation in vector-based buffer analysis. *International Journal of Geographical Information Science*, 17(3), pp. 251–271.

U.S. Bureau of the Budget, 1947. United States National Map Accuracy Standards. U.S. Bureau of the Budget, Washington, DC.

Van Groenigen, J. W. and Stein, A. 1998. Constrained optimization of spatial sampling using continuous simulated annealing. *Journal of Environmental Quality*, 27(5), pp. 1078–1086.

Wiens, D. P., 2005. Robustness in spatial studies I: Minimax design. *Environmetrics*, 16(2), pp. 191–203.

Zhu, Z. and Stein, M., 2005. Spatial sampling design for parameter estimation of the covariance function. *Journal of Statistical Planning and Inference*, 134(2), pp. 583–603.

Zhou, S. J., Lu, T. D., Guan, Y. L., and Zang D. Y, 2003. The analytical collocation of the measurement errors distribution. *Acta Metrologica Sinica*, 24(3), pp. 250–253.

7 Predictive Risk Mapping of Water Table Depths in a Brazilian Cerrado Area

Rodrigo Manzione, Martin Knotters, Gerard Heuvelink, Jos von Asmuth, and Gilberto Câmara

CONTENTS

7.1 INTRODUCTION

The Brazilian Cerrados extend from the northern margins of the Amazon rain forests to outliers on the southern borders of the country with extensions into Paraguay and Bolivia. Before recent human disturbances, Cerrados probably covered well over 2 million km^2, equivalent to 23% of the Brazilian territory (Jepson, 2005). The Cerrados are woody savannas, which vary from nearly treeless grasslands to what is almost a woodland of semideciduous trees (Furley, 1999). The structure and physiognomy of the Cerrados reflect climatic and other environmental variables. The total annual rainfall over the central area of Brazilian savannas ranges from 1300 to 1600 mm, concentrated in six to seven months between October and April. The rest of the year is characterized by a pronounced dry season. So, natural vegetation has developed adaptations to the seasonal wet rainfall, acid soils, and aluminium toxicity, and features protective devices against fire. Plants metabolize throughout the year, drawing on soil water reserves, and can withstand short-lived fires.

The expansion of modern agriculture over the past three decades and the establishment of large-scale ranches have affected the characteristic landscape and ecosystem of the Cerrados. These areas became Brazil's most important grain belt, facing deforestation rates much higher than in the Amazon rain forest (Oliveira et al., 2005). A complex wood/grass ecosystem was substituted by shallow-rooted monocultures, which are less well adapted to drought. Their need of water supply by irrigation techniques is likely to change the hydrological system (Klink and Moreira, 2002). With irrigation increasing, lowering of the water table occurs, and so risks of water shortage appear. The preservation of these resources is important because the Cerrado presents itself as an important receiver, keeper, and disperser of water to the three main hydrographic basins of the country (Amazon Basin, Prata Basin, and São Francisco Basin). The negative impact to its environment might be reflected in other Brazilian biomes.

Knowledge about the spatio-temporal dynamics of the water table is important to optimize and balance the interest of economical and ecological land use purposes (Von Asmuth and Knotters, 2004). In hydrology, water table dynamics are modeled in several ways. Many authors refer to transfer function–noise (TFN) models to describe the dynamic relationship between precipitation and water table depths (Box and Jenkins, 1976; Hipel and McLeod, 1994; Tankersley and Graham, 1994; Van Geer and Zuur, 1997). Basically, these models can be seen as multiple regression methods, where the system is seen as a black box that transforms series of observations on the input (the explanatory variables) into a series of output variables (the response variables). The parameters of time series models address the temporal variation of the water table depths, while the spatial component can be accessed by regionalizing the outputs using ancillary information related to the physical basis of these models (Knotters and Bierkens, 2000, 2001). This approach can be used to describe the spatio-temporal variation of the water table depths. It is assumed that the spatial differences in water table dynamics are determined by the spatial variation of the system properties, while its temporal variation is driven by the dynamics of the input into the system.

To link the response characteristics of the water table system to the dynamic behavior of the input, Von Asmuth et al. (2002) presented a method based on the use of a transfer function–noise model in continuous time, the so-called PIRFICT model. An important advantage of the PIRFICT model as compared to discrete-time TFN models is that it can deal with input and output series that have different observation frequencies and irregular time intervals. Using a time series model, it is possible to simulate over periods without observations, as long as data on explanatory series are available. For instance, long series (say 30 years' length) on precipitation and evapotranspiration can be assumed to represent the prevailing climatic conditions (Knotters and Van Walsum, 1997). Alternatively, series generated by climatic models might be applied as inputs to the PIRFICT model. From the simulated realizations, statistical characteristics of future water table dynamics can be calculated, such as mean, standard deviation, and limits of prediction.

The aim of this study is to estimate and map the expected water table depths in a watershed located at the Brazilian Cerrados. These estimations are made for a specific date in any future year, given the prevailing hydrological and climatic conditions, in order to support decision making in long-term water policy and indicate areas with potential risks of future water shortage and shallow water table depths. In addition, the uncertainty associated with the estimated water table depth is quantified simulating realizations of the stochastic processes.

7.2   MATERIALS AND METHODS

7.2.1   Study Area—The Jardim River Watershed

The Jardim River watershed is a representative Cerrado area in the eastern part of the Brazilian Federal District, latitudes 15°40′S and 16°02′S and longitudes 47°20′W and 47°40′W. The dry and wet seasons are well defined, with the rainfall concentrated between October and April. During the past years, almost all natural vegetation of the area was replaced by agricultural crops, and the use of irrigation systems has substantially increased in this region during the past years. The main cultivations in the area are soybeans, cotton, and corn crops, as well as pasture and horticultural crops. To monitor the water table depths, 37 wells were drilled (Figure 7.1). The locations were selected purposively, to cover the range of soil types in the area (Lousada, 2005). The water table was observed semimonthly from October 11, 2003, until October 06, 2006, resulting in series of 50 more or less regularly spaced semimonthly observations, during a period of 1092 days. Series of 33 years' length of precipitation and potential evapotranspiration were available from a climate station close to the basin. The period covered is from 1974 until 1996 with a monthly frequency, and from 1996 until March 2007 with a daily frequency. Ancillary information related to local geomorphology was derived from a digital elevation model (Figure 7.1) with 15 m resolution (Lousada, 2005).

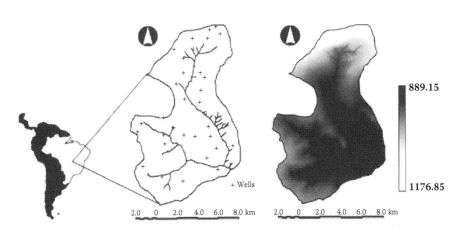

FIGURE 7.1   Jardim River watershed and location of the observation wells (+) (left) and the Digital Elevation Model (right).

7.2.2   MODELING WATER TABLE DEPTHS—THE PIRFICT MODEL

The behavior of linear input-output systems can be completely characterized by their impulse response (IR) function (Ziemer et al., 1998; Von Asmuth et al., 2002). The response of water table depth to impulses of precipitation series can be modeled by a transfer function–noise (TFN) model (Box and Jenkins, 1976; Hipel and McLeod, 1994; Von Asmuth and Knotters, 2004). For water table depths, the dynamic relationship between precipitation and water table depth can also be described using physical mechanistic groundwater flow models. However, by using much less complex TFN models, predictions of the water table depth can be obtained that are often as accurate as those obtained by physical mechanistic modeling (Von Asmuth and Knotters, 2004).

The basic idea behind TFN modeling is to split the observed series (output) into a sum of transfer components related to known causes (inputs) that influence the temporal variation of the output and an unknown noise component. TFN models are often applied to distinguish between natural and man-induced components of groundwater series (Van Geer and Zuur, 1997). In TFN models one or more deterministic transfer components and a noise component are distinguished. These components are additive. Each transfer component describes the part of the water table depth that can be explained from an input by a linear transformation of a time series of this input. The noise component describes the autoregressive structure of the differences between the observed water table depths and the sum of the transfer components. The input of the noise model is a series of independently and identically distributed disturbances with zero mean, and finite and constant variance, i.e., white noise. The PIRFICT model, introduced by Von Asmuth et al. (2002), is a specific type of TFN model and an alternative to discrete-time TFN models. In the PIRFICT model a block pulse of the input is transformed into an output series by a continuous-time transfer function. The coefficients of this function do not depend on the observation frequency. The following single-input continuous TFN model can be used to model the relationship between water table dynamics and precipitation surplus/deficit. For the simple case of a linear, undisturbed phreatic system that is influenced by precipitation surplus/deficit only (Von Asmuth et al., 2002),

$$h(t) = h^*(t) + d + r(t) \tag{7.1}$$

$$h^*(t) = \int_{-\infty}^{t} p(\tau)\theta(t - \tau)\partial\tau \tag{7.2}$$

$$r(t) = \int_{-\infty}^{t} \phi(t - \tau)\partial W(\tau) \tag{7.3}$$

where:
$h(t)$ = observed water table depth at time t $[T]$;
$h^*(t)$ = predicted water table depth at time t attributed to the precipitation surplus/deficit, relative to d $[L]$;

d = level of $h^*(t)$ without precipitation, or in other words the local drainage level, relative to ground surface $[L]$;

$r(t)$ = residuals series $[L]$;

$p(t)$ = precipitation surplus/deficit intensity at time t $[L/T]$;

$\theta(t)$ = transfer Impulse Response (IR) function $[-]$;

$\phi(t)$ = noise IR function $[-]$;

$W(t)$ = continuous Wiener white noise process $[L]$, with properties $E\{dW(t)\} = 0$, $E[\{dW(t)\}^2] = dt$, $E[dW(t_1)dW(t_2)] = 0$, $t_1 \neq t_2$.

The local drainage level d is obtained from the observations as follows:

$$d = \frac{\sum_{i=0}^{N} h(t_i)}{N} - \frac{\sum_{i=0}^{N} h^*(t_i)}{N} - \frac{\sum_{i=0}^{N} r(t_i)}{N} \tag{7.4}$$

with N the number of water table depth observations.

TFN models are identified by choosing mathematical functions that describe the IR and the autoregressive structure of the noise. This identification can be done in two ways: first, iteratively, using correlation structures in the available data and model diagnostics, and, second, physically, based on insight into the behavior of the analyzed system. Here, the second approach is followed. $\theta(t)$ is a Pearson type III distribution function (PIII df; Abramowitz and Stegun, 1964). Because of its flexible nature, this function adequately models the responses of a broad range of ground-water systems. Under the assumption of linearity, the deterministic part of the water table dynamics is completely determined by the IR function moments. In this case, based on Von Asmuth et al. (2002), the parameters can be defined as

$$\theta(t) = A \frac{a^n t^{n-1} e^{-at}}{\Gamma(n)}$$

$$\phi(t) = \sqrt{2\alpha\sigma_r^2}\, e^{-\alpha t} \tag{7.5}$$

where A, a, n, are the parameters of the adjusted curve; $\Gamma(n)$ is the Gamma function; α determines the decay rate of $\emptyset(t)$; and σ_r^2 is the variance of the residuals.

Equation 7.5 and its parameters have a physical meaning that is described in Von Asmuth and Knotters (2004). The physical basis of the PIII df lies in the fact that it describes the transfer function of a series of linear reservoirs (Nash, 1958). The parameter n denotes their number and a equals the inverse of the reservoir coefficient normally used. As Knotters and Bierkens (2000) explain, a single linear reservoir (a PIII df with $n = 1$) equals a simple physical model of a one-dimensional soil column, discarding lateral flow and the functioning of the unsaturated zone. The extra parameter A is necessary because in the case of Equation 7.5, where a precipitation and evapotranspiration series are transformed into a water table depths series, the law of conservation of mass does not apply.

The PIII df has been shown to be able to model fluctuations of water table closely and comparably to the Box-Jenkins TFN models with many more parameters (Von Asmuth et al., 2002). The parameter A is related to the local drainage resistance (the area of the IR function equals the ratio of the mean height of the water table to the mean water table recharge), while Aa is determined by the storage coefficient of the soil and n as the convection and dispersion time of the precipitation through the unsaturated zone. However, care should be taken when interpreting the parameters of the PIII df, or any other time series model for that matter, in the physical sense, because of their lumped and empirical nature (Von Asmuth and Knotters, 2004).

After the selection of an IR function that represents the underlying physical process, the available time series have to be transformed to continuous series. First, in order to characterize the variability of precipitation and evaporation, we rely on a simple but effective method to estimate the average precipitation surplus intensity and its annual amplitude. When precipitation surplus data are only available at discrete intervals, the continuous series $p(\tau)$ cannot be reconstructed exactly, but it can be approximated by assuming that the distribution of $p(\tau)$ is uniform during the period t_{pb} to t_{pe} (Ziemer et al., 1998). The average level of the precipitation surplus is obtained as

$$\bar{p} = \frac{\displaystyle\int_{pb}^{pe} p(\tau)\partial\tau}{t_{pe} - t_{pb}} \tag{7.6}$$

with t_{pb} and t_{pe} denoting the start and end of the period over which the meteorological characteristics are calculated. Next, time is split into year Y and the Julian day D, and the precipitation surplus is averaged over Y, which effectively filters out its yearly course:

$$\tilde{p}(D) = \frac{\displaystyle\sum_{Y_{pb}}^{Y_{pe}} p(Y,D)}{Y_{pe} - Y_{pb}}, \quad 1 \le D \le 365 \tag{7.7}$$

Because the temperature largely determines the annual evaporation cycle and is more or less harmonic, so is the precipitation surplus, and the annual amplitude can be obtained by matching a sine to the yearly course (Von Asmuth and Knotters, 2004).

Using Equation 7.6, the transfer model (Equation 7.2) can be evaluated using a block response (BR) function $\Theta(t)$. The BR function can be obtained by convoluting the IR function with a block pulse of precipitation surplus with unit intensity over a period Δt, as follow:

$$\Theta(t) = \int_{t-\Delta t}^{t} \theta(\tau)\partial\tau \tag{7.8}$$

Because $\Theta(t)$ is a continuous function, $h*(t)$ itself is also continuous, and for every observation of $h(t)$ a sample of the residual series $r(t)$ can be obtained. Next, the noise model (Equation 7.3) is evaluated in order to obtain a series of innovations $v(t)$. Following Von Asmuth et al. (2002), to evaluate the noise model without having to use a Kalman filter (which is computationally expensive) we will derive a direct relation between the residuals $r(t)$ and the innovations $v(t)$. Consider the series $v(t)$ as the nonequidistantly sampled changes in the solution to the stochastic integral describing the residual series

$$v(t) = \int_{t-\Delta t}^{t} \phi(t-\tau)\partial W(\tau)$$

(7.9)

with $\varnothing(t)$ from Equation 7.5 as the noise IR function, we can rewrite Equation 7.3 as

$$r(t) = e^{-\alpha\Delta t}r(t-\Delta t) + \int_{-\infty} \sqrt{2\alpha\sigma_r^2}e^{-\alpha(t-\tau)}\partial W(\tau)$$

(7.10)

which is known as an Ornstein-Uhlenbeck process (Uhlenbeck and Ornstein, 1930; Gardiner, 1994). Combining Equation 7.9 and Equation 7.10, we obtain the innovation series calculated from the available data:

$$v(t) = r(t) - e^{-\alpha\Delta t}r(t-\Delta t)$$

(7.11)

Subsequently, an estimative of the model parameters set $\beta = (A, a, n, \alpha)$ is made with the aid of a Levenberg-Marquardt algorithm, which numerically minimizes a weighted least-squares criterion based on the likelihood function of the noise model. Finally, the accuracy and validity of the model are checked using the auto- and cross-correlation functions of the innovations, the covariance matrix of the model parameters, and the variance of the IR functions. For a complete overview of the PIRFICT model formulation, applications, and study cases, refer to Von Asmuth and Maas (2001), Von Asmuth et al. (2002), Von Asmuth and Knotters (2004), and Von Asmuth and Bierkens (2005).

The PIRFICT model was applied in this study because the model can describe a wide range of response times with differences in sampling frequency between input series and output series. For the Cerrado situation, it is particularly interesting because different behaviors of water tables can be found even in small catchments. Being the most important driving forces of water table fluctuation, precipitation and evapotranspiration are incorporated as exogenous variables into the model.

7.2.3 UNCERTAINTY MEASURES—SIMULATING WATER TABLE DEPTHS

Time series models using precipitation surplus/deficit as the input variables, calibrated on time series of water table depths with limited years, enable us to simulate series of extensive length (Knotters and Van Walsum, 1997). From extensive series, statistics of WTD can be estimated. These will represent the prevailing hydrological and climatic conditions rather than specific meteorological circumstances during the monitoring period of water table depths.

The simulation of water table depths presented here is based on a time frequency filtering of the PIRFICT model performed as a convolution in the time frequency domain. This operation considers the shape of the PIII df adjusted from the parameters of each model. Since water table depths indicate the output of a time varying system, the interaction of the precipitation input signal and the system can be regarded as an operation in the time frequency domain between the time frequency expansion of the signal and the time frequency response of the system.

These models contain a dynamic component, describing the dynamic relationship between the input and the output, either physically or empirically. But variation of the water table cannot be completely explained from the precipitation and evapotranspiration series. So, the models must contain a noise component, which describes the part of water table fluctuation that cannot be explained with the used physical concepts or empirically from the input series. The unexplained part (noise component) has to be taken into account in the simulation procedure, since we are interested in the statistics of extremes, like the probabilities that critical levels are exceeded. Details about simulation can be found in Hipel and McLeod (1994). Here, we evaluate the uncertainty of the estimations of water table depths simulating 1000 realizations of the PIRFICT model in order to calculate probability distribution functions (PDFs) of the target variable. The uncertainty is taken into account by the probability thresholds established to risk management, generated from the PDFs. The following steps are followed:

- After modeling the relationship between precipitation surplus/deficit and water table depths using the PIRFICT model, series of water table depths are extrapolated to a length of 30 years. It is assumed that the average weather conditions during the last 30 years represent the prevailing climate. As a result, deterministic series of predicted water table depths are generated.
- Realizations of the noise process are generated by stochastic simulation and next added to the deterministic series, resulting in realizations of series of WTD. Realizations of the noise process can be generated either by random sampling from a normal distribution with zero mean and residual variance, or by resampling from the fitted residuals.
- From the previous step, N realizations of the stochastic simulation are generated. With a probability density function of the distribution of water table depths for each t instant, the statistics representing the prevailing hydrologic conditions can be calculated.

In this study, we applied random sampling from a normal distribution. We decided to calculate statistics about WTD for a t that equals October 1. This is a reasonable date at which cultivations in the Cerrado region start. The rainy season usually starts around this period. Farmers start cultivations just after the first rains of the season.

To enable risk management of water table depths for October 1, we calculated two levels of probability. First, a 5% probability level was considered as a measure for risk of water shortage. With this results, we can say that the area has just 5% probability to have water levels deeper than the values of the resulting map, and 95% higher water levels. The limits established for risks of water shortage at October 1 were the depths of the wells, with dry wells characterizing a scenario of water shortage in the area. It can be a problem during the beginning of the plants' development, affecting water availability and resulting in production losses.

Second, a 95% probability level was considered as a measure for risk of shallow water tables. Shallow water tables can be a problem in the beginning of the rainy season because it can make machinery impossible, affecting plowing and planting operations. It can also influence soil conditions, decreasing soil redox potential, increasing pH in acid soils, and decreasing in alkaline soils and increasing conductivity and ion exchange reactions. These modifications in the system might influence plant growth, by affecting the availability on toxicity of nutrients, regulating uptake in the rhizosfere. With these results, we can say that the area has just 5% probability of having water levels higher than the values of the resulting map, and 95% deeper water levels. The limits established for risks of shallow water table depths at October 1 were 0.5 m below the ground surface.

7.2.4   RISK MAPPING—REGIONALIZING SIMULATED WATER TABLE DEPTHS

The results of WTD simulations are interpolated spatially using universal kriging (Matheron, 1969; Pebesma, 2004). The use of exhaustive information on elevation is interesting, because it can decrease the variance and the uncertainty in the spatial prediction model. Also, when the ancillary information is physically related to the target variable, it can incorporate physical meaning to the predictions. In our case, areas with relatively low elevation and close to drainage devices present relatively shallow water tables, whereas in areas with relatively high elevation and far from drainage devices, the water table is relatively deep (Furley, 1999).

Incorporating DEM as drift (Odeh et al., 1994; Knotters et al., 1995) in the spatial prediction model works as follows. Let the simulated water table depths be given as $z(x_1)$, $z(x_2)$, ..., $z(x_n)$, where x_i is a (two-dimensional) well location and n is the number of observations (i.e., $n = 37$). At a new, unvisited location x_0 in the area, $z(x_0)$ is predicted by summing the predicted drift and the interpolated residual (Odeh et al., 1994; Hengl et al., 2004):

$$\hat{z}(x_0) = \hat{m}(x_0) + \hat{e}(x_0) \tag{7.12}$$

where the drift m is fitted by linear regression analysis, and the residuals e are interpolated using kriging:

$$\hat{z}(x_0) = \sum_{k=0}^{p} \hat{\beta}_k \cdot q_k(x_0) + \sum_{i=1}^{n} w_i(x_0) \cdot e(x_i); \tag{7.13}$$
$$q_0(x_0) = 1$$

Here, the β_k are estimated drift model coefficient $q_k(x_0)$ is the kth external explanatory variable (predictor) at location x_0, p is the number of predictors, $w_i(x_0)$ are the kriging weights, and $e(x_i)$ are the zero-mean regression residuals. In this case, for WTD, the model was formulated as follows:

$$WTD(x_0) = \beta_0 + \beta_1 \cdot EV(x_0) + e(x_0) \tag{7.14}$$

where EV is the elevation value for each location and e is a zero-mean spatially correlated residual. Its spatial correlation structure is characterized by a semivariogram.

7.2.5 SUMMARY OF THE METHOD

The methods described in Subsection 7.2.2 to Subsection 7.2.4 are now summarized as follows:

1. Calibrate the PIRFICT model.
2. Stochastically simulate the WTD series by using the PIRFICT model ($N = 1000$) and input series of 30 years' length.
3. Pick all selected date (October 1) values of WTD generated by stochastic simulation.
4. Create a probability distribution function (PDF) of these values.
5. Select the percentile values (5th and 95th) for the WTD.
6. Repeat steps 1–5 for all wells.
7. Model the spatial structure of the percentile values with geostatistics techniques.
8. Finally, use these values of the WTD to create risk maps of water levels that could be exceeded at the selected date with 5 and 95% probability.

7.3   RESULTS

7.3.1 TIME SERIES MODELING

Due to spatially varying hydrological conditions, a wide range of calibration results was found for the 37 observed wells. Table 7.1 summarizes the results of the time series modeling.

The percentage of variance indicated a good fit of the PIRFICT model to the data. Low percentages might be caused by errors in the data or lack of data, or possibly because other inputs that affect the groundwater dynamics are not incorporated into the model (Von Asmuth et al., 2002). The parameters of the PIRFICT model are summarized in Table 7.2. Some problems with the calibration were diagnosed by checking the impulse response function for each well. After several calibrations, the RMSE and RMSI values were the minimum founded for each well.

The physical plausibility of the results of a TFN model can be judged, for instance, by checking the IR functions. It is equivalent to the cross-correlation function. We check if the memory of the hydrological system, indicated by the time lag where the IR function approximates to zero, is covered by the monitoring period (De Gruijter

TABLE 7.1

Summary of the Statistics of PIRFICT Model Calibration

	Min	1st Q	Med	3rd Q	Max	Mean	SD
R^2_{adj}	57.68	76.16	82.91	88.04	95.45	81.85	9.21
RMSE	0.080	0.350	0.701	0.904	1.886	0.680	0.41
RMSI	0.072	0.313	0.555	0.750	1.433	0.552	0.32

Note: R^2_{adj} = percentage of explained variance; RMSE = root mean squared error (meters); RMSI = root mean squared innovation (meters); Min = minimum; 1st Q = first quartile; Med = median; 3rd Q = third quartile; Max = maximum; SD = standard deviation.

TABLE 7.2

Summary of Calibrated Parameters of the PIRFICT Model

	Min	1st Q	Med	3rd Q	Max	Mean	SD
A	54.8	770.9	1250	1955	6160	1711.3	705.5
a	0.001	0.004	0.007	0.011	0.162	0.013	0.01
n	0.49	1.05	1.30	1.72	2.86	1.43	0.24
E	−5.85	−0.94	0.88	1.65	2.58	0.24	0.93
α	5.92	22.03	32.24	47.69	95.50	38.60	14.73
IR	0	218	600	900	2200	656.58	570.6
LD	−50.6	−13.9	−7.37	−1.39	0.0	−10.16	11.97

Note: A = drainage resistance (days); a = decay rate (1/days); n = convection time (days); E = reduction factor (–); α = decay or memory of the white noise process (–); IR = impulse response (days); LD = local drainage base (meters); Min = minimum; 1st Q = first quartile; Med = median; 3rd Q = third quartile; Max = maximum; SD = standard deviation.

et al., 2006). A lack of relationship between the input series and the observed water table depths was found for three wells. The monitoring period apparently was not long enough to characterize the long memories of the hydrological system in these sites at the Jardim river area.

Parameter A is related to the shape of the IR function. Large values of A were calibrated to series at sites where a large fluctuation of the water table level was observed. From a mathematical point of view, these sites have large pulses from the signal of the input in the system. Some attention should be paid to parameter E, the reduction factor of evapotranspiration. This value should be between 0 and 1. For some wells we found estimates of E that are not realistic, like negative values. One reason could be that the climate station, located around 10 km outside the study area, does not represent the meteorological circumstances at all well locations. Another reason might be in the large temporal variation of land use, which makes these parameters difficult to estimate.

Stochastic simulation with the PIRFICT model was performed for the 34 wells, which remained after inspection of the results. For seven wells the simulation results indicated that the stationary conditions were not met. The distribution functions of the simulated WTD for these wells were bimodal. These wells were excluded from interpolation. Possibly the relatively short length of the water table time series did not completely cover the response time of the hydrological system. For example, the dry years of 2001, 2002, and 2003 might have a long-term effect on water tables in systems with long memories. The precipitation in these years was 24.4, 41.02, and 33.2% less than the annual average over the last 30 years, respectively. This effect acts different over the basin, due the presence of different geological systems (Lousada, 2005). Continuing monitoring of the WTD would enable us to clarify these questions (De Gruijter et al., 2006). For the remaining 27 wells, the distribution function of the WTD for October 1 was created from the simulated data. The WTDs that are expected to be exceeded with 5 and 95% probability at October 1 were used for spatial interpolation.

7.3.2 SPATIAL INTERPOLATION

The spatial dependence of the WTDs that are exceeded with 5% and 95% probability at October 1 was modeled by semivariograms (Figure 7.2). The alternative to using ancillary information on spatial prediction was taken, once the number of estimation points was sensibly reduced. Including elevation as a spatial drift into the geostatistical model caused a decrease in the semivariance.

The spatial dependence at short distances is poorly estimated because of the small number of observation wells that are fairly uniformly spread across the area. The nugget parameter of the semivariogram reflects the measurement precision of the WTD and the short-distance spatial variation in the WTDs.

The semivariogram fitted for WTDs that are exceeded with 95% probability, including a trend that depends on elevation, was used on UK estimation. The resulting map shows WTDs that could be exceeded with 95% probability at October 1. The probability of having lower (deeper) values than these in the maps is just 5%. Figure 7.3 gives these results for October 1 of any future year and the corresponding kriging variance.

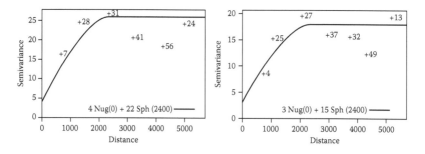

FIGURE 7.2 Semivariograms fitted for WTDs that are exceeded at October 1 with 95% (left) and 5% (right) probability, including a trend that depends on elevation.

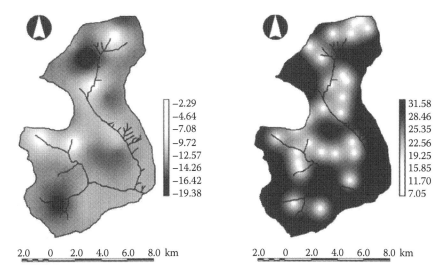

FIGURE 7.3   Map of WTDs (meters) that will be exceeded with 95% probability at Oct 1 (right) and the corresponding kriging variance (left).

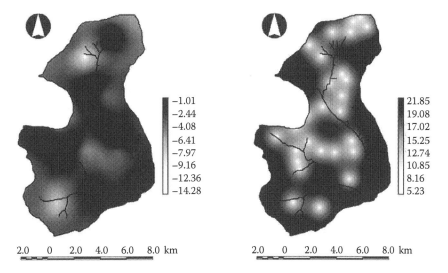

FIGURE 7.4   Map of WTDs levels (meters) that will be exceeded with 5% probability at Oct 1 (right) and the corresponding kriging variance (left).

Figure 7.4 gives the results of mapping WTDs with a confident level of 95%, including a trend that depends on using elevation and the corresponding kriging variances.

The UK resulted in maps with a physical meaning related to the local drainage. These maps were applied to quantify the risks on water management.

We compared the map of WTDs that will be exceeded with 95% probability with a map of estimated depth of wells as limit for water shortage. It indicated that no problems with deep water levels will occur at October 1, because no well can be dry with these levels. The same analysis was made for the map of WTDs that will be exceeded with 5% probability with a 0.5-m-limit map for shallow water table depths. The risk that these occur at October 1 is negligible.

7.3.3 Cross-Validation

The results of spatial interpolation were evaluated by cross-validation. Table 7.3 and Table 7.4 give the results. The SD values of the observations are much higher than those of predictions, indicating that the interpolation values are smooth (e.g., 4.86 vs. 1.79 in Table 7.3 or 4.16 vs. 1.83 in Table 7.4). The errors can be explained from

TABLE 7.3

Cross-Validation for the Spatial Interpolations of WTDs That Are Exceeded with 95% Probability at Oct 1

	Obs.	Pred.	Pred.–Obs.	Pred. SD	Z–Score
Min	−21.36	−14.80	−10.25	3.99	−2.24
1st Q	−11.83	−10.22	−1.99	4.54	−0.42
Med	−8.80	−9.28	1.13	4.77	0.25
3rd Q	−7.79	−8.18	2.67	5.04	0.59
Max	−0.72	−5.91	12.35	5.55	2.44
Mean	−9.37	−9.43	−0.05	4.78	−0.006
SD	4.86	1.79	5.49	0.42	1.12

Note: Pred. = predicted; Obs. = observed (meters); Z-score = (Pred.–Obs.)/ kriging variance (-); Min = minimum; 1st Q = first quartile; 3rd Q = third quartile; Max = maximum; SD = standard deviation.

TABLE 7.4

Cross-Validation for the Spatial Interpolations of WTDs That Are Exceeded with 5% Probability at Oct 1

	Obs.	Pred.	Pred.–Obs.	Pred. SD	Z–Score
Min	−16.04	−10.73	−7.04	3.36	−1.84
1st Q	−8.42	−6.51	−3.16	3.79	−0.86
Med	−5.62	−5.82	0.78	3.98	0.21
3rd Q	−2.62	−4.84	2.24	4.20	0.55
Max	0.14	−1.61	10.23	4.62	2.43
Mean	−5.75	−5.71	0.04	3.99	0.005
SD	4.16	1.83	4.57	0.34	1.12

Note: Pred. = predicted; Obs. = observed (meters); Z-score = (Pred.–Obs.)/ kriging variance (-); Min = minimum; 1st Q = first quartile; 3rd Q = third quartile; Max = maximum; SD = standard deviation.

uncertainty about the calibrated models, a poor relationship between elevation and WTD, and from a poor spatial correlation structure in both kriging models.

Predictions on Table 7.3 and Table 7.4 had the mean WTD value respected and indicate small mean interpolation errors (−0.05 and 0.04 m, respectively). The mean and standard deviation of the Z-score had values close to zero and one, respectively, which indicates a good performance of the kriging systems.

7.4   CONCLUSION

Time series modeling using the PIRFICTmodel was efficient to model a wide range of different responses of the hydrological system presented over the basin. Policy makers can and should use these results to optimize water use and to regulate the competing claims for water resources that often occur between small farmers, big farmers with irrigated crops, and water withdrawal for human use. However, the results reflect uncertainties from different sources: uncertainties related to the data (observed WTD, climatic database, DEM), and uncertainty associated with time series modeling and with the model of spatial variation.

The quality of the map was restricted by the effects of the relatively short time series that did not satisfactorily characterize the long memory systems. The quality of the time series models depends on both sampling frequency and the length of the series. The quality of the model of spatial structure depends on the number and the configurations the of well locations. The use of DEM as ancillary information slightly improved the quality of the final risk maps.

For the chosen date, October 1, there is a negligible risk of water shortage and shallow water table depths that could affect agriculture in some way. The analysis should be extended to other dates and periods that are critical to water supply. The method presented in this study enables this extension.

Given the long memories in the hydrological system of the study area, we recommend continued monitoring of water table depths in order to obtain more reliable results in the future.

ACKNOWLEDGMENTS

The first author is grateful to CAPES Foundation/Brazil (Foundation for the Coordination of Higher Education and Graduating Training) and to WIMEK (Wageningen Institute for Environment and Climate Research) for their financial support during his studies at ALTERRA, Wageningen, The Netherlands. We are also grateful to Dr. Suzana Druck (EMBRAPA/Cerrados—Brazilian Agricultural Research Corporation) for the use of this database (PRODETAB—Agricultural Technology Development Project for Brazil).

REFERENCES

Abramowitz, M. and I. A. Stegun. 1964. *Handbook of mathematical functions.* New York: Dover Publications Inc.
Box, G. E. P. and G. M. Jenkins. 1976. *Time series analysis: forecasting and control.* San Francisco: Holden-Day.

De Gruijter, J. J., D. J. Brus, M. F. P. Bierkens, and M. Knotters. 2006. *Sampling for natural resource monitoring*. Berlin: Springer-Verlag.

Furley, P. A. 1999. The nature and diversity of neotropical savannah vegetation with particular reference to the Brazilian Cerrados. *Global Ecol. Biogeogr.* 8:223–241.

Gardiner, C. W. 1994. *Handbook of stochastic methods*. New York: Springer-Velag.

Hengl, T., G. B. M. Heuvelink, and A. Stein. 2004. A generic framework for spatial prediction of soil properties based on regression-kriging. *Geoderma* 120:75–93.

Hipel, K. W. and A. I. McLeod. 1994. *Time series modelling of water resources and environmental systems*. Amsterdam: Elsevier.

Jepson, W. 2005. A disappearing bioma? Reconsidering land-cover change in the Brazilian savanna. *Geographical J.* 2:99–111.

Klink, C. A. and A. G. Moreira. 2002. Past and current human occupation and land-use. In *The Cerrado of Brazil: ecology and natural history of a neotropical savannah*, P. S. Oliveira and R. J. Marquis, Eds., pp. 69–88. New York: Columbia University Press.

Knotters, M. and M. F. P. Bierkens. 2001. Predicting water table depths in space and time using a regioinalised time series model. *Geoderma* 103:51–77.

Knotters, M. and M. F. P. Bierkens. 2000. Physical basis of time series models for water table depths. *Water Resour. Res.* 36:181–188.

Knotters, M. and P. E. V. Van Walsum. 1997. Estimating fluctuation quantities from time series of water-table depths using models with a stochastic component. *J. Hydrol.* 197:25–46.

Knotters, M., D. J. Brus, and J. H. Oude Voshaar. 1995. A comparison of kriging, co-kriging and kriging combined with regression for spatial interpolation of horizont depth with censored observations. *Geoderma* 67:227–246.

Lousada, E. O. 2005. *Hydrogeologic and isotopic studies in the Distrito Federal: conceptual flow models*. PhD diss., Brasília Univ.

Matheron, G. 1969. *Le krigeage universel. Cachiers du Centre de Morphologie Mathematique*. Fontainebleau: Ecole des Mines de Paris.

Nash, J. E. 1958. Determining runoff from rainfall. *Proc. Inst. Civ. Eng.* 10:163–184.

Odeh, I., A. McBratney, and D. Chittleborough. 1994. Spatial prediction of soil properties from landform attributes derived from a digital elevation model. *Geoderma* 63:197–214.

Oliveira, R. S., L. Bezerra, E. A. Davidson, F. Pinto, C. A. Klink, D. C. Nepstad, and A. Moreira. 2005. Deep root function in soil water dynamics in cerrado savannas of central Brazil. *Functional Ecology* 19:574–581.

Pebesma, E. J. 2004. Multivariable geostatistics in S: the Gstat package. *Comput. Geosci.* 30: 683–691.

Tankersley, C. D. and W. D. Graham. 1994. Development of an optimal control system for maintaining minimum groundwater levels. *Water Resour. Res.* 30:3171–3181.

Uhlenbeck, G. E., and L. S. Ornstein. 1930. On the theory of Brownian motion. *Physical Review* 36:823–841.

Van Geer, F. C. and A. F. Zuur. 1997. An extension of Box-Jenkins transfer/noise models for spatial interpolation of groundwater head series. *J. Hydrol.* 192:65–80.

Von Asmuth, J. R. and M. F. P. Bierkens. 2005. Modelling irregularly spaced residual series as a continuous stochastic process. *Water Resour. Res.* 41:W12404.

Von Asmuth, J. R. and M. Knotters. 2004. Characterising groundwater dynamics based on a system identification approach. *J. Hydrol.* 296:118–34.

Von Asmuth, J. R. and C. Maas. 2001, The method of impulse response moments: a new method integrating time series, groundwater and eco-hydrological modelling. In *Impact of human activity on groundwater dynamics*, J. C. Geherls, N. E. Peters, E. Hoehn, et al., Eds., pp. 51–58. Wallingford: IAHS Publication.

Von Asmuth, J. R., M. F. P. Bierkens, and C. Maas. 2002. Transfer function noise modelling in continuous time using predefined impulse response functions. *Water Resour. Res.* 38(12):23.1–23.12.

Ziemer, R. E., W. H. Tranter, and D. R. Fannin. 1998. *Signals and systems: continuous and discrete.* Upper Saddle River, NJ: Prentice-Hall.

8 Modeling Data Quality with Possibility Distributions

Gerhard Navratil

CONTENTS

8.1 INTRODUCTION

The amount of available data has increased with the development of new technologies. The availability of data and the capability of processing more data than before have led to new applications like online route planning or visualizations in landscape planning and architecture. The outcome of the application depends on the adequate selection of datasets. However, data quality varies with the source. Data quality descriptions have been defined to cope with that problem (Guptill and Morrison, 1995).

The automation of data processing requires the automatic handling of data quality. Fitness for use describes if a specific dataset is suitable for a specific task (Chrisman, 1984). Automatic preselection of the datasets can simplify the selection process for the user but requires automatic determination of the fitness for use. Byrom (2003) pointed out the necessity for the discussion of user needs. Grum and Vasseure (2004) and Pontikakis and Frank (2004) presented approaches to define the user needs and to use them to specify the fitness for use. A first step in that process is the suitable description of data quality.

Datasets are often collected over long periods. Graphs of street networks, for example, are determined once and updated periodically to reflect changes in the network. Since the quality of the determination may change due to improved technology, the quality of the data varies within the dataset. A worst-case scenario may be used for providing a number for the quality. However, this solution may give a wrong impression if a small part of the data is of significantly worse quality than the rest of the dataset. Quality should thus not be described by a single value.

The use of fuzzy numbers is a solution for varying data quality. Fuzzy numbers specify a range for the value. This allows showing the range of quality within the dataset. Fuzzy numbers are defined using distributions. Probability distributions specify probabilities for the different possible values, whereas possibility distributions only describe the possibility of the outcome. In this chapter I discuss an approach using possibility distributions. I use the Austrian cadastre as an example to specify possibility distributions. Different aspects of quality are modeled with different distributions. I then assume user requirements and present a method to compare the data quality with the user requirements. This allows assessing the fitness for use.

8.2   DATA QUALITY

How can we describe data quality? ISO 19113 "Quality Principles" (ISO 19113, 2002) defines the framework for a quality model. The data quality elements are completeness, positional accuracy, temporal accuracy, logical consistency, and thematic accuracy. Each of these elements describes a specific aspect of geographic data. Determination of positional accuracy provides an example where the use of precise numbers is not sufficient for complex datasets. The other elements can be treated in a similar way.

The determination of data quality must consider the creation process. The creation process is influenced by three different aspects, which influence the quality of the resulting data (Navratil and Frank, 2005): technology, legality, and usability.

Technological possibilities limit the achievable quality. In general, there is a maximum quality as well as a minimum quality. The usual method to create a terrain model, for example, is either laser scanning or aerial photogrammetry. In both cases the quality of the terrain model depends on the expenditure. Lower flight height will improve the quality of the model. Different measurement equipment will result in different precisions. The precision will not be arbitrarily high since there are always small changes in the terrain, e.g., footsteps, which should not influence the terrain model. Reduction of quality is limited, too. Less than a single height point in a terrain model is useless.

Laws have an impact on the quality of available datasets, too (Navratil, 2004). Laws may prohibit the use of data with higher quality than specified due to data protection laws or security reasons. In contrast to the technological limitations, legal influences are not "hard." It is possible to specify the maximum technical quality of a distance measure. This is not true for laws. The Austrian law stipulates a minimum precision of 15 cm for boundary points in the coordinate-based cadastre. Since it is difficult to prove the quality of coordinates, a point with a precision of 20 cm may be accepted, too. Thus legal rules on data quality can be seen as guidelines to develop technical solutions. It is then assumed that the results of the process meet these rules.

Usability may affect data quality. Data used more often may have higher quality since they produce more revenue and thus more money is available for collecting additional data. Nautical maps, for example, have higher quality in the areas where it is needed. Users only need details in coastal areas where the danger of hitting the ground is high. Thus more money is spent on mapping coastal areas than on mapping the ocean.

8.3 IMPRECISE NUMBERS

Many real-world situations cannot be described precisely. Statistics on the number of cars waiting at a red traffic light seems to be a simple task, but the definition of a "waiting car" is difficult. A stopped car is definitely waiting, but how about a car rolling slowly toward the traffic light? What is the maximum speed that a rolling car may have to be labeled "waiting"? Questions like that led to the development of mathematics with imprecise numbers.

Reasoning can be defined as testing the correspondence of a specified hypothesis with given statements. The statements can be data stored in a database and the hypothesis is a query on these data. A typical example is a database containing the heights of persons and the question of whether a specific person is "tall." Four different situations can be determined (Dubois and Prade, 1988a):

- Both the data in the database and the definition of "tall" are crisp. The entry in the database for the person could be 1.7 m and "tall" is defined as ">1.65 m." This leads to traditional, two-valued logic.
- The data are vague and the definition of "tall" is crisp. Here the definition for "tall" is the same as above but the entry in the database is expressed with a possibility distribution. This leads to possibility theory as published by Zadeh (1978, 1979) and expanded by Dubois and Prade (1988b).
- The data are crisp and the definition of " tall" is vague. The entry in the database could be 1.7 m, but the concept of "tall" is uncertain. This leads to many-valued logic.
- Both the data and the definition of "tall" are vague. This leads to fuzzy logic (Zadeh, 1975).

Which of these types of logic shall we use for modeling data quality? Data describe the world. Since the world changes, the data must change, too. Thus the data acquisition is a continuous process. Data quality parameters shall describe the quality of this data. It will not be possible to use a crisp description because the quality will vary throughout the dataset, and this variation should be reflected by the data quality description. Thus we deal with uncertain data.

The questions are crisp or can be made crisp. Users have two different questions:

- I need a dataset with a specific quality. Is it available?
- There is a dataset with a specific quality. Can I use it for the purpose at hand?

Both questions are crisp. In the first case, there may be several parameters for the data quality. All of these parameters must be fulfilled. Thus a dataset either fulfills the quality specification or it does not. This gives a crisp answer to the question. The second question is more complex. Again data quality issues must be considered, but in addition a cost-benefit analysis is necessary. According to Krek (2002), the value of a dataset emerges from better decisions. The value can be compared to the costs of acquisition and processing of the data. The dataset is applicable if the costs are lower than the benefits and there is no other possible outcome than using or not using the dataset. Thus both questions are crisp and we must use possibility theory.

8.4   POSSIBILITY DISTRIBUTIONS

A discussion of processes requires a method to describe the outcome of the processes. Possibility distributions (Zadeh, 1978) are such a method. In general, the use of fuzzy methods is suitable for the results of precise observation processes, and they can be used for statistical analysis (Viertl, 2006). Viertl uses probability distributions, which assign probabilities to each possible outcome. Determination of probabilities requires detailed knowledge. Possibility distributions avoid that problem. Possibility distributions only specify the possibility of the result: The value 0 shows impossibility and 1 shows possibility. Values between 0 and 1 provide information on the plausibility of the outcome. Thus, a result with value 0.4 is possible but less plausible than a result with 0.8. However, a result with 0.8 is not twice as probable as a result with 0.4.

The use of a set Θ of mutually exclusive and exhaustive possibilities is the most common way to express propositions (Wilson, 2002). A possibility distribution π assigns a value of possibility to each element of the set. If there is an element with value 1, then the function is said to be normalized:

$$\pi: \Theta \to [0,1]$$

8.5   QUALITY OF CADASTRAL DATA

The Austrian cadastral data are used as an example for a large dataset collected over an extended period. The dataset includes parcel identifiers, parcel boundaries, and current land use. Details on the Austrian cadastral system can be found in different publications (e.g., Twaroch and Muggenhuber, 1997). An important aspect is the definition of boundary. Whereas evidence in reality (like boundary marks, fences, or walls) defines the boundary in the traditional Austrian cadastre, the new, coordinate-based system uses coordinates to specify the position of the boundary. This change allows the creation of datasets reflecting reality since the data provide the legal basis for the boundaries.

The elements of data quality as listed in Section 8.2 must be defined in order to specify the quality of the Austrian cadastre. Positional accuracy connects to the elements defining the boundary lines. The Austrian cadastre uses boundary points to define the boundary. Thus the positional accuracy of the boundary points stipulates the positional accuracy of the dataset.

8.6 MODELING DATA QUALITY WITH POSSIBILITY DISTRIBUTIONS

8.6.1 Technological Influence

Positional accuracy for cadastral boundaries depends on the accuracy of boundary points, which depends on the precision of the point determination and the point definition itself. Thus the accuracy of the points will be used in the following discussion. Modern technical solutions for point determination use GPS and high-precision measurement equipment. This results in a standard deviation of 1–5 cm for the points based, e.g., on Helmert's definition, $\sigma_H^2 = \sigma_x^2 + \sigma_y^2$. This can be reached if the whole dataset is remeasured to eliminate influences of outdated measurement methods. Reduction of quality is possible, e.g., by using cheaper equipment. The lower limits are reached if the topology described by the dataset is influenced by random deviations. These effects may start with an accuracy of approximately 1 m and the dataset will become unusable in large parts of Austria with an accuracy of 10 m. Figure 8.1 shows the possible positional accuracy for boundary points.

8.6.2 Legal Influence

The positional accuracy of boundaries depends on the cadastral system used, the coordinate-based cadastre or the traditional cadastre. The traditional cadastre allows adverse possession. A person acquires ownership of land by using the land for 30 years in the belief that the person is the lawful owner. This is only detected during boundary reconstruction or in case of dispute. Thus parts of the dataset will

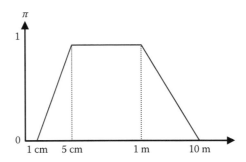

FIGURE 8.1 Possibility distribution for technological influence on positional accuracy.

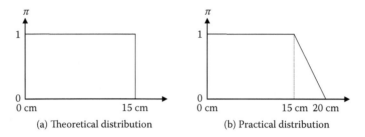

FIGURE 8.2 Possibility distribution for legal influence on positional accuracy in the coordinate-based system.

not be describing the correct boundaries, and even points with high internal accuracy may be incorrect. Therefore, the overall accuracy is low, but it is impossible to specify precise numbers. An estimate of the percentage of affected points cannot be provided, but it seems plausible that the number is not high because many boundaries are fixed by walls or fences. The possibility distribution will be similar to Figure 8.2b, but the values will be in the range of meters.

Accuracy is better defined for boundary points in the coordinate-based cadastre. The decree for surveying (Austrian Ministry for Economics, 1994) stipulates a minimum positional accuracy of 15 cm for boundary points. This value determines the standard deviation for the boundary points. Thus, theoretically, the possibility distribution for the positional accuracy looks like the one in Figure 8.2a. This rule is strict, as the law disregards statistical measures like standard deviation for decision making (Twaroch, 2005). However, it is difficult to control the actual accuracy of a boundary point. The existence of points with lower accuracies is possible. This is modeled in Figure 8.2b. Accuracies of less than 20 cm should not be possible since they should have been detected.

8.7 MODELING USER NEEDS

Two different groups of users of cadastral data are considered:

- Users of the boundary itself: Owners of land need data on their parcel and the neighboring parcels with high accuracy.
- Users of the positional reference in general: The cadastre is the only large-scale map available for the whole area of Austria, and thus it is often used to provide spatial reference.

These two groups have different requirements. The differences will show in the possibility distributions. In contrast to the technological and legal influences, the possibility distributions are not based on the specifications of the dataset but on the intended application. The possibility distribution shows if it is possible to use the dataset for the specific application.

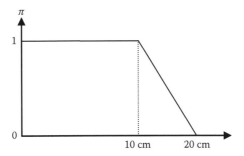

FIGURE 8.3 Positional accuracy for users of boundary.

8.7.1 USERS OF THE BOUNDARY

Positional accuracy is important for land owners. Land owners want to use their land, e.g., by creating a building. In Austria buildings must comply with legal rules specifying, for example, the maximum building height or the distance from the parcel boundary. The last point requires high positional accuracy to fit the strategies of courts. Thus, although an accuracy of 20 cm may be sufficient for some tasks of land owners, most tasks require a positional accuracy of 10 cm at maximum (compare Figure 8.3).

8.7.2 USERS OF THE SPATIAL REFERENCE

Spatial reference has limited demands for positional accuracy. Assuming a scale of 1:10.000 and accuracy on the map of 1/10 mm, then the accuracy of the points should be 1 m. Higher mapping accuracy leads to higher accuracy demands, but accuracy better than 0.5 m is not needed for positional reference. The lower limit of accuracy depends on the type of visualization. Accuracy of less than 10 m in built-up areas may result in less plausible datasets because it will not be possible to determine on which side of a street a point is (compare Figure 8.4).

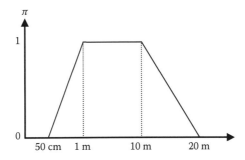

FIGURE 8.4 Positional accuracy for users of boundary.

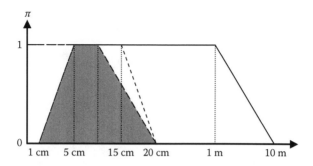

FIGURE 8.5 Combination of possibility distributions for users of the boundary.

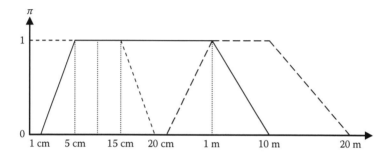

FIGURE 8.6 Combination of possibility distributions for users of the spatial reference.

8.8 COMBINATION OF POSSIBILITY DISTRIBUTIONS

In many cases data quality must meet several conditions. These conditions can be combined by a logical "and"-relation. The minimum-function provides this for possibility distributions. The "or"-relation would lead to the maximum-function (Viertl, 2006; Viertl and Hareter, 2006). Figure 8.5 shows the combination of the possibility distributions for positional accuracy. The gray area marks the overlap of the possibilities. Figure 8.6 shows the same combination for users of the spatial reference. This combination has no solution.

The example shows that the technical solutions and legal rules for cadastral systems do meet the demands of owners of land. Other types of use have different demands and thus the possibility distribution is different. Users who need spatial reference only require a different technical solution and different legal rules.

8.9 CONCLUSIONS

As we have seen, it is possible to model the influences on data quality with possibility distributions. It was possible to specify all necessary possibility distributions. The combination of influences produced a result that can be verified by practical experience. The method thus can be used to assess the correspondence of the influences on data quality.

Left for future investigation is the application for dataset selection. The chapter showed how to model possibility distributions for influences on data quality. It showed a simple method of combination. A general method will be needed to create the possibility distribution for more general examples. These distributions might require more sophisticated methods of combination.

REFERENCES

Austrian Ministry for Economics, 1994. Verordnung des Bundesministers für wirtschaftliche Angelegenheiten über Vermessung und Pläne (VermV). BGBl.Nr. 562/1994.

Byrom, G. M., 2003. Data Quality and Spatial Cognition: The Perspective of a National Mapping Agency. In: *International Symposium on Spatial Data Quality*, The Hong Kong Polytechnic University, pp. 465–473.

Chrisman, N. R., 1984. The Role of Quality Information in the Long-Term Functioning of a Geographical Information System. *Cartographica* 21: 79–87.

Dubois, D. and H. Prade, 1988a. "An Introduction to Possiblistic and Fuzzy Logics." Chapter in *Non-Standard Logics for Automated Reasoning*. P. Smets, E. H. Hamdani, D. Dubois, and H. Prade, Eds., London, Academic Press Limited, pp. 287–326.

Dubois, D. and H. Prade, 1988b. *Possibility Theory: An Approach to Computerized Processing of Uncertainty*. New York, NY, Plenum Press.

Grum, E. and B. Vasseure, 2004. How to Select the Best Dataset for a Task? In: *International Symposium on Spatial Data Quality*, Vienna University of Technology, pp. 197–206.

Guptill, S. C. and J. L. Morrison, Eds., 1995. *Elements of Spatial Data Quality*. Oxford, U.K., Elsevier Science, on behalf of the International Cartographic Association.

ISO 19113, 2002. Geographic Information—Quality Principles.

Krek, A., 2002. An Agent-Based Model for Quantifying the Economic Value of Geographic Information. PhD thesis, Vienna University of Technology.

Navratil, G., 2004. How Laws Affect Data Quality. In: *International Symposium on Spatial Data Quality*, Vienna University of Technology, pp. 37–47.

Navratil, G. and A. U. Frank, 2005. Influences Affecting Data Quality. In: *International Symposium on Spatial Data Quality*, Peking.

Pontikakis, E. and A. U. Frank, 2004. Basic Spatial Data according to User's Needs—Aspects of Data Quality. In: *International Symposium on Spatial Data Quality*, Vienna University of Technology, pp. 13–21.

Twaroch, C., 2005. Richter kennen keine Toleranz. In: *Intern. Geodätische Woche*, Obergurgl, Wichmann.

Twaroch, C. and G. Muggenhuber, 1997. Evolution of Land Registration and Cadastre. In: *Joint European Conference on Geographic Information*. Vienna, Austria.

Viertl, R., 2006. Fuzzy Models for Precision Measurements. In: *Proceedings 5th MATHMOD*, Vienna, ARGESIM/ASIM.

Viertl, R. and D. Hareter, 2006. *Beschreibung und Analyse unscharfer Information*. Vienna, Springer.

Wilson, N., 2002. A Survey of Numerical Uncertainty Formalisms, with Reference to GIS Applications. Annex 21.1 to REV!GIS Year 2 Task 1.1 deliverable.

Zadeh, L. A., 1975. Fuzzy Logic and approximate Reasoning. *Synthese* 30: 407–428.

Zadeh, L. A., 1978. Fuzzy Sets as a Basis for a Theory of Possibility. *Fuzzy Sets and Systems* 1: 3–28.

Zadeh, L. A., 1979. "A Theory of Approximate Reasoning." Chapter in *Machine Intelligence*, Vol. 9. J. E. Hayes, D. Michie and L. I. Mikulich, Eds. New York, Elsevier, pp. 149–194.

9 Kriging and Fuzzy Approaches for DEM

Rangsima Sunila and Karin Kollo

CONTENTS

9.1 INTRODUCTION

Models that have benefited from a strong mathematical background are best able to support the ability to form pictures of geographical information. In modern geodesy, existent computer-based programs are used, as well as models created for data computation. Digital elevation data are essential products derived from geodetic measurements. The effort needed to achieve high-quality data relating to vertical network needs is very time consuming and expensive, mainly because of the method involved. This method is known as *leveling*. There are various applications designed to present topographic information; the DEM (digital elevation model), used in geodesy and cartography, is one well-known model of this kind.

The DEM is often based on GRID (regular raster model) or TIN (triangulated irregular network). Seemingly, TIN is often preferable. Nevertheless, TIN has some disadvantages, such as each prediction depends on only three data, it makes no use of data further away, and there is no measure of error. The resulting surface has abrupt changes in gradient at the margins of the triangles, so the resulting surface is discontinuous, which makes a map with nonsmooth isolines (Webster and Oliver, 2001). As a result, alternative models based on possibility theories were constructed in order to introduce a new approach to viewing coordinate information. Fuzzy concepts were then brought into focus.

The fuzzy approach is based on the premise that key elements in human thinking are not just numbers but concepts that can be approximated to tables of fuzzy sets or, in other words, classes of objects in which the transition from membership to nonmembership is gradual rather than abrupt (Sankar and Dwijesh, 1985). Unlike crisp sets that allow only true or false values, fuzzy sets allow membership functions with boundaries that are not clearly defined. The grade of membership is expressed in the scale of 0 to 1 and is a continuous function (Sunila et al., 2004). To whatever extent, the use of fuzzy methods simplifies the mathematical models that are usually used in geodetic applications. Such mathematical models make use of polynomials simplified to a very high degree, but these cannot be computed and visualized by means of a simple method (Kollo and Sunila, 2005).

Geostatistics is a subject concerned with spatial data. That is, each data value is associated with a location, and there is at least an implied connection between the location and the data value (Jesus, 2003). There are various methods for doing this in geostatistics, all of which have different approaches suitable to various kinds of data and environments of model design. The geostatistical methods also provide error measures. Unlike the TIN model, geostatistical models—kriging, for example— provide a great flexibility of interpolation, which yields a smooth surface. Where a fuzzy model based on possibility theories may not be suitable for solving all problematic cases, probability theories may arise to provide a better alternative for reasonable modeling.

The aims of this chapter are to present alternatives in modeling digital elevation data using a geostatistical method, such as kriging, in order to compare results derived from using different methods in DEM and also to provide the possibility of establishing choices to obtain height data for use in different lower-accuracy geodetic and cartographic applications. The expected results will be the gaining of insights into using geospatial methods in geodesy and geostatistical models, and the analysis of the use of geostatistical methods as alternatives for providing height information. The results of the kriging technique from this research and of fuzzy DEM and TIN from the previous research will be compared and discussed.

9.2   KRIGING

9.2.1   THEORETICAL BACKGROUND

Kriging is a term coined by G. Matheron in 1963 after the name of D. G. Krige. It is based on a statistical model of a phenomenon instead of an interpolating function.

It uses a model for a spatial continuity in the interpolation of unknown values based on values at neighboring points (Sunila et al., 2004). Kriging is regarded as an optimal method because the interpolation weights are chosen to provide the best linear unbiased estimate (BLUE) for the value at a given point (Jesus, 2003).

There are several kriging techniques for different purposes such as ordinary kriging, simple kriging, universal kriging, indicator kriging, cokriging, point kriging, block kriging, disjunctive kriging, Bayesian kriging, and so on. In this chapter, we focus on ordinary kriging as, in practice, it is by far the most common type.

Ordinary kriging is a variation of the interpolation technique, one that implicitly estimates the first-order component of the data and compensates for this accordingly. This technique enables interpolation without the necessity of explicitly knowing the first-order component of the data a priori (GIS Dictionary, 1999). The basic equation used in ordinary kriging is as follows. (Note: variables denoted in the equations in the text below are defined for the first equations and then used with these definitions through the rest of the chapter.)

$$\hat{Z}\left(x_0\right) = \sum_{i=1}^{n} \lambda_i z\left(x_i\right) \tag{9.1}$$

where

n = number of sample points
λ_i = weights of each sample point
$z(x_i)$ = values of the points

When the estimate is unbiased, the weights are made to sum to 1 or

$$\sum_{i=1}^{n} \lambda_i = 1.$$

The prediction variance is given by

$$\sigma^2\left(x_0\right) = \sum_{i=1}^{n} \lambda_i \gamma\left(x_i, x_0\right) + \phi \tag{9.2}$$

where

σ^2 = variance
$\gamma(x_i, x_0)$ = semivariance between sample point x_i and unvisited point x_0
ϕ = Lagrange multiplier

9.2.2 VARIOGRAM MODELS

A variogram is a geostatistical technique that can be used to examine the spatial continuity of a regionalized variable and how this continuity changes as a function of distance and direction. Computation of a variogram involves plotting the relationship between the semivariance ($\gamma(h)$) and the lag distance (h) (Iacozza and Barber, 1999). The variogram is an essential step on the way to determining the optimal weights for interpolation (Burrough and McDonnell, 1998).

The most commonly used variogram models are spherical, exponential, linear, and Gaussian. Other models are those such as the pentaspherical model and Whittle's elementary correlation and Pure nugget; these are omitted from this chapter. The four commonly used models that were mentioned earlier are described and explained in Section 9.2.2.1 through Section 9.2.2.4.

9.2.2.1 Spherical Model

The spherical function is one of the most frequently used models in geostatistics (Webster and Oliver, 2001). The spherical model is a good choice when the nugget variance is important but not too large, and when there is also a clear range and sill (Burrough and McDonnell, 1998):

$$\gamma(h) = \begin{cases} c_c + c_1 \left\{ \dfrac{3h}{2a} - \dfrac{1}{2}\left(\dfrac{h}{a}\right)^3 \right\} & \text{for} \quad 0 < h < a, \\ \\ c_0 + c_1 & \text{for} \quad h \geq a \end{cases} \tag{9.3}$$

$$\gamma(0) = 0$$

where
$\quad\quad \gamma(h)$ = semivariance
$\quad\quad\quad h$ = lag
$\quad\quad\quad a$ = range
$\quad\quad\quad c_0$ = nugget variance
$\quad\quad c_0 + c_1$ = sill

9.2.2.2 Exponential Model

The exponential model is a good choice when there is a clear nugget and sill but only a gradual approach to the range:

$$\gamma(h) = c_0 + c_1 \left\{ 1 - \exp\left(-\frac{h}{a}\right) \right\} \tag{9.4}$$

9.2.2.3 Linear Model

This is a nontransitive variogram as there is no sill within the area sampled and typical attributes vary at all scales:

$$\gamma(h) = c_0 + bh \tag{9.5}$$

where b = the slope of the line.

9.2.2.4 Gaussian Model

If the variance is very smooth and the nugget variance is very small compared to the spatially dependent random variation, then the variogram can often be best fitted with the Gaussian model (Burrough and McDonnell, 1998):

$$\gamma(h) = c_0 + c_1 \left\{ 1 - \exp\left(-\frac{h^2}{a^2} \right) \right\} \tag{9.6}$$

9.2.3 Variance and Standard Deviation

To measure the variability, or how spread out the data values are, the variance is computed as the summation of the square of the difference of each data value from its mean divided by the number of points or number of points − 1 for an unbiased estimate of variance. The formula for computing variance is

$$S^2 = \frac{\sum_{j=1}^{n} (x_j - \bar{x})^2}{n-1} \tag{9.7}$$

where

S^2 = variance
x_j = data value
$\bar{x}$ = mean of the data values

The standard deviation (S) is simply calculated by the square root of the variance:

$$S = \frac{\sqrt{(x - \bar{x})^2}}{n-1}$$

In the unbiased estimator case, one can say that the minimum value of the mean square error is the variance. Hence, the standard deviation is the minimum of the root mean square error.

9.2.4 Cross-Validation

Cross-validation is a model evaluation method for checking the validity of the spatial interpolation method used. There are several validation techniques, e.g., the hold-out set method, and the 5-fold, 10-fold, K-fold, and N-fold methods. In this study, we selected the N-fold method, which is also known as the leave-one-out cross-validation method. In N-fold cross-validation, the dataset is split into N subsets of roughly equal size. The classification algorithm is then tested N times, each time training with $N-1$ of the subsets and testing with the remaining subset (Weiss and Kulikowski, 1991).

9.3 CASE STUDY

In this chapter, the coordinates and absolute heights of initial points are used. The area called Rastila in the eastern suburb of Helsinki was chosen. The whole area, in which about 2000 laser-scanning points are situated, is about 2 km² in area. Some points were eliminated due to errors in the data collecting process. Summary statistics of the sample points are shown in Table 9.1.

TABLE 9.1
Summary Statistics of the Sample Points in Rastila

Number of Observations	Min	Max	Mean
1985	0	18	4.926

The plotting of sample points is shown in Figure 9.1. The grayscale range presents the different heights from 0 to 18 m.

The histogram in Figure 9.2 shows a frequency of different heights in our sample points.

9.3.1 MODELING THE VARIOGRAM

In the elevation data we used in this study, there seem to be no directional differences, so the separate estimates can be averaged over all directions and yield the isotropic variogram. In Section 9.2.2, the four most commonly used variogram models are mentioned; we limited the scope of this study to focus on analyzing these four models (spherical, exponential, linear, and Gaussian). One question we had to answer was what model was most suitable for modeling our sample data. In order to answer this, we decided to construct potential variogram models based on different parameters and compare them, and then select the best-fit one for our analysis.

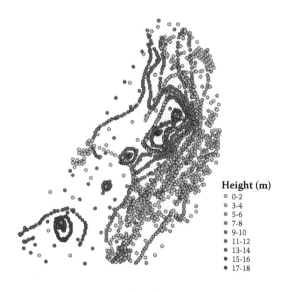

FIGURE 9.1 Plotted sample points of the study area in Rastila.

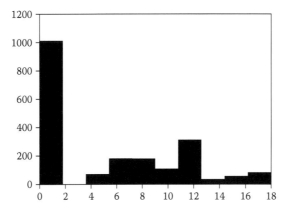

FIGURE 9.2 Histogram of the elevation sample points in the study area.

TABLE 9.2
Parameters of Variogram Models

Variogram Model	Range	Nugget	Sill	Length
Exponential	0.5	0.10847	1.1232	0.18064
Linear	0.5	0.50754	1.806	—
Gaussian	0.5	0.4546	1.0447	0.22811
Spherical	0.75	0.35265	1.0666	0.48576
Exponential	0.75	0.13161	1.1388	0.19144
Linear	0.75	0.58671	1.5711	—
Gaussian	0.75	0.48089	1.0728	0.24717
Spherical	0.95	0.54285	1.9597	1.8513
Exponential	0.95	0.36872	1.4082	0.40794
Linear	0.95	0.55676	1.6424	—
Gaussian	0.95	0.57831	1.1922	0.34484

The model parameters with different ratios of nugget, sill, length, and range that are used in the computation of candidate variogram models are shown in Table 9.2.

It is noticeable that, during the fitting model computational process, the spherical model at range 0.5 gave an error due to the spherical curve certainly not fitting the model. This was therefore subsequently ignored.

9.3.2 SELECTING THE BEST-FIT MODEL

To select the model most suitable for modeling our observations, the standard deviation or the sum of square error guided us as to how well the model displaying the specified parameter values fit the empirical semivariogram. Using the parameters in Section 9.3.1, the resulting standard deviation for each estimated model is shown in Table 9.3.

From Table 9.3, it can be seen that the best-fit variogram model for our study is the exponential variogram model with the range 0.75, nugget = 0.13161, sill = 1.1388, and length = 0.19144. The chosen variogram model, exponential, was then drawn as shown in Figure 9.3.

After choosing the variogram model, the kriging method was implemented to estimate and interpolate the data. As was stated in Section 9.2.1, that ordinary kriging is our focus in this study. This technique was then used in the estimation and interpolation computations. The search radius is 0.3, where the minimum number of kriging points is 10 and the maximum is 30. As a result, the kriging interpolation result from the chosen exponential model is displayed by means of the krig map in Figure 9.4.

TABLE 9.3
The Standard Deviation Values Calculated from the Various Variogram Models

Variogram Model	Std
Exponential (0.5)	3.281
Linear (0.5)	3.367
Gaussian (0.5)	3.431
Spherical (0.75)	3.307
Exponential (0.75)	3.272
Linear (0.75)	3.397
Gaussian (0.75)	3.439
Spherical (0.95)	3.381
Exponential (0.95)	3.302
Linear (0.95)	3.386
Gaussian (0.95)	3.466

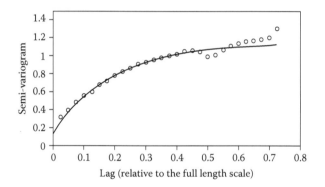

FIGURE 9.3 The exponential variogram mode.

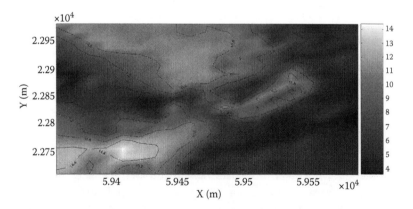

FIGURE 9.4 Kriging interpolation map of the study area.

The variance map displays the krig variance normalized by the variance from the covariance function. The resulting variance map is shown in Figure 9.5.

The leave-one-out cross-validation technique is used to present observation values and prediction values that are plotted in the cross-validation graph in Figure 9.6. The validation also reveals the standard deviation or the sum of square error.

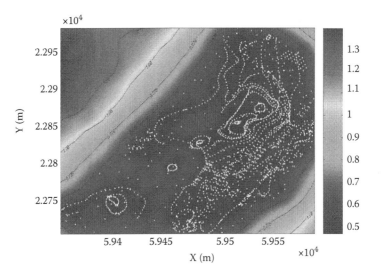

FIGURE 9.5 Variance map.

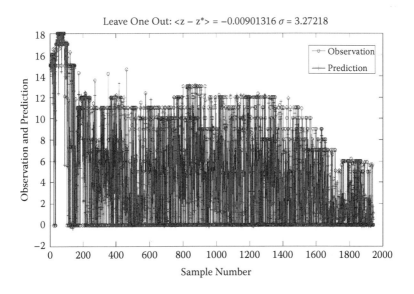

FIGURE 9.6 Cross-validation of exponential variogram model.

9.3.3 Comparison with Models from the Previous Study

9.3.3.1 Fuzzy DEM from the Previous Study

The digital elevation model (DEM) is often used when referring to topographic information. As topographic information contains uncertainty, the methods applied for should allow overcoming these issues. The fuzzy method is said to be a good tool for DEM construction, because it allows users to model the imprecision naturally and different methods of defuzzifications also give a better understanding of the results. The use of fuzzy methods might not be appicable in geodesy. To whatever extent, the use of fuzzy methods simplifies the mathematical models that are usually used in geodetic applications. These mathematical models may be very complified mathematical polynomials up to very high degree, which cannot be computed and visualized with a simple method.

Fuzzy set theory and fuzzy logic were invented by Lotfi Zadeh in the 1960s. They provide an intelligible basis for coping with imprecise entities (Niskanen, 2004). Unlike crisp sets, which allow only true or false values, fuzzy sets allow membership functions with unclear defined boundaries. The grade of membership is expressed in the scale of 0 to 1 and is a continuous function (Sunila et al., 2004). The most popular models of fuzzy systems are the Mamdani models and the Takagi-Sugeno-Kang (TSK) models. In this research, the TKS model gives better possibilities to construct fuzzy digital elevation models based on the hypothesis without demanding complexity.

To compute fuzzy DEM models, we used the fuzzy toolbox in MatLab. It provides opportunities to model fuzzy DEMs with varied methods. Two methods were chosen for this study—grid partition and subclustering. To test the hypothesis, several models were created, using different functions like triangular, trapezoidal, and Gaussian with varied numbers of membership functions. For the data analysis, several statistical quantities were used, namely, training error, standard deviations, and RMSE (root mean square error). In the second stage of the analysis, three models were compared with the TIN model. For statistical analysis, the standard deviation, the mean difference, and the max and min differences were computed for the models; histograms and correlation lines also were constructed by checking the model data against the real data. Based on the statistical analysis, the subclustering method was chosen, as it gives the best suitability for the fuzzy modeling in this research. To find the best compatible fit to the real data using fuzzy modeling, one more model was constructed. The RMSE for this model was about 40 cm, which is suitable for many geodetic applications, like cadastral surveying, engineering, control measurements, height determination, etc.

9.3.3.2 Comparison of Three Methods

From the previous study (Kollo and Sunila, 2005), three different fuzzy models were computed: firstly, the constant grid partition method with Gaussian membership function and 12 membership functions, referred to as Model 1; secondly, the linear grid partition method with trapezoidal membership function and nine membership functions, referred to as Model 2; and thirdly, the subclustering method with the range of influence being 0.2, referred to as Model 3. The RMSE values for the fuzzy DEM models were 4.45, 4.38, and 4.21 for Models 1, 2, and 3, respectively. From these three models, Model 3 (subclustering method) was chosen for the final comparison with the TIN model.

From this research, ordinary kriging using the exponential variogram model gives us the lowest RMSE, 3.27. The result is then compared with the results from the previous fuzzy DEM study. The comparison of RMSEs of both methods is shown in Table 9.4.

The results from Table 9.4 obviously show that the kriging model gives us a lower RMSE value than the fuzzy model. This can be explained by reference to the fact that our data contain sufficient points that benefit geostatistical modeling, in this case, kriging.

It should be mentioned here that the resulting map from the TIN model was constructed by using ArcGIS as a completed product. The TIN map is, however, not in our main area of interest in this research, but simply used as a comparison for map visualization to give us a better view of the different results obtained from various modeling methods.

TABLE 9.4

The RMSE Comparison between the Fuzzy and Kriging Models

Method/Model	RMSE
Fuzzy/subclustering	4.21
Ordinary kriging/exponential	3.27

The visualization of the resulting interpolation maps constructed from three different methods based on TIN, fuzzy, and ordinary kriging is shown in Figure 9.7.

9.4   DISCUSSION

The study provided different alternatives to modeling height surface. Three different kinds of methods, i.e., triangulated irregular network, fuzzy theories and techniques, and the geostatistical method of ordinary kriging were presented. As a result of this research, the advantages and disadvantages of these techniques were separated out and listed, before being gathered again for the purpose of discussion.

The triangulated irregular network (TIN) has the advantage that it is simple, so it does not require much time for constructing the TIN map. However, some difficulties are found with TIN. Although the surface of the resulting map is continuous, the map is presented with nonsmooth isolines. TIN uses only three data, so no more-distant data are included. TIN contains no measure of error.

The fuzzy technique, using concepts of possibility, allows us to visualize and model the data in a new way. The advantages of this technique in DEM relate to it not requiring the dataset to be dense. Based on the data we have, the fuzzy approach gave us very promising results. The computational algorithm does not require much work, but nevertheless can be time consuming. A disadvantage of fuzzy DEM that we come across is that the resulting map either has no continuous extrapolation or it is vague.

The kriging technique based on probability theories provides us many advantages in modeling elevation data. It is possible to use this technique to calculate missing locations. The result gave us an estimate of potential error. Kriging interpolation is smooth and visualization of the surface is good. Fortunately, our sample

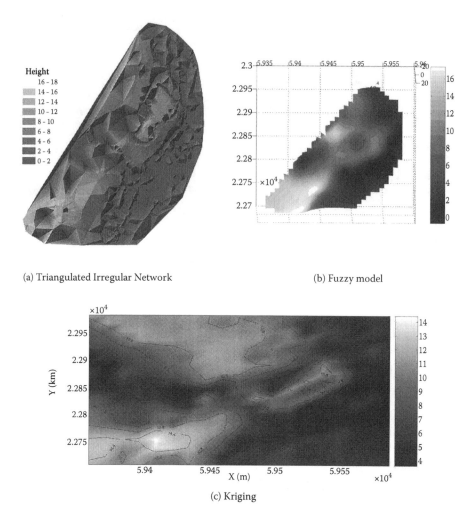

(a) Triangulated Irregular Network

(b) Fuzzy model

(c) Kriging

FIGURE 9.7   Comparison of resulting maps from different techniques used in DEM: TIN, fuzzy model, and kriging.

data are efficient; the resulting map looks very reliable. A disadvantage of kriging is, for example, that fitting variogram models can be time consuming because doing so requires skill and judgment. The computational process is also demanding. Kriging will not work well if the number of sample points is too small. Fitting a model can be difficult if the behavior of the data points is extremely noisy. Table 9.5 shows the advantages and disadvantages of both fuzzy and kriging methods.

9.5   CONCLUSIONS

This study aimed to present an alternative method for constructing DEM models to be used in geodetic and cartographic applications. The kriging method associated

TABLE 9.5

Advantages and Disadvantages of Fuzzy and Kriging Methods

	Fuzzy	Kriging
Advantages	+ Data need not be dense. + Algorithm is ready and easy to use. + Visualization is smooth and interpolation looks real.	+ It is possible to calculate missing points. + Potential error measure can be estimated. + Visualization is smooth and interpolation looks real.
Disadvantages	− Possibly requires more time for model construction. − The models are dependent on data structure. − The extrapolation is vague.	− Demands time and effort in computational process. − Fitting variogram models can be time consuming. − Fitting variogram model requires skill and effort if data points are inadequate. − Does not work well with too few data points. − Can be expensive regarding data-collection method.

with the geostatistical approach was chosen to test our hypothesis. From various kriging methods, ordinary kriging was chosen. Then, four commonly used variogram models—spherical, exponential, linear, and Gaussian—were brought into our variogram model construction. To try to search for the best-fit variogram model, we generated various models based on different parameters and assumptions to provide a wider range of alternatives in selecting the final model. Next, the best-fit model was selected. Our computations showed that the exponential model was suitable for our data, giving us fewer statistical error measures than the other models. Then, the ordinary kriging technique was applied to the construction of the krig interpolation map. The variance map was also constructed, and a cross-validation analysis was carried out as well to derive the finalized RMSE value.

We can conclude that, among the three interpolation techniques in our research, the kriging approach gives us a better error measure, so the height surface produced by means of kriging also gives a better visualization of the output. When the number of observations is adequate, this technique seems to provide a better fit in elevation surface modellng in terms of realistic visualiztion of the results than the fuzzy approach.

9.6 FURTHER RESEARCH

The authors have found that research in this field is very interesting. Comparisons of various methods based on different theories and approaches are useful for model construction. Each method or technique has its own advantages and disadvantages. It gives us alternatives for choosing suitable models for data analysis. Yet, a best-fit model for one kind of data may not fit with another kind of data. In their future research, the authors aim to study a new model construction or selection of regions with different terrain forms.

ACKNOWLEDGMENTS

The authors would like to thank Professor Kirsi Virrantaus for supporting our ideas and research, Mr. Matias Hurme from the Department of Real Estate, Division of City Mapping, Helsinki City provided the sample data and relevant information. Sincere thanks also go to the 5th ISDQ committee for giving us an opportunity to present our work. The second author sincerely thanks the Kristjan Jaak Scholarship Foundation and Estonian Land Board for providing his scholarship.

REFERENCES

Burrough, P. and McDonnell, R., 1998. *Principles of Geographical Information Systems.* Oxford University Press, pp. 132–161.

GIS Dictionary, 1999. Association for Geographic Information. http://www.geo.ed.ac.uk/agidexe/term?292 (accessed 20 Jan. 2007).

Iacozza, J. and Barber, D., 1999. An Examination of the Distribution of Snow on Sea-Ice. *Atmosphere-Ocean*, 37(1), pp. 21–51.

Jesus, R., 2003. Kriging: An Accompanied Example in IDRISI. GIS Centrum University for Oresund Summer University, Sweden.

Kollo, K. and Sunila, R., 2005. Fuzzy Digital Elevation Model. In: *Proceedings of the 4th International Symposium on Spatial Data Quality '2005*, Beijing, China, pp. 144–151.

Niskanen, V., 2004, *Soft Computing Methods in Human Sciences,* Springer-Verlag.

Sankar, K. P. and Dwijish, K. D. M., 1985, *Fuzzy Mathematical Approach to Pattern Recognition.* Wiley Eastern Limited.

Sunila, R., Laine, E., and Kremenova, O., 2004. Fuzzy Modelling and kriging for imprecise soil polygon boundaries. In: *Proceedings 12th International Conference on Geoinformatics—Geospatial Information Research: Bridging the Pacific and Atlantic*, Gävle, pp. 489–495.

Webster, R. and Oliver, M., 2001. *Geostatistics for Environmental Scientists.* John Wiley & Sons.

Weiss, S. and Kulikowski, C., 1991. *Computer Systems That Learn: Classification and Prediction Methods from Statistics, Neural Nets, Machine Learning, and Expert Systems*, Morgan Kaufmann.

Section III

Error Propagation

INTRODUCTION

Spatial data quality has as an important component the fitness for use. In order to do so, computer-based modeling is applied time in many fields of science. To quantify the fitness for use, error propagation is an important tool. All spatial information has its uncertainties. Error propagation relates input uncertainty to output uncertainty. It is usually possible as well to define the contributions of the different sources to the final error uncertainty. This may then in turn lead to focused actions, like additional sampling, better measuring, and thus reducing (or exploring) the input uncertainties. Error propagation is usually, but not exclusively, divided into two broad approaches: the analytical approach, e.g., based upon Taylor series, and Monte Carlo approaches, based upon randomization. The first approach is particularly useful when the operations are based upon simple analytical functions, calculations are rapid, and the results can then usually be simply visualized and understood. Monte Carlo methods are computer-intensive, are able to deal with modeling approaches of a large complexity, and require some skills in an appropriate visualization.

This section of the book deals with various modern approaches to error propagation. It first addresses error propagation of measurement errors of positional data. This is done by means of two papers on agricultural operations. It concerns field boundaries and precision agriculture. In modern agricultural practices, where information systems are transported on agricultural devices, the recognition of field boundaries requires determining the relevant information for agricultural operations. Environmental models, in addition, play a role in precision agriculture, the scientific development that aims to better integrate agriculture with the environment, taking farmers' interests and crop requirements into account. Spatial modeling with associated spatial data quality issues is also important in environmental and landscape studies, such as when simulating ice sheets. A study along these lines is included in the book. Finally, error propagation is also of importance in geodetic networks, where, on the basis of a limited set of stations, relevant geodetic statements are to be made. Geodesy, traditionally, has a very strong mathematical component, thus extending the approach to error propagation clearly broader.

The final two chapters contain intrinsic aspects of error propagation. First, they consider remote sensing data, by comparing different MODIS time series. As remote sensing data are mainly of a spatial nature, the inclusion of a time series analysis emphasizes the multitemporal character of such imagery. Second, modeling required in environmental studies can be done by applying a multicriteria fusion approach. This is shown here for a geographical data matching.

10 Propagation of Positional Measurement Errors to Field Operations

Sytze de Bruin, Gerard Heuvelink, and James Brown

CONTENTS

10.1 INTRODUCTION

The use of GIS and GPS in agriculture has increasingly moved from research to practical application. For example, in the Hoeksche Waard in The Netherlands, farmers are ready to invest in real-time kinematic (cm-accuracy) GPS technology (RTK-GPS) and peripheral devices to support optimal allocation of field margins, vehicle path planning, variable rate application, and other agricultural operations. A few years ago, the equipment required for these operations was very expensive and often required extensive customization of machinery (Keicher and Seufert, 2000). Nowadays, standard solutions are becoming available and costs are decreasing (Hekkert, 2006), thus improving the feasibility of GPS-assisted farming (Nijland, 2006). In addition to these applications, GPS is used to validate the agricultural subsidies claimed by farmers who, based on topographic field boundaries, apply for money under the European Union Common Agricultural Policy (CAP). For this purpose, less accurate hand-held receivers are being used (Bogaert et al., 2005).

It is widely agreed that site-specific management, also known as precision agriculture, requires detailed information on soil and environmental attributes, such as texture; organic matter content; nutrient concentrations; and incidence of diseases, weeds, and pests (Atherton et al., 1999). Recently, however, the importance of accurate geometric positioning for the development of field operation maps has also been recognized (Earl et al., 2000; Gunderson et al., 2000; Choset, 2001; Fountas et al., 2006). It has been claimed that the upcoming targeted approach to managing field operations requires field boundaries to be measured with cm-level accuracy, thus avoiding losses such as wasted inputs, unharvested crops, and inefficient use of the area.

The aim of this chapter is to demonstrate a method for experimental verification of such claims. It employs positional error models and random sampling from these models (Monte Carlo) to assess error propagation from GPS measurements or digitized vertices along field boundaries through the planning procedure. We demonstrate the approach using error models based on three measurement scenarios, namely: (1) using hand-held GPS with EGNOS (European Geostationary Navigation Overlay Service) correction; (2) using RTK-GPS measurements; and (3) based on a topographic vector product. The simulations were performed using the reference geometry of an irregularly shaped field of approximately 15 hectares located in the Hoeksche Waard (see Figure 10.1).

Note that our analysis only considers the positional uncertainty of mapped fields; semantic differences between topographic fields (which may have boundaries in the center of ditches) and cultivated fields were not accounted for.

10.2   METHODS

10.2.1   ERROR MODELS

The (x_i, y_i) coordinates in the Dutch grid system of the $n = 14$ corner points ($i = 1, \ldots, n$) of the agricultural field shown in Figure 10.1 were measured by a professional surveyor using RTK-GPS equipment. The resulting coordinates and mapped field boundaries were used as the reference geometry in the present work. By construction, any observation error in these locations is of no consequence for our results, because the reference geometry constitutes our "true" geometry in all subsequent calculations.

Under a measurement scenario, however, the coordinates of vertices (e.g., corner points) are subject to observational error, which can be represented by the random variables X and Y, with marginal cumulative probability distribution functions (mpdfs) F_X and F_Y:

$$F_X(x) = P(X \leq x) \quad \text{and} \quad F_Y(y) = P(Y \leq y) \tag{10.1}$$

where x and y are real numbers and P denotes probability.

0 50 100 150 Meters

FIGURE 10.1 Potato field in the Hoeksche Waard; the black boundaries represent reference geometry; the greyish stripes are spray paths.

The random variables X and Y typically have means (expected values) μ_X and μ_Y providing information on positional bias and standard deviations σ_X and σ_Y, which are measures of spread in the x and y directions, respectively. In the two-dimensional case, description of the positional uncertainty of a deformable object composed of n vertices requires a $2n$-dimensional joint probability density function (jpdf) that describes all mpdfs of the individual vertices together with all (cross-) correlations:

$$F_{X_1 Y_1 \cdots X_n Y_n}(x_1, y_1, \ldots, x_n, y_n) =$$
$$P(X_1 \leq x_1, Y_1 \leq y_1, \ldots, X_n \leq x_n, Y_n \leq y_n) \qquad (10.2)$$

Estimation of Equation 10.2 typically relies on the assumption of second-order stationarity, and on assumptions regarding the shape of the bivariate distribution and the function of statistical dependence (Heuvelink et al., 2007).

While geostatistical error models usually consider spatial correlation among observations, temporal variations in satellite clock errors, orbit errors, atmospheric delays, and filtering by the GPS receiver itself may result in the temporal correlation of observed positional errors (Olynik et al., 2002; Tiberius, 2003). Likewise, manual digitization of polygons is a sequential procedure that is likely to result in temporal correlations among errors of vertices. Therefore, our error model considers the temporal correlation of positional errors, which can be described by semivariograms (see below). Clearly, these temporal correlations will lead to spatial correlations in the positional errors, but are better modeled as temporal correlations.

Similar to Bogaert et al. (2005), we assumed the errors to be normally distributed. However, unlike that work, we allowed for different variances for the GPS errors in the x and y directions. In this context, the GPS satellite orbits cross the equator with an angle of 55°, which reduces the signal availability from the northern (y) direction in The Netherlands (52°N latitude).

10.2.1.1   EGNOS

EGNOS provides GPS correction data and satellite integrity messages that improve the accuracy of GPS code-phase positioning (European Space Agency, 2004). In October 2005, during the GPS workshop of the EU Joint Research Centre, EGNOS was still in the test bed phase (European Space Agency, 2006).

The teams operating Thales MobileMapper Pro receivers provided us with time series of EGNOS-augmented positional data, which were acquired to determine the area of three agricultural fields (Joint Research Centre, 2005). The positions were acquired at 1-s intervals, while the operator walked along pickets for which accurate RTK-GPS coordinates had been recorded. Some observations were removed by automatic filtering within the receiver. Each field was measured 10 to 14 times, but only the EGNOS-augmented data were used in our analysis. Depending on the size of the field and the speed of the operator, a time series of GPS positions comprised 225 to 613 s of data. The errors in the x and y directions were defined as the differences between the EGNOS positions and the nearest point on the line segments connecting the pickets. The final dataset consisted of 10,839 points.

Temporal dependence of the x and y errors was assessed by semivariogram analysis using Gstat (Pebesma and Wesseling, 1998). We used MobileMapper product information (Magellan, 2007) to set reasonable sills for the semivariograms. The data of all EGNOS augmented measurements were pooled, but temporal dependences between repeated measurements of the same field and between different fields were not analyzed. We also assessed possible bias in the errors.

The thus parameterized error model was used to illustrate the positional uncertainty in field measurements in the CAP. To apply the model on the field represented in Figure 10.1, we increased the number of vertices to 1 per 1.4 m ($n_{Egnos} = 1{,}258$), which represents a measurement rate of 1 Hz by an operator walking around the field (common practice for verification CAP).

10.2.1.2   RTK-GPS

RTK-GPS is a real-time surveying method with cm-level accuracy that employs correction signals from a (virtual) base station to solve the integer ambiguities, i.e., the number of integer cycles of 19 cm of the GPS carrier signal that fit along the path between the GPS receiver and the satellite. Several providers, including 06-GPS in

The Netherlands, provide correction signals obtained from a network of fixed base stations (Henry and Polman, 2003).

The company 06-GPS provided us with a time series of 17,570 RTK-GPS positions acquired at a 1 Hz sample rate from their control station in Sliedrecht (almost 5 hours of data). Temporal dependences of the x and y positions were assessed by semivariogram analysis using Gstat. Because of the nature of the data, we assumed no bias in the x and y coordinates (zero-mean errors).

The thus parameterized error model was used to illustrate the positional uncertainty in the RTK-GPS measurements. We used the original ($n = 14$) vertices and assumed that the operator walks between the individual measurements at corner points. Each measurement was assumed to take 1 min.

10.2.1.3 Topographic Map

The BRP (Basis Registatie Percelen) is a Dutch registry of agricultural fields and nature areas. It is largely derived from the Top10Vector digital topographic dataset (Hoogerwerf et al., 2003). Based on Van Buren et al. (2003) we assumed zero-mean positional errors in the x and y direction ($\mu_X = \mu_Y = 0$) for the original ($n = 14$) vertices, with a standard deviation of 2 m in each direction ($\sigma_X = \sigma_Y = 2$ m) and no cross-correlation. We further assumed that the vertices were digitized by hand at a speed of one per second and that the temporal dependence has a spherical structure with a range of 12 s.

10.2.2 SIMULATION OF MEASURED FIELD BOUNDARIES

The Data Uncertainty Engine version 3.0 (DUE) (Brown and Heuvelink, 2007) was used for generating 250 realizations of each of the above-described error models (parameterized mpdfs and temporal [cross-] correlations). The agricultural field was classified as a deformable object, i.e., the relative positions of the vertices along its boundary can vary under uncertainty.

The coordinates of the vertices were read from a simple time series data file, i.e., one header line specifying the variable names, another line giving the no data values, and the next 14 (RTK-GPS and topographic map) or 1258 (EGNOS) records each listing date/time and x and y data. DUE 3.0 includes an option to read and write ESRI Shape files, but this functionality currently does not support time series analysis.

In DUE, sampling from the joint-normal distribution is first attempted by factorizing the covariance matrix Σ, giving L, such that $\Sigma = LL^T$, where T represents the transpose. Secondly, a vector of samples is obtained from the standard normal distribution $N(0, I)$, with covariance matrix equal to the identity matrix I. Sampling from the pdf then involves rescaling by L and adding the vector of means μ:

$$x = \mu + Lz \tag{10.3}$$

where z is a random sample from $N(0, I)$ and x is a random sample from the required distribution $N(\mu, \Sigma)$.

TABLE 10.1

Parameter Values (*m*) for the Marginal

Probability Distribution Functions *FX* and *FY*

Scenario	μ*X*	μ*Y*	σ*X*	σ*Y*
EGNOS	0.508	0.230	1.16	1.63
RTK-GPS	0	0	0.0061	0.011
Topographic map	0	0	2.0	2.0

If Σ is too large to store in memory, or to factorize directly, a sequential simulation algorithm is called from Gstat within DUE (Brown and Heuvelink, 2007).

The parameterized error models were entered as expert judgement on the model page of DUE 3.0. The standard deviation or spread of normally distributed errors (σ) was defined by the square root of the sill of the semivariograms (see Table 10.1). The normal distributions of the coordinates were either centered on the reference coordinates (in case $\mu_X = \mu_Y = 0$) or an offset was added to model bias (otherwise; see Table 10.1). The semivariograms modeled in Gstat were transformed into correlograms, because DUE employs these as the single option of the dependence model. This allows σ to vary for each location while the correlogram (ρ) remains a simple function of the absolute (temporal) distance (Brown and Heuvelink, 2007). In case of cross-correlations between the *x* and *y* errors, the linear model of co-regionalization was used to ensure a valid bivariate covariance structure (Goovaerts, 1997).

10.2.3 Effects on Field Operations

In this work, we did not consider individual field operations but assumed that a farmer would optimize all field operations (e.g., ploughing, seeding, spraying, harvesting, etc.) based on the mapped field geometry. In this case two types of error may occur: (1) the farmer plans field operations outside the true field; and (2) the farmer subutilizes his field because he leaves parts uncultivated. The first type of error may severely harm the environment because agrochemicals may be sprayed into ditches, for example. Both types of errors reduce income. We assessed the two types of error by their area by overlaying the realized geometry according to the three error models with the reference geometry depicted in Figure 10.1. Other losses that may result from suboptimal planning within the field were not considered.

The 750 realizations of the uncertain time series data produced by DUE were converted to ESRI-generated files to create ArcInfo coverages. The topology of the polygons was postprocessed to eliminate any sliver polygon caused by self-intersection of the field boundaries. Next, the realized polygons were intersected with the reference polygon and the statistics of the two types of errors were obtained by querying the area attribute from the associated tables. All geo-processing was done in ArcGis 9.1 and Python scripts to allow for looping over the realizations.

10.3   RESULTS AND DISCUSSION

10.3.1   ERROR MODEL

10.3.1.1   EGNOS

The EGNOS sample data had biases of $\mu_X = 0.508$ m and $\mu_Y = 0.230$ m. Figure 10.2 shows the semivariograms of the EGNOS residuals in the x and y directions. The plots were cut at 400 s because of relatively few data pairs at larger temporal distances. For temporal distances exceeding 300 s, the model fits are poor, but this was assumed to be of little consequence in the subsequent simulations because adjacent vertices are at 1 s distance. Note also that toward the right the semivariograms are based on fewer data pairs. Based on the semivariograms, the spread of the x and

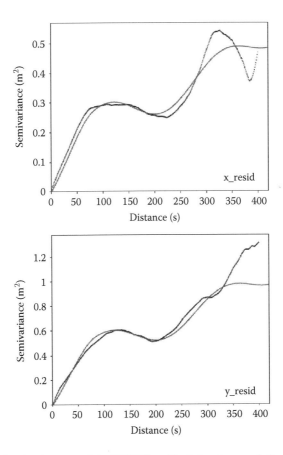

FIGURE 10.2   Semivariograms of the EGNOS residuals in x (upper plot) and y (lower plot) direction. Fitted models are indicated by gray lines; the dots represent experimental data.

the y residuals were set at $\sigma_X = 1.16$ m and $\sigma_Y = 1.63$ m. We found no evidence for cross-correlation. The correlogram (ρ) for both the x and the y residuals was modeled as follows :

$$1-\rho = 0.132 Sph(85)+0.792 Gau(720)+0.076 Per(220) \qquad (10.4)$$

where $Sph(85)$ = spherical structure with range 85 s, Gau = Gaussian structure, and $Per(220)$ = periodic structure with period 220 s.

The Gaussian structure with a long range was added to improve the fit at larger temporal distances and to bring the standard deviations close to documented values. The cause of the periodic structure is not clear. Periodicity has been observed over short time spans with a static receiver owing to multipath effects (Van Willigen, 1995; Amiri-Simkooei and Tiberius, 2007), but in our case time series were based on a moving receiver. Multipath effects happen when a GPS unit receives both the direct GPS signal and signals reflected by, e.g., buildings. It differs from site to site and from time to time as it depends on the azimuth and elevation of the satellites and local geometry (Amiri-Simkooei and Tiberius, 2007). Atmospheric signal path delay variations have also been identified as a potential cause of periodic variations in GPS time series (Poutanen et al., 2005).

The semivariograms are notably different from the structure reported by Bogaert et al. (2005). The latter only comprised a single Gaussian structure with an effective range of 30 s and a rather high sill in comparison to the accuracy that is claimed to be possible with EGNOS augmentation. Obviously, there is no single semivariogram that can be used for all EGNOS-enabled GPS receivers under all circumstances. Therefore, our parameterizations of the error model should not be used uncritically beyond the scenarios presented here.

10.3.1.2   RTK-GPS

Figure 10.3 shows the semivariograms and cross-variogram of the RTK-GPS data in the x and y directions. The spreads of the x and y errors were set at $\sigma_X = 0.0061$ m and $\sigma_Y = 0.011$ m, respectively, and, as indicated above, biases were ignored ($\mu_X = \mu_Y = 0$). These figures are consistent with RTK-GPS accuracies reported elsewhere. The correlogram for x and y errors was modeled by

$$1-\rho = 0.041 Nug(0)+0.581 Sph(500)+0.378 Per(1250) \qquad (10.5)$$

where $Nug(0)$ = nugget component.

Under the linear model of coregionalization, cross-correlation was modeled by

$$1-\rho_{xy} = 0.0385 Nug(0)+0.552 Sph(500)+0.36 Per(1250) \qquad (10.6)$$

There is an even more pronounced periodic component in the spatial dependence structures. While such periodicity over short time spans may be attributed to multipath

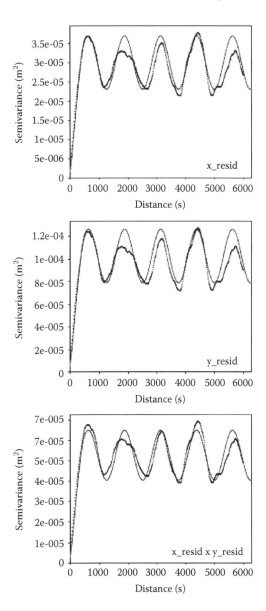

FIGURE 10.3 Semivariograms of the x and y RTK-GPS coordinates (upper and middle plots) and cross variogram (lower plot). Fitted models are indicated by gray lines; the dots represent experimental data.

effects (Amiri-Simkooei and Tiberius, 2007), its prolonged presence in these data is remarkable. Possibly, the periodicity was produced by multipath effects somewhere in the RTK network. However, we observed a similar periodic structure but with smaller amplitude in semivariograms of a 3-hour time series of RTK measurements acquired with a Trimble 4700 and 4800 base station/rover (short baseline) configuration.

10.3.2 SIMULATION WITH DUE

Table 10.1 summarizes the parameter values for the marginal cumulative probability distribution functions, which were entered as expert judgement on the model page of DUE 3.0. All simulations were performed using the full jpdf, by factorization of the covariance matrix. In the case of the EGNOS scenario (with 1258 vertices), this involved factorisation of a 1258 × 1258 matrix (cross-correlations in the EGNOS data were not accounted for).

10.3.3 EFFECTS ON FIELD OPERATIONS

Figure 10.4 shows an example realization obtained under the topographic map scenario. If the farmer would rely on this topographic map to plan field operations, in the North he would leave a large strip uncultivated while in the South his plans would cover a ditch. Realizations obtained under the EGNOS scenario (detail shown in Figure 10.5) give comparable results, but with more irregular field boundaries because of the increased number of vertices. Conversely, on the maps resulting from the RTK-GPS scenario, the errors cannot be discerned by the eye unless displayed at a very large scale.

Table 10.2 lists several summary statistics of erroneously mapped areas under our three scenarios. The corresponding histograms are shown in Figure 10.6. Not

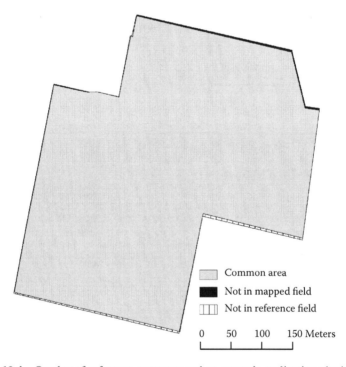

Common area
Not in mapped field
Not in reference field

0   50   100   150 Meters

FIGURE 10.4 Overlay of reference geometry and an example realization obtained under the topographic map scenario.

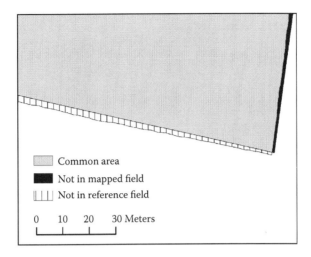

FIGURE 10.5 Detail of an overlay of reference geometry and an example realization obtained under the EGNOS scenario, i.e., an operator walking around the field while EGNOS augmented GPS positions are recorded at 1 Hz.

TABLE 10.2
Summary Statistics of Areas in Error (m²) under
Three Scenarios

				Percentile	
Scenario	Error[a]	Mean	SD	0.10	0.90
EGNOS	1	980	435	469	1566
	2	1002	432	485	1561
RTK-GPS	1	6.02	3.54	2.08	10.9
	2	5.99	3.01	3.25	9.38
Topographic map	1	1348	685	489	2299
	2	1230	642	447	2134

[a] 1 = area included in mapped field, but outside reference field; 2 = area outside mapped field, but inside reference field.

surprisingly, the expected incorrectly mapped areas are approximately proportional to the standard deviations of the positional errors for each type of error (see Table 10.1 and Table 10.2). On the other hand, the non-Gaussian distributions shown in Figure 10.6 are symptomatic of the nonlinear operation performed on the data. This demonstrates the utility of Monte Carlo simulation, which enables incorporation of operations of any complexity in an error propagation study.

Data such as those presented in the percentile columns of Table 10.2 may be used to assess risks. For example, under the EGNOS scenario there is a probability of 90% that the area with error type 1 exceeds 469 m². Whether such risks are

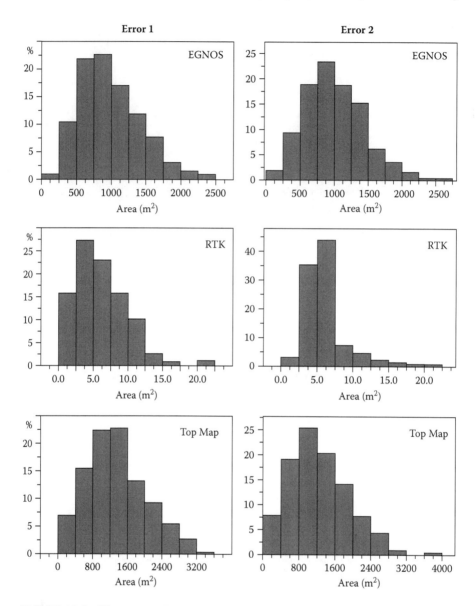

FIGURE 10.6 Histograms of areas in error under three scenarios. Types of errors are explained below Table 10.2.

acceptable depends on the environmental, financial, and other consequences of the errors. In practice, there is some uncertainty surrounding these probabilities, including uncertainty originating from sampling effects and modeling of the joint pdf. While confidence intervals could be computed for these estimates, we do not present them here.

10.4   CONCLUSIONS

We have demonstrated a general error propagation method that can be used for experimental verification of claims regarding the positional accuracy required for planning field operations based on digital field maps. In our current example we did not plan any field operations (e.g., ploughing, seeding, spraying, harvesting, etc.) directly, but rather assessed the areas where a farmer would erroneously plan activities (while they are outside his field) and areas that would be left without cultivation (while they could be used). The method can easily be adopted to compute error propagation through more complex applications such as path planning for field operations.

We observed periodic components in the temporal dependence structures of the GPS errors. Such periodicity has been attributed to multipath effects over short time spans, but its presence in our experiments and relevance beyond our scenarios require further study.

A definite answer to the question of whether agricultural fields should be measured with cm-level accuracy depends on the environmental and financial consequences of the above-described errors and other costs that may occur within the fields. It also depends on the level and type of automation employed by the farmer. Our scenario analysis, nevertheless, showed that for planning and executing field operations a farmer should not blindly rely on approximate field geometry as this would leave ample room for accidents (e.g., 90% chance that an erroneously cultivated area adjacent to the studied field is larger than 469 m^2, under the EGNOS scenario).

ACKNOWLEDGMENTS

This work was partly carried out within the project "Geo-information Requirement for Agri-environmental Policy," which is co-financed from the program "Space for Geo-information" (Project RGI-017). We gratefully acknowledge Thales/Magellan and 06-GPS for providing GPS time series, Kemira GrowHow for the aerial photograph used in Figure 10.1, and Van Waterschoot Landmeetkunde for the reference geometry. Special thanks go to Aad Klompe (farmer in the Hoeksche Waard and chairman of H-WodKa), who introduced us to the problem setting.

REFERENCES

Amiri-Simkooei, A. R. and C. C. J. M. Tiberius 2007. Assessing receiver noise using GPS short baseline time series. *GPS Solutions* 11(1): 21–35.

Atherton, B. C., M. T. Morgan, S. A. Shearer, T. S. Stombaugh, and A. D. Ward 1999. Site-specific farming: A perspective on information needs, benefits and limitations. *Journal of Soil and Water Conservation* 54(2): 455–461.

Bogaert, P., J. Delinc, and S. Kay 2005. Assessing the error of polygonal area measurements: a general formulation with applications to agriculture. *Measurement Science & Technology* 16(5): 1170–1178.

Brown, J. D. and G. B. M. Heuvelink 2007. The Data Uncertainty Engine (DUE): A software tool for assessing and simulating uncertain environmental variables. *Computers & Geosciences* 33(2): 172–190.

Choset, H. 2001. Coverage for robotics—a survey of recent results. *Annals of Mathematics and Artificial Intelligence* 31(1–4): 113–126.

Earl, R., G. Thomas, and B. S. Blackmore 2000. The potential role of GIS in autonomous field operations. *Computers and Electronics in Agriculture* 25(1–2): 107–120.

European Space Agency 2004. EGNOS. http://esamultimedia.esa.int/docs/br227_EGNOS_2004.pdf (accessed January 28, 2007).

European Space Agency 2006. EGNOS system test bed. http://esamultimedia.esa.int/docs/egnos/estb/esaEG/estb.html (accessed January29, 2007).

Fountas, S., D. Wulfsohn, B. S. Blackmore, H. L. Jacobsen, and S. M. Pedersen 2006. A model of decision-making and information flows for information-intensive agriculture. *Agricultural Systems* 87(2): 192–210.

Goovaerts, P., 1997. *Geostatistics for natural resources evaluation*. New York: Oxford University Press, pp. 108–123.

Gunderson, R. W., M. W. Torrie, N. S. Flann, C. M. U. Neale, and D. J. Baker 2000. The collective—GIS and the computer-controlled farm. *Geospatial Solutions* July 2000: 2–6.

Hekkert, G. 2006. Volautomaten sturen goed—Een prima stuurautomaat is er al voor €7500. *Boerderij* 91: 18–23.

Henry, P. J. A. and J. Polman 2003. GPS-netwerk operationeel in heel Nederland. *Geodesia* 3: 108–114.

Heuvelink, G. B. M., J. D. Brown, and E. E. Van Loon 2007. A probabilistic framework for representing and simulating uncertain environmental variables. *International Journal of Geographical Information Science* 21(5): 497–513.

Hoogerwerf, M. R., J. D. Bulens, J. Stoker, and W. Hamminga 2003. Verificatie Kwaliteit BRP gegevensbank. Wageningen, Alterra, Research Instituut voor de Groene Ruimte.

Joint Research Centre, 2005. 2005 GPS Workshop—5th and 6th October 2005 (Wageningen). http://agrifish.jrc.it/marspac/LPIS/meetings/2005-10-5NL.htm (accessed January 29, 2007).

Keicher, R. and H. Seufert 2000. Automatic guidance for agricultural vehicles in Europe. *Computers and Electronics in Agriculture* 25(1–2): 169–194.

Magellan 2007. MobileMapper Pro Specifications. http://pro.magellangps.com/en/products/product_specs.asp?PRODID=1043 (accessed January 29, 2007).

Nijland, D. 2006. Futuristisch boeren wordt haalbaar, want betaalbaar. *VI Matrix* 108: 6–7.

Olynik, M., M. Petovello, M. Cannon, and G. Lachapelle 2002. Temporal impact of selected GPS errors on point positioning. *GPS Solutions* 6(1): 47–57.

Pebesma, E. J. and C. G. Wesseling 1998. Gstat: A program for geostatistical modelling, prediction and simulation. *Computers & Geosciences* 24(1): 17–31.

Poutanen, M., J. Jokela, M. Ollikainen, H. Koivula, M. Bilker, and H. Virtanen 2005. Scale variation of GPS time series. *A Window on the Future of Geodesy—Proceedings of the International Association of Geodesy IAG General Assembly*, Sapporo, Japan, June 30–July 11, 2003. Berlin: Springer-Verlag, pp. 15–20.

Tiberius, C., 2003. Handheld GPS receiver accuracy. *GPS World* 14(2): 46–51.

Van Buren, J., A. Westerik, and E. J. H. Olink 2003. Kwaliteit TOP10vector—De geometrische kwaliteit van het bestand TOP10vector van de Topografische Dienst, Kadaster—Concernstaf Vastgoedinformatie en Geodesie.

Van Willigen, D. 1995. Integriteit van (D)GPS signalen. Paper presented at Precisie plaatsbepaling met DGPS in Nederland—Kwaliteit, netwerken en toepassingen, Rotterdam, Nederlands Instituut voor Navigatie.

11 Error Propagation Analysis Techniques Applied to Precision Agriculture and Environmental Models

Marco Marinelli, Robert Corner, and Graeme Wright

CONTENTS

11.1 INTRODUCTION

The way in which the uncertainty in input data layers is propagated through a model depends on the degree of nonlinearity in the model's algorithms. Consequently, it can be shown (Burrough and McDonnell, 1998) that some GIS operations in environmental modeling are more prone to exaggerate uncertainty than others, with exponentiation functions being particularly vulnerable. Also of influence are the magnitude of the input values and the statistical distribution of the datasets. It is generally assumed, often through lack of information, that the uncertainty in a data layer is normally distributed (Gaussian).

Many environmental and agricultural models have either been derived from an understanding of the biophysical processes involved, or empirically as a result of long-term trials. They are therefore not usually designed or assessed with regard to the possible effects of error propagation on accuracy. If the error is propagated in such a way as to be exaggerated, then clearly the usefulness of the recommendations made (by the model) may be compromised. Other parameters of the error, such as nonnormality in the input data layers (and their associated errors), may further influence the result. This is especially important as often the error distribution in inputs has to be estimated due to lack of data. An example of this is when a spatial data layer (for input into a model) is generated using an interpolation technique. This is considered important as interpolation methods are often used in GIS software with the default "black box" settings. This is especially the case with end users who are unfamiliar with the limitations of the interpolation method, and/or the system being studied.

The aim of this work is to test which error results calculated using Monte Carlo simulation and a range of assumed distributions best replicated the Taylor method results, and hence which may be most accurate in assessing error propagation through a model. This will help in best assessing a model's accuracy and its limitations. However, an assumption is made in this case that the Taylor method is best for assessing the propagation of error in a model, but this may not always be the case for a number of reasons such as nonnormality in the input variable error distribution and noncontinuity. This work therefore also aims to illustrate the influence of nonnormal input distributions on a model's final results and associated statistics. In particular, the error and skew of the synthesized results are investigated relative to the synthesized results to see if these results can give an insight into how a model and its inputs may influence the final result.

11.1.1 THE MODELS INVESTIGATED FOR THIS CHAPTER

The models used in this study are the nitrogen- (N-) availability component of the SPLAT model (Adams et al., 2000) and the Mitscherlich precision agricultural model. The N-availability (in soil) model is linear whereas the Mitscherlich is not, a key factor effecting the shape and size of the propagated error. Therefore, these are ideal for the study and assessment of error propagation analysis techniques. The details of the models are as follows.

11.1.1.1 N-Availability

$$N \text{ (available)} = \left(RON * RONDep\left(T-1\right) * RONEff\right) + 10000 *$$
$$\left(OC * \left(1 - GravProp\right) * SONEff\right) + \left(15 \times FerTeff\right) \tag{11.1}$$

where the input data layers are the residual organic nitrogen (RON), organic carbon in the soil (OC), and the gravel proportion in the soil (Grav Prop). The other parameters, RONEff, SonEff, and FertEff, are fixed values for all of the area studied, but do have some uncertainty. The other variable is time (T) in years, since the last lupin crop. The N-available result is in kilograms per hectare (Kg/Ha).

11.1.1.2 The Mitscherlich Model

An inverted form of this model (Edwards, 1997) has been proposed (Wong et al., 2001) as a method of determining the spatially variable potassium fertilizer requirements for wheat. This relationship, which describes the response of wheat plants to potassium, is shown in

$$Y = A - B \times e^{-CR} \tag{11.2}$$

where Y is the yield in tonnes per hectare, A is the maximum achievable yield with no other limitations, B is the response to potassium, C is a curvature parameter, and R is the rate of applied fertilizer.

It has been shown (Edwards, 1997) that the response, B, to potassium fertilizer for a range of paddocks in the Australian wheat belt may be determined by

$$B = A\left(0.95 + 2.6 \times e^{\left(-0.095 K_\theta\right)}\right) \tag{11.3}$$

where K_0 is the soil potassium level

Substituting Equation 11.2 into Equation 11.1 and inverting provides a means of calculating the potassium requirements for any location with any given soil potassium value. This is shown in

$$R = \frac{-1}{C} \times LN\left[\frac{\left(Y_t - A\right)}{-A\left(0.95 + 2.6 \times e^{\left(-0.095 K_0\right)}\right)}\right] \tag{11.4}$$

where R is the fertilizer requirement (Kg/Ha) to achieve a target yield of Y_t tonnes per hectare.

11.2   DATA USED

The layers required for the N-availability equation are from a 20-ha paddock in the northern wheat belt of Western Australia.

The data for the Mitsherlich model are from an 80-ha paddock in the central wheat belt, from an area where potassium fertilization is often required. Achievable yield was calculated by aggregating NDVI representations of biomass, derived from Landsat 5 images over a period of 3 years and estimating water limited achievable yield using the method of French and Schultz (1984). This method of deriving achievable yield is described more fully in Wong et al. (2001). Soil potassium was determined at 74 regularly spaced sample points using the Colwell K test (Rayment & Higginson et al., 1992). These values were interpolated into a potassium surface using inverse distance weighting. All data were assembled as raster layers with a spatial resolution of 25 m. For the work described here, a target yield of 2 tonnes/ha was set. This is within the achievable yield value for 97% of the paddock.

11.2.1   SKEWED SPATIAL ERROR PATTERNS

In certain cases the method by which the error distribution of a data layer is calculated does not result in a pattern with a normal error distribution. For example, Figure 11.1 shows the error distribution of an interpolated digital elevation map (DEM) surface from (a) randomly spaced and (b) equally spaced samplings. These random points were sampled from a DEM that covered a region in Western Australia: latitude 116.26–117.23 East, longitude 27.17–27.14 South. The interpolation methods used to generate the six DEM surfaces were inverse distance weighting (IDW), spline, and ordinary kriging. The data points were sampled at random and equally spaced positions and equaled to 0.1% of the original DEM surface. The actual error (per point) in the input layer is unknown and therefore was not included in the calculation. Arc-GIS (ESRI, 2006) software was used to generate the interpolated surfaces, and in each case default settings were used. The generated layers were subtracted from the original DEM to generate the error layers for each method. For comparison, a normal distribution is also shown, with a standard deviation equaling the mean standard deviation of the interpolation results.

For the spline and kriging techniques, the error results with the lowest skew (and hence higher normality) occurred when the sampled points were equally spaced (see Table 11.1). The exception was the result for the IDW, which was less skewed (but not by a great degree when compared with the other changes observed). It is also noted that the greatest agreement between the three methods occurs when the data are evenly spaced. A general statement that can be made from these results is that most of the results appear relatively normally distributed, but there are some points in the generated data layers where the difference from the original DEM is considerably higher in the positive range. This in turn is reflected in the skew of the error results, which in turn suggests that the Monte Carlo method is appropriate in this case for studying the propagation of this error.

Several questions relating to the accuracy of a interpolated data layer arise from these results: (1) What interpolation method gives me the most accurate results (and

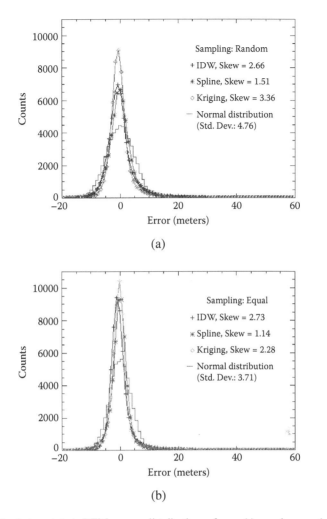

FIGURE 11.1 Interpolated DEM error distributions from (a) random and (b) equal sampling.

TABLE 11.1

Skew and Kurtosis of Error in Interpolated DEM Layers

	Randomly Spaced Samples		Equally Spaced Samples	
	Skew	**Kurtosis**	**Skew**	**Kurtosis**
IDW	2.66	14.22	2.73	14.26
Spline	1.51	9.50	1.14	12.36
Kriging	3.36	21.34	2.28	15.68

hence the least error)? (2) Is the skew a true representation of the distribution of the error? (3) Can the skew be used in a data simulation to generate a valid random data set from which error propagation can be investigated? These questions may easily be answered if the interpolated data layer can be compared to on-field measurements, but as is often the case in environmental studies, this is not possible due to lack of data. These are also key questions that must be asked when choosing the method of investigating error propagation.

11.3   METHODS USED IN ERROR PROPAGATION ANALYSIS

The error propagation methods used in this study are the first- and second-order Taylor series methods and Monte Carlo simulation. These will be assessed to determine how the estimated error propagated through the model varies between the different methods.

11.3.1   THE TAYLOR SERIES METHOD

11.3.1.1   Theory

The method of error propagation analysis referred to as the Taylor series method relies on using either the first, or both the first and second, differentials of the function under investigation. In the case when the error is normally distributed and the algorithm is continuous, it is effectively considered a "gold standard" and widely used. Its main limitation is that it can only be used in the analysis of the parts of an algorithm that are continuous. Since the function in Equation 11.4 is differentiable, that is not a constraint here. For a detailed description of the theory and how this method is used, refer to Heuvelink (1998) and Burrough and McDonnell (1998).

11.3.1.2   Implementation

This method was carried out using a procedure written in the Interactive Data Language (IDL) (Research Systems, 2006). The variables in the N-availability model are the residual organic nitrogen (RON); the organic carbon fraction (OC); the gravel proportion (Grav Prop); and the RON, SON, and fertilizer coefficients. Equation 11.1 was partially differentiated with respect to each of these inputs, to the first order. The resulting equations (not shown here) were converted to spatially variable data layers and combined with the absolute error layers for the inputs. The error layers for the inputs were generated as follows:

1. For the RON, OC, and Grav Prop a relative error of ±10% was assumed for each data point. The error was therefore calculated by first multiplying the data by 0.1. This value was assumed to represent the full width of the error distribution. In order to provide the same approximate error magnitude as was being used in the Monte Carlo simulations (described below) the error was represented by 3.33% being equivalent to 1 standard deviation.
2. The RON, SON, and fertilizer coefficients are not spatially variable, but are known to contain errors of ±0.4, 0.025, and 0.025, respectively, which were divided by their respective coefficients to obtain an absolute error ratio.

Using the same logic as above, the absolute error was regarded as being one third of the difference between the mean and the extreme values quoted.

3. The output generated using the Taylor series method is an absolute error surface for N-availability.

The input variables in the Mitsherlich model are the achievable yield (A_y), the soil potassium level (K_0), and the curvature term (C). Equation 11.4 was partially differentiated with respect to each of these inputs to both the first and second order. Error for these layers were generated as follows:

1. For the A and the K_0 data layers, a relative error of ±10% was assumed for each data point.
2. The curvature term C is not spatially variable but is known to contain uncertainty. In this case, the value is derived from a series of regional experiments on potassium uptake by wheat crops and is quoted in the literature as having a value of between 0.011 and 0.015 for Australian Standard Wheat (Edwards, 1997). The work described here used the mean of those two values as the "true value" for C. Using the same logic as above, the absolute error was regarded as being one third of the difference between the mean and the extreme values quoted.
3. The output generated using the Taylor series method is an absolute error surface for R. The error surface produced incorporated any correlation that exists between the data layers. Correlation was only able to be determined between the A and K_0 input surfaces, with a ρ value of 0.53.

11.3.2 MONTE CARLO METHOD

11.3.2.1 Theory

The Monte Carlo method of error propagation assumes that the distribution of error for each of the input data layers is known. The distribution is frequently assumed to be Gaussian with no positive or negative bias. For each of the data layers an error surface is simulated by drawing, at random, from an error pool defined by this distribution. Those error surfaces are added to the input data layers and the model is run using the resulting combined data layers as input. The process is repeated many times with a new realization of an error surface being generated for each input data layer. The results of each run are accumulated, and both a running mean and a surface representing deviation from that mean are calculated. Since the error surfaces are zero centered, the stable running mean may be taken as the true model output surface and the deviation surface as an estimate of the error in that surface. Another important point is that the Monte Carlo method can be used in the analysis of disjoint functions, whereas the Taylor method cannot. Again the reader is referred to Heuvelink (1998) for a full description.

It reality, uncertainties in input data layers are not always evenly distributed. Therefore, for the Mitsherlich model, error simulations were drawn from distributions that were skewed to differing degrees. The skewed distribution was generated using the "RANDOMN" command in IDL with the "Gamma" option set to differing

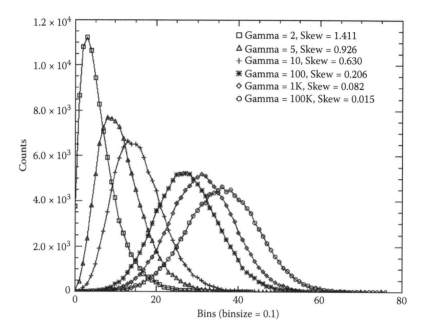

FIGURE 11.2   Gamma distributions.

levels. This produces a family of curves with a variety of skews; a selection of these and an unbiased normal distribution are compared in Figure 11.2.

11.3.3   IMPLEMENTATION

A procedure was written in IDL to perform the process described above. Simulated random datasets were generated for the appropriate inputs with the incorporation of appropriate error realizations. For each run 100,000 simulated data points were generated for each valid grid cell in each of the input data layers and coefficients. From these, the mean, absolute error, and relative error for R were calculated for each location.

For the Mitsherlich model, in some cells either the achievable yield (A_y) is less than the target yield (Y_t) or the soil K values are adequate for the achievable yield and hence a calculation of the fertilizer requirement (R) returns a negative value. Where this happened the result was classified as invalid and the cell value set to null. For the N-availability, the same was implemented if the values in the input layers or the simulated results where less than zero.

The level of agreement between the calculated values of N-availability and fertilizer recommendation (R) and their associated error surfaces calculated by the error propagation methods was determined by performing pairwise liner regressions between the various outputs. Two error surfaces that agree completely should have a slope of 1 and a correlation coefficient of 1.

11.4   RESULTS AND DISCUSSION

11.4.1   Calculated N-Availability Results and Associated Statistics

There is a high agreement between the N-available results calculated from the input layers and the synthesized input layers (correlation: 0.999, slope: 0.999). There is also a high agreement in the calculated error, even though the number of simulations investigated varied significantly (2,000 to 100,000). The curves in all cases do follow a slightly nonlinear upward slope, possibly suggesting that the N-availability algorithm is influencing these results.

This is further reflected in the skew of the synthesized results (per point; see Figure 11.3). At first it would appear that there is a significant difference in these skew results. However, a closer inspection shows that for the low and high values of N, the center of the skew is approximately the same (~0.4 and 0.6, respectively). The major difference is the width of the skew results, which is lower for the greater number of simulations, suggesting that a higher number of simulations is required for more accurate and easily interpreted results; e.g., as seen in Figure 11.3b, the increase in skew with higher N-availability is more easily seen.

Also of note is that the skew is not centered on zero. As the skew of the synthesized input layers are centered on zero, this suggests that the model itself is influencing not only the propagated error results but also the shape of the synthesized results. This influence is most likely to be greater in the more complex nonlinear Mitsherlich

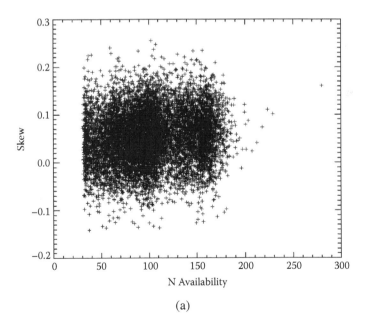

(a)

FIGURE 11.3   A comparison of N-availability and Skew for (a) 2,000 and (b) 100,000 simulations.

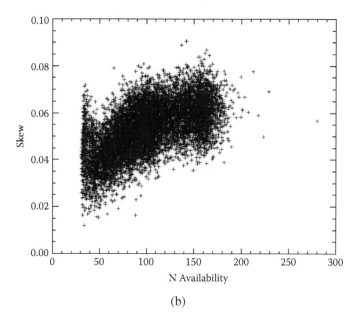

(b)

FIGURE 11.3   (continued).

model (as may also be the case for nonnormal inputs) and both are investigated in the following sections.

There is a good linear fit between the Taylor and Monte Carlo simulated error results, with a slope of 1.0 and correlation of 0.999. The relative error is also small, with a minimum and maximum of 0.048 and 0.078, respectively, which suggests that the N-availability component of the SPLAT model does not propagate error to any large degree.

11.4.2   MITSHERLICH MODEL RESULTS AND ASSOCIATED STATISTICS

Figure 11.4 shows the error of the Monte Carlo synthesized results verses the Taylor method's results, for both a Gaussian and Gamma distribution (+ and – distribution, Gamma = 2 and 100,000). The maximum number of simulations (per point) is 100,000 for all the following results.

Clearly, the greatest agreement between the methods is when the error calculated is low. The greatest agreement (with the Taylor method) is with the Gaussian distribution. Closer inspection of the results shows that the best agreement occurs at points where R is less than or equal to 100 (calculated errors < 30; regression analysis in this range gives a slope of 0.93). Also, in this error range, the Gamma distribution of 100,000 gives a similar result of 0.93. However, as is clearly seen, at higher values of R, the Taylor method results increase significantly.

The heavily skewed distributions (Gamma = 2) clearly are in even less agreement with the Taylor method results. Furthermore, in this case, the positive and negative Gamma distributions are not in agreement. This is reflected (to a lesser degree) in Figure 11.5, which shows fertilizer recommendation values (R) plotted against

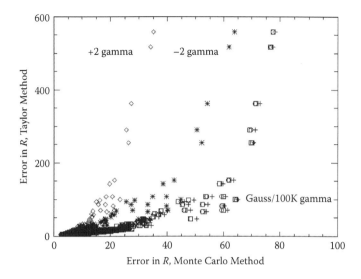

FIGURE 11.4 A comparison of the error of R, simulated (Monte Carlo) and Taylor methods.

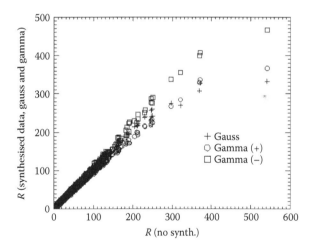

FIGURE 11.5 Comparison of simulated and directly calculated R values (Gamma = 2).

Gaussian and Gamma distribution results. As one might expect, for the most part, the best agreement occurs with the mean R calculated from the Gaussian distribution (slope of 0.935). However, above a value of 250 there is greater agreement with the negative (and to a lesser degree with the positive) Gamma distribution. The reason for this is due to the Gaussian distribution R results that are invalid and filtered out, e.g., where A is less than Y_t. This weights the calculated mean in a negative direction. More importantly it highlights how biased results may occur depending on the structure of the model and the skew of the input variables. This is further discussed in the following sections.

11.4.3   ERROR RELATIVE TO R

Figure 11.6 compares the calculated mean and standard deviations of R per point from the Guassian and Gamma distribution synthesized inputs. It can be seen that there appears to be a similar pattern for all three distributions, with notable changes occurring in the R vs. error relationship at approximately 100–200 Kg/ha and then at 250–400 Kg/ha. The second of these changes is most likely due to the bias in

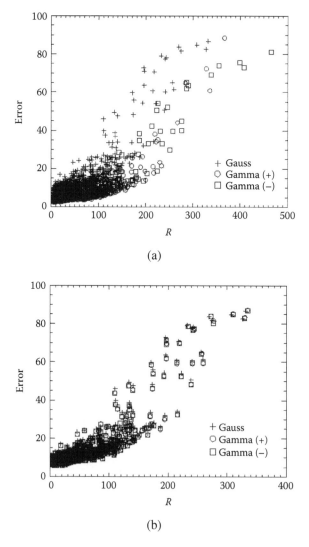

(a)

(b)

FIGURE 11.6   Mean synthesized R vs. error: (a) Gamma = 2; (b) Gamma = 100,000. For comparison, a Gauss distribution is included in both plots.

the results due to the decrease in the number of valid data points. However, the first change suggests that at a point where one or more of the inputs contribute to a higher output R, a significant increase occurs in the error associated with that result. Also notable is that both positive and negative skew Gamma inputs generally have lower error. This is most likely due to the concentration of the simulated inputs into a smaller range than occurs in a Gaussian distribution.

Figure 11.6b shows the results for a skewed distribution for which the gamma value is 100,000. The R vs. error relationship is essentially the same as when Gamma is set to 2, but notably smoother in the curve (as R increases). There is also very good agreement between the Gaussian and Gamma distribution results. This is expected as the Gamma distribution of 100,000 is equally biased (and hence the skew is very close to 0).

11.4.4 SKEW RELATIVE TO R

The skew in the R results calculated from the synthesized datasets show three features. (1) As in the error results, the skew values appear to remain approximately the same when R is equal to or less than 100, but then increases (see Figure 11.7). Three of the four Gamma distributions investigated eventually peaked and then fell. However, the negatively skewed Gamma = –2 distribution continues to increase. This mirrors the "valid results pattern" discussed earlier. (2) The heavily nonnormal distribution in the inputs is reflected in the position of the skew results relative to the Gaussian skew results, easily seen in Figure 11.7a. (3) As shown in Figure 11.7b, the skew of the Gaussian and equally weighted Gamma distribution is not centered on 0, even when the values of R are low (and hence considered valid) and the skew of the input layers is insignificant. Analysis of the Mitsherlich model shows that this

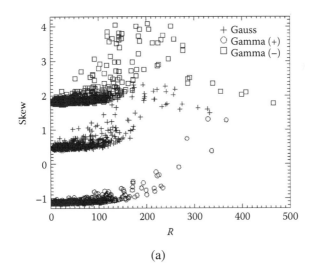

(a)

FIGURE 11.7   Mean synthesized R vs. skew: (a) Gamma = 2; (b) Gamma = 100,000. For comparison, a Gauss distribution is included in both plots.

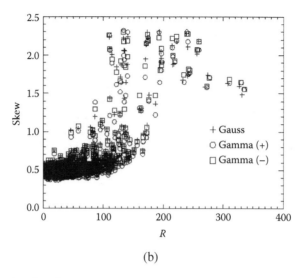

(b)

FIGURE 11.7   (continued).

is due to the mathematical structure of the model and is important as it may bias R and its associated error.

11.5   CONCLUSIONS

The values of N-available and R (K requirement) calculated from the given data layers are in close agreement with the mean values calculated from the Monte Carlo synthesized datasets under a Gaussian assumption. The exception occurs at extreme values of R and is an artifact of the nonlinearity of this model.

As is generally the case for linear models, error propagation in the linear N-available model is negligible. However, the model structure did influence the skew of the N-available results (calculated from the synthesized input layers). For the Mitsherlich model, error propagation increases as R increases, and the rate of this increase can vary significantly and abruptly. In the case of this nonlinear model, there appears to be several reasons for this that are dependent on how the input/model interaction can change as the input values change.

The closest agreement in the absolute error trends is seen between the combined first- and second-order Taylor series results and the Monte Carlo Gaussian distribution for calculated R values of less than or equal to 100. Above this value there is considerably less agreement.

For the skewed Gamma distribution, the best in the calculated R agreement is seen when the synthesized dataset has little positive or negative bias (within a given valid range for R). However, the heavily negatively skewed distribution produces results that are less prone to the model bias at higher R values.

Both the error and skew statistical results (for the calculated R) can give an insight into how a model and/or its inputs may influence the validity of the final results.

ACKNOWLEDGMENTS

This work has been supported by the Cooperative Research Centre for Spatial Information, whose activities are funded by the Australian Commonwealth's Cooperative Research Centres Programme. DEM data were obtained from the Department of Land Information, Western Australia.

REFERENCES

Adams, M. L., Cook, S. E., and Bowden, J. W. (2000), Using yield maps and intensive soil sampling to improve nitrogen fertiliser recommendations form a deterministic model in the Western Australian wheatbelt. *Australian Journal of Experimental Agriculture*, 40, No 7. 959–968.

Burrough, P. A. and McDonnell, R. A. (1998), *Principles of Geographical Information Systems*, Oxford: Oxford University Press.

Edwards, N. K (1997). Potassium fertiliser improves wheat yield and grain quality on duplex soils. In *Proceedings of the 1st Workshop on Potassium in Australian Agriculture*, Perth, Western Australia: UWA Press.

ESRI (2006), ArcGIS 9.1, Redlands, CA: Environmental Systems Research Institute.

French, R. J. and Shultz, J. E. (1984), Water-use efficiency of wheat in a Mediterranean-type environment. II. Some limitation to efficiency. *Australian Journal of Agricultural Research*, 35, 765–775.

Heuvelink, G. B. M. (1998), *Error Propagation in Environmental Modelling with GIS*, London: Taylor & Francis.

Rayment G. E. and Higginson F. R. (1992), *Australian Laboratory Handbook of Soil and Water Chemical Methods*, Melbourne: Inkata Press.

Research Systems Inc, (2006), IDL 6.2. Boulder, CO: Research Systems Inc.

Wong, M. T. F., Corner, R. J., and Cook, S. E. (2001), A decision support system for mapping the site-specific potassium requirement of wheat in the field. *Australian Journal of Experimental Agriculture*, 41, 655–661.

12 Aspects of Error Propagation in Modern Geodetic Networks

Martin Vermeer and Karin Kollo

CONTENTS

12.1 INTRODUCTION

In geodesy we produce and manage highly precise coordinate data. Traditionally we do this by successive, controlled propagations of precise measurements down a *hierarchy* of progressively more localized and detailed network densifications: working "from the large to the small."

Compared to the market for geographic information used for mapping applications, where precision is less critical and often in the range ±0.1 to 1 m, geodetically precision-controlled coordinate data form a much smaller field of application. However, this field is vitally important, including the precise cadastral, urban planning, and construction surveys that make modern society possible. Bringing this area of activity within the scope of geographic information services would require adapting these to the management of the spatial precision *structures* found in these network hierarchies, codifying traditional geodetic practice.

One of us (KK) has studied in detail the technical aspects of coordinate Web services for geodesy (Kollo, 2004).

In geodesy, the complexity of describing the precision of point sets is often handled by defining simple *criterion functions* that model the point coordinates' overall variance behavior as a function of relative point location, without having to specify a detailed covariance matrix.

Next we first briefly present the current state of spatial information services for the World Wide Web, including coordinate transformation services. Then we discuss geodetic networks, network hierarchies, error propagation, and criterion matrices.

We propose to bring sets of geodetic coordinate data upon a globally unique realization of WGS84 by a two-step procedure:

1. Perform an overdetermined tie of the given geodetic network (which may be a traditionally measured, pre-GPS, one) by a triangle-wise affine (bilinear) transformation to a given set of GPS-positioned points. This technique is currently in use in Finland, cf. Anon. (2003, appendix 5); after this, the network will be in the national realization EUREF-FIN of WGS84, i.e., the locally canonical realization for the territory of Finland.
2. Perform a three-dimensional Helmert transformation of the result to a single, globally unique WGS84 realization. In this operation, the given set of GPS points, which could be considered errorless in the EUREF-FIN datum, will acquire a nonzero variance structure again.

We derive criterion functions modeling the variance propagation behavior of both steps.

12.2   SPATIAL DATA WEB SERVICES

Geographic information services, as they exist today, supply spatial information over the World Wide Web. They are commonly based upon standards established by the Open Geospatial Consortium (OGC), an international nonprofit geospatial information standards group. Using these standards, one may extract geographic data from a variety of conforming data sources, which may all be in different datums or coordinate reference systems.

Services of this kind can be classified as Web Map Services (WMS; OGC, 2001), Web Feature Services (WFS; OGC, 2002), and many others.

Web standards are based on the XML (Extensible Modeling Language) description language; OGC has defined the GML (Geographical Mark-up Language) for this (OGC, 2003). The language provides for specifying position precisions of points and point sets, either as individual point position precisions or as between-points relative position precisions. Additionally, it allows specification of a full variance-covariance matrix. The standard speaks of "data quality" (dataQuality.xsd).

More recent work on data quality is going on in ISO, the International Standards Organization, e.g., ISO 19113 "Quality Principles" and ISO 19138 " Data Quality Measures" (A. Jakobsson, personal comm.). Clearly coordinate precision is only a small part of what the concept of "quality" covers when applied to spatial information. Conversely, however, there is much more to coordinate quality than is often understood, about which more later.

In practical implementations such as GeoServer (Anon., 2005d) or MapServer (Anon., 2005b) we tend to see a limited set of predefined datums (e.g., the European Petroleum Survey Group (EPSG) set (cf. European Petroleum Survey Group, 2005) and projections and transformations (e.g., the PROJ.4 set;, cf. Anon., 2005c) being included. A more scalable approach is using a coordinate service specification for the Web. Both standardization and implementation work in this direction are now being done in a number of places. There exist a WCTS (Web Coordinate Transformation Server) specification and experimental implementations (see, e.g., Anon., 2005a).

Spatial data Web services, as they are currently designed, are aimed at the large, complex market of users of various map products for a broad range of applications. These products are often of limited resolution and precision, coordinate precision not being their focus. To some extent this is also a cultural difference, cf., e.g., Jones and Winter (2005/2006).

12.3   CHOOSING A "ROSETTA FRAME"

A well-known spatial data application like PROJ.4, e.g., does not distinguish between the various realizations of WGS84, such as the different international ITRF and European ETRF frames (F. Warmerdam, email). As long as we work within a domain where there is only one canonical realization, like EUREF-FIN in Finland, this is a valid procedure. PROJ.4 uses WGS84 as the common "exchange datum" to which all other datums are transformed, typically by applying (after, if necessary, transformation to three-dimensional Cartesian using a reference ellipsoid model) either a three-parameter shift or a seven-parameter Helmert transformation. For geodetic use, it is not enough to consider the various realizations of WGS84 as representing the same datum. The differences between the various regional and national "canonical realizations"—as well as between the successively produced international realizations of ITRS/ETRS—are on the several-centimeter level. To illustrate this, we mention a recent report (Jivall et al., 2005) that derives the transformation parameters between the various Nordic national realizations of ETRS 89, and a common, truly geocentric system referred to as ITRF2000 epoch 2003.75. This allows the combination of coordinate data from these countries in an unambiguous way.

12.4   NETWORK HIERARCHY IN THE GPS AGE

Some claim that in the GPS age the notion of network hierarchy has become obsolete. We can measure point positions anywhere on Earth, using the satellite constellation directly, without referring to higher-order terrestrial reference points. In reality, if again robustly achieving the highest possible precision is the aim, this isn't quite true.

Measurements using the satellite constellation directly violate the "from the large to the small" principle. If we measure, e.g., independently absolute positions in a terrestrial GPS network on an area of 1000×1000 km using satellites at least 20,000 km away, we will not obtain the best possible *relative* positions between these terrestrial points. Rather, one should measure *vectors* between the terrestrial points, processing measurements made simultaneously from these points to the same satellites, to obtain coordinate *differences* between the points. This *relative GPS measurement* is the standard for precise geodetic GPS.

In relative GPS positioning within a small area, one point may be kept fixed to its conventionally known coordinates, defining a local datum. From this datum point outward, precision deteriorates due to the various error contributions of geodetic GPS. For covering a larger area, one should keep more than one point fixed. These points are typically taken from a globally adjusted point network, like the well-known ITRF or ETRF solutions. In Finland, e.g., one uses points in the EUREF-FIN datum, a national realization of WGS84 providing a field of fixed points covering Finland. To bring a geodetic network into the EUREF-FIN datum, it must be attached to a number of these points, which formally, in the EUREF-FIN datum, are "errorless."

12.5   VARIANCE BEHAVIOR UNDER DATUM TRANSFORMATION

Any realistic description of a geodetic network's precision should capture its *spatial structure*, the fact that interpoint position precision between adjacent points is the better, the closer together the two points are. For points far apart, precision may be poorer, but that will be of no practical consequence. What matters is the *relative* precision, e.g., expressed in parts per million (ppm) of the interpoint distance.

The precision structure of a network depends on its *datum*, the set of conventionally adopted reference points that are used to calculate the network points' coordinates. For example, in the plane, two fixed points may be used to define a coordinate datum; the coordinates of those points, being conventionally agreed, will be errorless. Plotting the uncertainty ellipses describing the coordinate imprecision of the other points, we will see them grow outward from the datum points in all directions.

Choosing a different set of datum points will produce a different-looking pattern of ellipses: zero now on, and growing in all directions outward from, these new datum points. Yet, the precision structure described is the same, and well-defined transformations exist between the two patterns: *datum transformations*, also called S-transformations.

12.6   CRITERION FUNCTIONS

We refer to the work of Baarda (1973) for the notion of criterion matrices, as well as the related notion of S- or datum transformations. The precision of a set of network points can be described collectively by a *variance-covariance matrix*, giving the variances and covariances of network point coordinates. If all point positions are approximately known, as well as the precision of all geodetic measurements made between them, this variance matrix is obtained as a result of the least-squares adjustment of the network.

In a three-dimensional network of n points there will be $9n^2$ elements to the variance matrix—or $\frac{3}{2}n \cdot (3n + 1)$ essentially different ones—so this precision representation doesn't scale very favorably. Also, the original measurements and their precisions may be uncertain or not readily available. For this reason, geodesists have been looking for ways of describing the precision structure of a geodetic network—realistically, if only approximately—using a small number of defining parameters. Such synthetic variance matrices are called *criterion matrices* and their generating functions *criterion functions*.

Criterion functions are an attractive and parsimonious way to describe the precision structure of geodetic point sets or corpora of spatial information. They offer a more complete description than point or interpoint coordinate precision, yet take less space than full variance matrices, while in practice being likely just as good.

A formal requirement to be placed upon criterion matrices is that they transform under datum transformations in the same way as real variance-covariance matrices would do. As this is known geodetic theory, we will not elaborate further.

12.7 GEOCENTRIC VARIANCE STRUCTURE OF A GPS NETWORK

Let us first derive a rough but plausible geocentric expression for the variance-covariance structure of a typical geodetic network. The true error propagation of GPS measurements is an extremely complex subject. Here, we try to represent the bulk coordinate precision behavior in a simple but plausible way.

Also, the full theory of criterion matrices and datum transformations is complicated (Baarda, 1973; Vermeer et al., 2004). Here we shall cut some corners. We assume that the interpoint position variance between two network points A and B, coordinates (X_A, Y_A, Z_A) and (X_B, Y_B, Z_B), is of the form

$$\text{Var}(\mathbf{r}_B - \mathbf{r}_A) = Q_0 \left((X_B - X_A)^2 + (Y_B - Y_A)^2 + (Z_B - Z_A)^2 \right)^{\frac{k}{2}} = Q_0 d_{AB}^k \quad (12.1)$$

with k and Q_0 as the free parameters (assumed constant for now) and $d_{AB} = \|\mathbf{r}_A - \mathbf{r}_B\|$ the $A - B$ interpoint distance.

For this to be meaningful, we must know what is meant by the variance or covariance of vectors. In three dimensions, we interpret this as

$$\text{Cov}(\mathbf{r}_A, \mathbf{r}_B) = \text{Cov}\left(\begin{bmatrix} X_A \\ Y_A \\ Z_A \end{bmatrix}, \begin{bmatrix} X_B \\ Y_B \\ Z_B \end{bmatrix} \right) = \begin{pmatrix} \text{Cov}(X_A, X_B) & & \\ & \ddots & \\ & & \text{Cov}(Z_A, Z_B) \end{pmatrix},$$

i.e., a 3×3 elements tensorial function. Also, Q_0 is in this case a 3×3 tensor. The approach is not restricted to three dimensions, however.

Equation 12.1 is fairly realistic for a broad range of geodetic networks: for (one-dimensional) leveling networks we know that $k = 1$ gives good results. In this case, $\sqrt{Q_0} = \sigma_0$, a scalar called the *kilometer precision* is expressed in mm/$\sqrt{\text{km}}$. For two-dimensional networks on the Earth's surface, we have due to isotropy $Q_0 = \sigma_0^2 I_2$, with I_2 the 2×2 unit matrix. This is valid in a small enough area for the Earth's curvature to be negligible, so that map projection coordinates (x, y) can be used.

Also for GPS networks an exponent of $k = 1$ has been found appropriate (e.g., Beutler et al., 1989). The 3×3 matrix contains the component variances and will, in a local horizon system (x, y, H) in a small enough area, typically be diagonal:

$$Q_{0,hor} = \begin{pmatrix} \sigma_h^2 & & \\ & \sigma_h^2 & \\ & & \sigma_v^2 \end{pmatrix},$$

where σ_h^2 and σ_v^2 are the separate horizontal and vertical standard variances. In a geocentric system we then get the location-dependent expression

$$Q_0(\mathbf{r}) = R(\mathbf{r})Q_{0,hor}R^T(\mathbf{r})$$

with $R(\mathbf{r})$ the rotation matrix from geocentric to local horizon orientation for location r.

Now if we choose the following expressions for the variance and covariance of absolute (geocentric) position vectors:

$$\text{Var}(\mathbf{r}_A) = Q_0(\mathbf{r}_A)R^k,$$

$$\text{Var}(\mathbf{r}_B) = Q_0(\mathbf{r}_B)R^k,$$

$$\text{Cov}(\mathbf{r}_A, \mathbf{r}_B) = \bar{Q}_{0,AB}\, d_{AB}^k,$$

with R the Earth's mean radius, then we obtain the following, generalized expression for the difference vector:

$$\text{Var}(\mathbf{r}_B - \mathbf{r}_A) = \bar{Q}_{0,AB}\left[R^k - \frac{1}{2}d_{AB}^k\right],$$

with

$$\bar{Q}_{0,AB} = \frac{1}{2}\left[Q_0(\mathbf{r}_A) + Q_0(\mathbf{r}_B)\right].$$

This yields a consistent variance structure.

In practice, the transformation to a common geocentric frame will be done using known parameters found in the literature (Boucher and Altamimi, 2001) for a number of combinations ITR-Fxx/ETRFyy, where xx/yy are year numbers. Our concern here is only the precision of the coordinates thus obtained. We need to know this precision when combining GPS datasets from domains having different canonical WGS84 realizations, requiring their transformation to a suitable common frame.

12.8   AFFINE TRANSFORMATION ONTO SUPPORT POINTS

Often, one connects traditional local datums to a global datum by an overdetermined Helmert transformation with least-squares estimated parameters. While this will work well in a small area, it doesn't yield geodetic precision over larger national or continental domains.

The PROJ.4 software models such transformations more precisely by augmenting the Helmert transformation by a regular "shift grid" of sufficient density describing a residual deformation field between the two datums. Unfortunately this technique obfuscates how these shifts were originally determined, usually by using a field of irregularly located "common points" known in both global and local systems.

We may derive a plausible variance structure for the current Finnish practice documented in Anon. (2003) of transforming existing old *kkj* network coordinates into the new EUREF-FIN datum by a per-triangle affine transformation applied to a Delaunay triangulation of the set of points common to both datums. The parameters of this transformation follow from the shift vectors in a triangle's corner points and produce an overall transformation continuous over triangle boundaries. We abstract from the actual process producing those local measurements and postulate a formal covariance structure.

Let a given network be transformed to a network of support points assumed exact, forming a (e.g., Delaunay-) triangulation. Let one triangle be ABC and the target point P inside it. The transformation takes the form

$$\mathbf{r}_P^{(ABC)} = \mathbf{r}_P - p^A\left(\mathbf{r}_A - \mathbf{r}_A^{(ABC)}\right) - p^B\left(\mathbf{r}_B - \mathbf{r}_B^{(ABC)}\right) - p^C\left(\mathbf{r}_C - \mathbf{r}_C^{(ABC)}\right)$$

where p^A, p^B, p^C are point P's *barycentric coordinates* within triangle ABC (cf. Vermeer et al., 2004 and Figure 12.1), with always $p^A + p^B + p^C = 1$. These are readily computable.

Then, if we postulate the a priori covariance function to be of the form

$$\mathrm{Cov}(\mathbf{r}_P, \mathbf{r}_Q) = g(\mathbf{r}_P - \mathbf{r}_Q) = g(d_{PQ})$$

with $d_{PQ} = \|\mathbf{r}_P - \mathbf{r}_Q\|$ the $P - Q$ interpoint distance, and assume the "given coordinates" $\mathbf{r}_A^{(ABC)}$, $\mathbf{r}_B^{(ABC)}$, $\mathbf{r}_C^{(ABC)}$ to be error free, we get, by propagation of variances, the a posteriori variance at point P as

$$\mathrm{Var}(\mathbf{r}_P^{(ABC)}) = \begin{bmatrix} 1 & -p^A & -p^B & -p^C \end{bmatrix} \cdot$$

$$\begin{bmatrix} g(0) & g(d_{PA}) & g(d_{PB}) & g(d_{PC}) \\ g(d_{PA}) & g(0) & g(d_{AB}) & g(d_{AC}) \\ g(d_{PB}) & g(d_{AB}) & g(0) & g(d_{BC}) \\ g(d_{PC}) & g(d_{AC}) & g(d_{BC}) & g(0) \end{bmatrix} \begin{bmatrix} 1 \\ -p^A \\ -p^B \\ -p^C \end{bmatrix}.$$

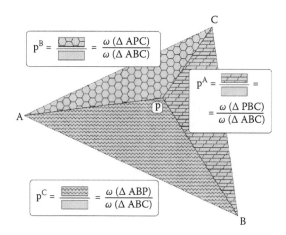

FIGURE 12.1 Barycentric coordinates illustrated. Every barycentric coordinate is the quotient of two triangle surface areas ω; e.g., p^C is the area of triangle ABP divided by the total surface area of *ABC*.

If we further postulate, implicitly defining *f*,

$$\text{Var}(\mathbf{r}_P) = \text{Var}(\mathbf{r}_Q) \quad = g(0) \quad\quad = \alpha^2$$
$$\text{Cov}(\mathbf{r}_P, \mathbf{r}_Q) \quad = g(d_{PQ}) \quad = \alpha^2 - \tfrac{1}{2} f(d_{PQ})$$

then substituting this into the above yields

$$\text{Var}(\mathbf{r}_P^{(ABC)}) = -\frac{1}{2}\begin{bmatrix} 1 & -p^A & -p^B & -p^C \end{bmatrix} \cdot$$

$$\begin{vmatrix} 0 & f(d_{PA}) & f(d_{PB}) & f(d_{PC}) \\ f(d_{PA}) & 0 & f(d_{AB}) & f(d_{AC}) \\ f(d_{PB}) & f(d_{AB}) & 0 & f(d_{BC}) \\ f(d_{PC}) & f(d_{AC}) & f(d_{BC}) & 0 \end{vmatrix} \begin{bmatrix} 1 \\ -p^A \\ -p^B \\ -p^C \end{bmatrix} \quad (12.2)$$

where the arbitrary α^2 (assumed only to make the variance positive over the area of study) has vanished. A plausible form for the function *f*, which describes the *inter-point* (a priori) variance behavior, i.e., that of the point difference vector $\mathbf{r}_Q - \mathbf{r}_P$, would be

$$\text{Var}(\mathbf{r}_Q - \mathbf{r}_P) = f(d_{PQ}) = Q_0 d_{PQ}^k, \quad (12.3)$$

with *k* and Q_0 as the free parameters.

Symbolically we can describe the above as

$$\text{Var}\left(\mathbf{r}_P^{(ABC)}\right) = \mathbf{P}_{P(ABC)} \mathbf{Q}_{P(ABC)}^{P(ABC)} \mathbf{P}_{P(ABC)}^{\text{T}}$$

where

$$\mathbf{p}_{P(ABC)} \equiv \begin{bmatrix} 1 & -p_A & -p_B & -p_C \end{bmatrix}$$

and

$$Q_{P(ABC)}^{P(ABC)} = -\frac{1}{2}\begin{vmatrix} 0 & f(d_{PA}) & f(d_{PB}) & f(d_{PC}) \\ f(d_{PA}) & 0 & f(d_{AB}) & f(d_{AC}) \\ f(d_{PB}) & f(d_{AB}) & 0 & f(d_{BC}) \\ f(d_{PC}) & f(d_{AC}) & f(d_{BC}) & 0 \end{vmatrix}.$$

In Figure 12.2 we give for illustration one example of the point variance behavior after tying to the three corner points of a triangle; cf. Vermeer et al. (2004).

Including the uncertainty of the given points, we can write

$$\mathrm{Var}\left(\mathbf{r}_P^{(ABC)}\right) = \mathbf{p}_{P(ABC)}\left[\mathbf{Q}_{P(ABC)}^{P(ABC)} + \mathbf{Q}_{ABC}^{ABC}\right]\mathbf{p}_{P(ABC)}^{\mathrm{T}}$$

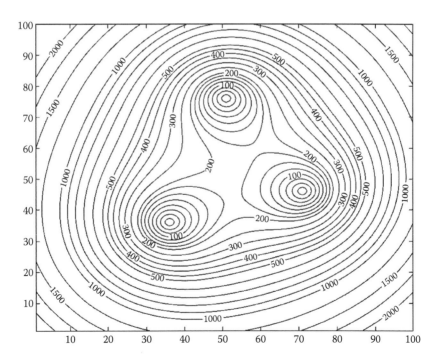

FIGURE 12.2 Example plot of points variance after transformation to support points (assumed errorless) within a single triangle. MatLab™ simulation, arbitrary units.

where we have denoted the a priori variance matrix of the given points by

$$
\mathbf{Q}_{ABC}^{ABC} \equiv
\begin{vmatrix}
0 & 0 & 0 & 0 \\
0 & Q_{AA} & Q_{AB} & Q_{AC} \\
0 & Q_{AB} & Q_{BB} & Q_{BC} \\
0 & Q_{AC} & Q_{BC} & Q_{CC}
\end{vmatrix}
$$

This represents the given points' variance-covariance information, computed geocentrically as described earlier, i.e., $Q_{AA} = \mathrm{Var}(\mathbf{r}_A) = Q_0(\mathbf{r}_A)R^k$, $Q_{AB} = \mathrm{Cov}(\mathbf{r}_A,\mathbf{r}_B) = \frac{1}{2}[Q_0(\mathbf{r}_A) + Q_0(\mathbf{r}_B)][R^k - \frac{1}{2}\, d_{AB}^k]$, etc. As a result, we will obtain the *total* point variances and covariances in a *geocentric, unified* datum.

12.9   INTERPOINT VARIANCES

It is straightforward if laborious to also derive expressions for the a posteriori inter-point variances:

$$
\mathrm{Var}\left(\mathbf{r}_Q^{(DEF)} - \mathbf{r}_P^{(ABC)}\right) = \mathrm{Var}\left(\mathbf{r}_Q^{(DEF)}\right) + \mathrm{Var}\left(\mathbf{r}_P^{(ABC)}\right) - 2\,\mathrm{Cov}\left(\mathbf{r}_Q^{(DEF)},\mathbf{r}_P^{(ABC)}\right), \quad (12.4)
$$

by application of variance propagation like in Equation 12.2: separately for the cases of P and Q within the same triangle, in different triangles, or in different but adjacent triangles sharing a node or a side. We obtain, for the general case of different triangles ABC and DEF:

$$
\mathrm{Cov}\left(\mathbf{r}_Q^{(DEF)},\mathbf{r}_P^{(ABC)}\right) = \mathbf{P}_{Q(DEF)}\left[\mathbf{Q}_{P(ABC)}^{Q(DEF)} + \mathbf{Q}_{ABC}^{DEF}\right]\mathbf{P}_{P(ABC)}^{\mathrm{T}}
$$

where

$$
\mathbf{Q}_{P(ABC)}^{Q(DEF)} = -\frac{1}{2}
\begin{vmatrix}
0 & f(d_{DP}) & f(d_{EP}) & f(d_{FP}) \\
f(d_{QA}) & 0 & f(d_{EA}) & f(d_{FA}) \\
f(d_{QB}) & f(d_{DB}) & 0 & f(d_{FB}) \\
f(d_{QC}) & f(d_{DC}) & f(d_{DC}) & 0
\end{vmatrix}
$$

and

$$
\mathbf{Q}_{ABC}^{DEF} \equiv
\begin{vmatrix}
0 & 0 & 0 & 0 \\
0 & Q_{DA} & Q_{DB} & Q_{DC} \\
0 & Q_{EA} & Q_{EB} & Q_{EC} \\
0 & Q_{FA} & Q_{FB} & Q_{FC}
\end{vmatrix}.
$$

From this we obtain the general relative variance expression by substitution into Equation 12.4. Note that for a datum of this type, the locations of the fixed points used become *part of the datum definition*, though for any single variance or covariance to be computed, only six point positions are needed at most.

When representing the spatial precision structure in this way, the representation chosen should also be *semantically valid*, in that it should be possible to extract both point and interpoint mean errors for specified points and *use them*, e.g., for detecting inconsistencies between different data sources by statistical testing. This is related to the topic of the Semantic Web and the use of ontologies for specifying integrity constraints (K. Virrantaus, personal comm.; Mäs et al., 2005).

12.10   THE CASE OF UNKNOWN POINT LOCATIONS

If the locations of the common fit points are not actually known, we may derive a *bulk covariance structure* that does not depend on them. Assume a mean point spacing D and a uniform triangle size. Formula 12.2 yields, with $P = A$,

$$\mathrm{Var}(\mathbf{r}_P^{(ABC)}) = -\frac{1}{2}\begin{bmatrix} 1 & -1 & 0 & 0 \end{bmatrix} \cdot \begin{bmatrix} 0 & f(d_{AA}) & f(d_{AB}) & f(d_{AC}) \\ f(d_{AA}) & 0 & f(d_{AB}) & f(d_{AC}) \\ f(d_{AB}) & f(d_{AB}) & 0 & f(d_{BC}) \\ f(d_{AC}) & f(d_{AC}) & f(d_{BC}) & 0 \end{bmatrix} \begin{bmatrix} 1 \\ -1 \\ 0 \\ 0 \end{bmatrix}$$

$$= -\frac{1}{2}\begin{bmatrix} 1 & -1 \end{bmatrix}\begin{bmatrix} 0 & 0 \\ 0 & 0 \end{bmatrix}\begin{bmatrix} 1 \\ -1 \end{bmatrix} = 0$$

and similarly for the other corner points. The a posteriori variance reaches its maximum in the center of gravity of the triangle, where the barycentric weights are $p^A = p^B = p^C = \frac{1}{3}$. Assuming furthermore that the triangle is equiangular, i.e., $d_{AB} = d_{AC} = d_{BC} \equiv D$, we have also

$$d_{PA} = d_{PB} = d_{PC} = \frac{D}{\sqrt{3}}$$

and

$$\mathrm{Var}(\mathbf{r}_P^{(ABC)}) = -\frac{1}{2}\begin{bmatrix} 1 & -\frac{1}{3} & -\frac{1}{3} & -\frac{1}{3} \end{bmatrix} \cdot \begin{bmatrix} 0 & f(\frac{D}{\sqrt{3}}) & f(\frac{D}{\sqrt{3}}) & f(\frac{D}{\sqrt{3}}) \\ f(\frac{D}{\sqrt{3}}) & 0 & f(D) & f(D) \\ f(\frac{D}{\sqrt{3}}) & f(D) & 0 & f(D) \\ f(\frac{D}{\sqrt{3}}) & f(D) & f(D) & 0 \end{bmatrix} \begin{bmatrix} 1 \\ -\frac{1}{3} \\ -\frac{1}{3} \\ -\frac{1}{3} \end{bmatrix}$$

$$= f(\frac{D}{\sqrt{3}}) - \frac{1}{3}f(D).$$

For power law 12.3 we obtain

$$\text{Var}(\mathbf{r}_P^{(ABC)}) = \sigma_0^2 D^k 3^{-k/2} - \frac{1}{3}\sigma_0^2 D^k = \sigma_0^2 D^k (3^{-k/2} - 3^{-1}).$$

For $k = 1$ this becomes

$$\text{Var}(\mathbf{r}_P^{(ABC)}) = \sigma_0^2 D \cdot \frac{1}{3}\left(\sqrt{3} - 1\right) \approx 0.244\sigma_0^2 D.$$

We can symbolically write

$$\Delta_k \equiv (3^{-k/2} - 3^{-1}).$$

We use the above derived upper bound for the single point variance and postulate the following replacement variance structure:

$$\text{Var}(\mathbf{r}_P^{(\Delta)}) = \Delta_k \sigma_0^2 D^k,$$

$$\text{Cov}(\mathbf{r}_P^{(\Delta)}, \mathbf{r}_Q^{(\Delta)}) = \Delta_k \sigma_0^2 D^k - \frac{1}{2}F(d_{PQ}).$$

Note that *here*, the constant $\Delta_k \sigma_0^2 D^k$, unlike α^2 above, is no longer arbitrary. It does similarly vanish, however, when we derive the interpoint variance:

$$\text{Var}\left(\mathbf{r}_Q^{(\Delta)} - \mathbf{r}_P^{(\Delta)}\right) = \text{Var}\left(\mathbf{r}_Q^{(\Delta)}\right) + \text{Var}\left(\mathbf{r}_P^{(\Delta)}\right) - 2\,\text{Cov}\left(\mathbf{r}_Q^{(\Delta)}, \mathbf{r}_P^{(\Delta)}\right) = F(d_{PQ}).$$

We wish to see a variance structure, in which these a posteriori interpoint variances behave in the following reasonable way:

- For P and Q close together (and often within the same triangle), we want the relative variance to behave according to the k-power law.
- For larger distances, and P and Q in different triangles, we want the relative variance to "level off" to a constant value. We know it can never exceed twice the posterior variance of a single point, which is $\Delta_k \sigma_0^2 D^k$ max (and never less than 0, which happens if both P and Q coincide with nodes of the triangulation).

Therefore we choose

$$F(d_{PQ}) = \frac{1}{\frac{1}{\sigma_0^2 d_{PQ}^k} + \frac{1}{2\Delta_k \sigma_0^2 D^k}} = \sigma_0^2 \frac{1}{\frac{1}{d_{PQ}^k} + \frac{1}{2\Delta_k D^k}} = \sigma_0^2 \frac{2\Delta_k d_{PQ}^k D^k}{d_{PQ}^k + 2\Delta_k D^k},$$

which behaves in this way, with a smooth transition between the two regimes.

12.11   FINAL REMARKS

We believe that geodesists and spatial information specialists should get better acquainted with each other's ideas. Precise geodetic information still commonly moves around as files of coordinates, processed by dedicated software to maintain the highest precision. Dissemination using standard Web services promises many practical benefits but is not well known in geodetic circles and currently used only for mapping-grade geographic information.

Now, also in geodesy, awareness is growing, e.g., in connection with the GGOS (Geodetic Global Observing System) initiative (cf. Neilan et al., 2005) that precise geodetic coordinate information should be seen and integrated as part of our spatial data infrastructure. Care should then be taken to properly represent and manage its spatial precision structure. In this chapter we have not addressed the issue of coordinate changes with time due to geodynamical processes. At the current precision of geodetic measurement, this issue must be taken into account as well.

12.12   CONCLUSIONS

We have derived criterion matrix expressions for modeling the variance-covariance behavior of the geocentric coordinates of a set of GPS-determined "fixed points," as well as coordinates in a local geodetic network that have been tied to a set of GPS-positioned points by a triangle-wise affine (bi-linear) transformation

We were motivated to present these derivations by their possible use in coordinate Web services for geodesy. They will allow proper coordinate precision modeling when bringing geodetic coordinate material from heterogeneous sources on a single common geocentric datum.

ACKNOWLEDGMENTS

Discussions with Reino Ruotsalainen and Antti Jakobsson of the Finnish National Land Survey, and with Kirsi Virrantaus of the TKK Dept. of Surveying, are gratefully acknowledged, as is the support of the National Committee for Geodesy and Geophysics making possible the participation of one of us (MV) in the Dynamic Planet 2005 Symposium in Cairns, Australia. Furthermore, one of us (KK) gratefully acknowledges the Kristjan Jaak Scholarship Foundation and the Estonian Land Board for travel support to the ISSDQ 2007 Symposium in Enschede, The Netherlands. Also, the detailed remarks by three anonymous *Journal of Geodesy* reviewers on a very rough precursor of this chapter are gratefully acknowledged.

REFERENCES

Anon., 2003. JHS154. ETRS89—järjestelmään liittyvät karttaprojektiot, tasokoordinaatistot ja karttalehtijako (Map projections, plane coordinates and map sheet division in relation to the ETRS89 system). Web site, Finnish National Land Survey. URL: www .jhs-suositukset.fi/intermin/hankkeet/jhs/home.nsf/files/JHS154/$file/JHS154.pdf, accessed August 30, 2005.

Anon., 2005a. Deegree – building blocks for spatial data infrastructures. URL: http://deegree .sourceforge.net/, accessed August 29, 2005.

Anon., 2005b. MapServer Homepage. URL: http://mapserver.gis.umn.edu/, accessed August 29, 2005.

Anon., 2005c. PROJ.4—Cartographic Projections Library. URL: http://www.remotesensing .org/proj/, accessed August 29, 2005.

Anon., 2005d. The GeoServer Project: an Internet gateway for geodata. URL: http://geoserver.sourceforge.net/html/index.php, accessed August 29, 2005.

Baarda, W., 1973. S-transformations and criterion matrices. Publications on Geodesy, Netherlands Geodetic Commission, Delft. New Series, Vol. 5, No. 1.

Beutler, G., Bauersima, I., Botton, S., Boucher, C., Gurtner, W., Rothacher, M., and Schildknecht, T., 1989. Accuracy and biases in the geodetic application of the global positioning system. *Manuscripta geodaetica* 14(1), pp. 28–35.

Boucher, C. and Altamimi, Z., 2001. Specifications for reference frame fixing in the analysis of a EUREF GPS campaign. Memo. December 4. URL: lareg.ensg.ign.fr/EUREF/memo.pdf.

European Petroleum Survey Group, 2005. EPSG Geodetic Parameter Dataset v. 6.7. URL: http://www.epsg.org/Geodetic.html, accessed August 19, 2005.

Jivall, L., Lidberg, M., Nørbech, T., and Weber, M., 2005. Processing of the NKG 2003 GPS Campaign. Reports in Geodesy and Geographical Information Systems LMV-rapport 2005:7, Lantmäteriet, Gävle.

Jones, B. A., 2005/2006. Where did that geospatial data come from? *ESRI ArcNews* 27(4), pp. 1–2.

Kollo, K., 2004. The coordinate management service. Internal report, TKK Surveying Dept., Inst. of Geodesy.

Mäs, S., Wang, F., and Reinhardt, W., 2005. Using ontologies for integrity constraint definition. In: *Proceedings, 4th Int. Symp. Spatial Data Quality*, Beijing 2005, pp. 304–313.

Neilan, R., 2005. Integrated data and information system for the Global Geodetic Observing System. In: *Dynamic Planet* 2005 *Symposium*, Cairns, Australia, IAG. Invited paper, unpublished.

OGC, 2001. Web Map Service Implementation Specification. Open GIS Consortium Inc., Jeff de La Beaujardière, Editor. URL: http://www.opengeospatial.org/docs/01-068r2. ppf, accessed April 27, 2005.

OGC, 2002. Web Feature Service Implementation Specification. Open GIS Consortium Inc., Panagiotis A. Vretanos, Editor. URL: https://portal.opengeospatial.org/files/?artifact_ id=7176, accessed April 27, 2005.

OGC, 2003. OpenGIS ® Geography Markup Language (GML) Implementation Specification. Open GIS Consortium Inc., Simon Cox, Paul Daisey, Ron Lake, Clemens Portele, Arliss Whiteside, Editors. URL: http://www.opengeospatial.org/docs/02-23r4.pdf, accessed April 27, 2005.

Vermeer, M., Väisänen, M., and Mäkynen, J., 2004. Paikalliset koordinaatistot ja muunnokset (local coordinate systems and transformations). Publication 37, TKK institute of Geodesy, Otaniemi, Finland.

13 Analysis of the Quality of Collection 4 and 5 Vegetation Index Time Series from MODIS

René R. Colditz, Christopher Conrad, Thilo Wehrmann, Michael Schmidt, and Stefan Dech

CONTENTS

13.1 INTRODUCTION

Time series provide the possibility to monitor interannual and intra-annual processes of the Earth's surface. Annual cycles of vegetative activity are used for phenological analysis (Asner et al., 2000), crop monitoring (Tottrup and Rasmussen, 2004), or estimating net primary productivity (Running et al., 2000). Changes or modifications of these cycles due to droughts (Tucker et al., 1994), El Niño events (Anyamba et al., 2002), or human impacts (de Beurs and Henebry, 2004) are observed with multiannual time series mostly using vegetation indices such as NDVI from the AVHRR sensor. Climate modeling, change detection studies, and other applications in the framework of global change require high-quality time series with a standardized, consistent, and reliable time series generation process (Sellers et al., 1996; Justice et al., 2002). Therefore, the quality of the time series determines its usability for long-term analysis.

The level of required data quality is highly dependent on the subsequent analysis. Hereby data quality is related to the level of uncertainty contained in the data and propagated to the results with a high influence on their accuracy (Atkinson and Foody, 2002). With regard to time series, in particular vegetation indices describing the phenological development, the consistency during the year and for multiple years is most important (Roy et al., 2002). Intra-annual variations of a vegetation index profile that cannot be attributed to actual changes on the Earth's surface are serious quality issues. For example, cloud coverage and other atmospheric particles have a substantial influence on the signal and need to be either corrected or at least indicated. Intra-annual comparisons such as trend estimations and mapping of subtle multiyear earth surface processes (e.g., bush encroachment in semi-arid environments) may yield misleading conclusions if either the data are not corrected for sensor degradations or sensors generations are not correctly intercalibrated.

Data from both MODIS instruments are used for a large suite of global, value-added products. The improved sensor design and the standardized radiometric, geometric, and atmospheric calibrations are suitable for high-quality time series (Justice et al., 1998). The MODIS data production put much emphasis on the data quality starting at raw level 1 data to level 4 modeled products. The innovative concept of a simultaneous generation of remote sensing products and quality assurance indicators facilitates standardized and consistent global products. This is particularly important for high temporal resolution products suitable for time series generation but should also be considered for other satellite datasets.

Several MODIS science teams are concerned with quality assurance and product validation (Roy et al., 2002; Morisette et al., 2002). The land data operational product evaluation facility (LDOPE) tests the accuracy and consistency of all MODIS land products. Additional quality assurances are computed by science computing facilities (SCF) for individual products. It is only possible to investigate a selection of all MODIS products for particular areas. Both, LDOPE and SCF ensure high data quality by visual inspections and a number of operational checks, e.g., using time series of summary statistics for globally distributed regions (Roy et al., 2002). General and product-specific quality information is provided for the user as metadata and at the pixel-level. The science quality flags of the metadata describe quality issues for the entire spatial extent and contain the informative quality indicators in seven levels for data ordering. The pixel-level information (quality assurance science data set; QA-SDS) can be used to assess the quality of each grid cell (Roy et al., 2002). This unique concept of product-specific pixel-level quality indicators provides full information, maximum flexibility, and leaves the decision about the sufficiency of data quality to the user.

Multiple versions, also called collections, of MODIS data have been released since the launch of MODIS onboard Terra and Aqua in 2000 and 2002, respectively. A new collection of MODIS products incorporates the most recent scientific findings into data processing and requires a complete reprocessing of the current data archive. This chapter analyzes the quality of time series of the present collection 4 (C4) and the currently released collection 5 (C5) for the vegetation index product MOD13. The Time Series Generator (TiSeG) was used to evaluate the pixel-level QA-SDS and to interpolate invalid data (Colditz et al., 2006a, 2008). The chapter describes the modifications in quality retrieval and changes in quality settings between both collections and shows the impacts on annual NDVI and EVI time series for selected natural regions and land cover types in Germany.

13.2   CHANGES OF THE VEGETATION INDEX PRODUCT IN COLLECTION 5

Two vegetation indices, NDVI and EVI, are included in the MODIS product (MOD13). The NDVI, also known as the continuity index, matches to long-term observations of the AVHRR instruments. The EVI has an improved sensitivity in high biomass areas and decouples the vegetation signal from soil background and atmospheric influences (Huete et al., 2002).

Considerable changes in science and structure were applied to C5 of the MODIS vegetation index product (Didan and Huete, 2006). Scientific modifications were made to (1) cloud and aerosol retrieval, (2) a different backup algorithm for EVI computation, and (3) an improved constrained view maximum value compositing for better spatial consistency. The analysis of cloudy pixels in C4 yielded residual pixels labeled clear, and vice versa. For example, insufficient cloud masking in C4 data was observed at the margins of clouds and for partly cloudy pixels. Furthermore, the aerosol retrieval was insufficient for heavy aerosols and if climatology parameters had to be used. Changes in C5 occurred in data filtering with emphasis on cloud shadows, pixels adjacent to clouds, and aerosols. The simpler maximum value compositing approach (Holben, 1986) is used in C5 if all pixels during the compositing period are cloudy, partly cloudy, or adjacent to clouds. Second, in C4, the soil-adjusted vegetation index (SAVI; Huete, 1988) was used as an EVI backup for cloudy pixels, snow- and ice-covered surfaces, or if the blue band was out of range. C5 uses a newly developed equation, called EVI2, for better continuity with the standard EVI (Huete et al., 2002):

$$EVI2 = 2.5 \frac{\rho_{NIR} - \rho_{RED}}{1 + \rho_{NIR} + \rho_{RED}}$$

The constrained view maximum value compositing approach (Huete et al., 2002) in C4 processing considers only the two highest values with a deviation of less than 30% from the maximum and selects the value with the lowest view zenith angle. Despite the ratio effect, the selection of different days for adjacent pixels resulted in a low spatial connectivity in the composite. The approach was modified to a deviation of less than 10% and contextual selection according to the temporal behavior of suitable pixels. This omits a high temporal variability of selected days for compositing, i.e. adjacent pixels are more likely to be selected from the same observation or another observation close to the neighbor.

Structural changes comprise (1) additional layers, (2) modifications of the QA-SDS specifications, and (3) phased production between Terra and Aqua data. A layer indicating the day selected for compositing and an indicator for pixel reliability were added. In particular, the actual day of each vegetation index will be helpful to many vegetation studies and will enable more accurate monitoring and timing of phenological key stages such as the beginning of green-up or senescence. In C4, the actual day of image acquisition within the 16-day composite period was unknown. The new reliability SDS includes important pixel-level information on cloud cover, snow and ice surfaces, general usability, and fill values for pixels that cannot be retrieved. Second, negligible differences between separate QA-SDS for NDVI and EVI of C4 data lead to a combined QA-SDS in C5. In addition to the reliability SDS, structural modifications of the QA-SDS involve a full three-bit land and water mask instead of a two-bit reduction. In exchange, the bit field of the compositing approach was eliminated because the algorithm does not use the BRDF compositing method. An 8-day phasing in the production of 16-day composites between Aqua and Terra allows the generation of a combined 8-day time series. Additional changes in C5 are an effective internal compression and additional metadata parameters.

A merge of C4 and C5 data is not recommended by the MODIS scientists (Didan and Huete, 2006). Changes in the generation of the QA-SDS will lead to rather different results, and changes in the algorithm, e.g., EVI2, contribute to a different absolute vegetation index value.

13.3   TIME SERIES GENERATION

Multiple approaches have been successfully used for time series production mainly focusing on AVHRR datasets (el Saleous et al., 2000; Viovy et al.,1992; Colditz et al., 2006b; Jönsson and Eklundh, 2002; Roerink et al., 2000). New sensor developments and dataset production systems e.g., for MERIS and MODIS, also create data quality indicators (Brockmann, 2004; Roy et al., 2002). These ancillary datasets have been successfully used for data analysis and time series generation (Leptoukh et al., 2005; Lobell and Asner, 2004; Landmann et al., 2005; Lunetta et al., 2006). The Time Series Generator (TiSeG; Conrad et al., 2005; Colditz et al. 2006a, 2008) evaluates the pixel-level QA-SDS available for all value-added MODIS land products and selects suitable pixels according to user-defined settings. The resulting gaps can be masked or interpolated by temporal or spatial functions. The freely available software package generates two indices of data availability for time series quality assessment: the number of invalid pixels and the maximum gap length. While the first indicates the total of useful data for the entire period, the latter is an important indicator for a feasible interpolation. In order to mitigate interpolation problems, quality settings can be modified spatially and temporally. A detailed description of TiSeG and examples of time series for various quality settings are described in Colditz et al. (2008). The software package has been extended to C5 data, and adjustments have been made to include the redefined QA-SDS and the additional reliability SDS.

C4 and C5 data of Germany with 500 m resolution and 16-day compositing period (MOD13A1) were downloaded for one year starting in mid-February: day 2000-049 (the earliest available day) to day 2001-033 (the completed period of C5 data production at the point of writing). The tiles h18v03 and h18v04 were mosaiced, reprojected to UTM zone 32N, and subset to the area of Germany using the MODIS reprojection tool. EVI and NDVI time series with three different quality settings were generated (Table 13.1) and interpolated using linear temporal interpolation. Specifics on the MOD13 data generation and quality assessment approach can be obtained from Huete et al., (1999, 2002). The usefulness index is a weighted score and consists of several other indicators including cloud coverage and shadow, adjacency and BRDF correction during surface reflectance processing, angular information, and aerosol

TABLE 13.1

Quality Settings of 16-Day 500 m Vegetation Index Data (MOD13A1) for Collections 4 and 5

Setting	Usefulness Index	Mixed Clouds	Snow/Ice	Shadow
C-S-S		No	No	No
UI3-C-S-S	Perfect–Acceptable	No	No	No
UI5	Perfect–Intermediate			

Note: The table only shows the quality settings used in this analysis. For a detailed description on quality settings for MODIS vegetation index products, see Huete et al. (1999) and Didan and Huete (2006).

quantity (Huete et al., 1999). It ranks from perfect to low quality. C5 data were generated without and with consideration of the newly introduced reliability SDS (indicated by C5+R), selecting good and marginal quality.

13.4 TIME SERIES ANALYSIS

Temporal plots of the number of invalid pixels and the maximum gap length, indicating data availability for time series generation, are depicted in Figure 13.1. With regard to the quality settings, UI3-C-S-S was the strictest, followed by C-S-S and UI5. Considerable differences are shown between C4 and C5 data for the number of invalid pixels with different trends for day 49 to 81, day 177 to 209, and day 305 to 1. While the first and last periods mark the end and beginning of wintertime and transitional seasons with both snow and cloud effects, the middle period of the lower data quality is related to a typical summer rain period in July. The average and maximum differences in data availability in percent between collections are shown in Table 13.2. Generally, the quality analysis of C5 data is stricter and therefore omits more pixels. Some reverse patterns are shown for lenient setting UI5, which excludes more data in C4 for days 81, 193, 353, and 17. This effect is due to changes in the masking of clouds and shadow as well as aerosol mapping, which contribute to a different score of the usefulness index. Furthermore, the additional use of the reliability SDS for C5 excludes more pixels from the analysis (see also Table 13.2).

The cumulative plot of the maximum gap length (Figure 13.1) is a suitable indicator of how many data can be interpolated with sufficient confidence (Colditz et al., 2008). For example, 95% of C4 data with lenient setting UI5 have a gap shorter or equal to three composites. On the other hand, at a gap of five consecutively missing observations, 92% of all C4 data with strict setting UI3-C-S-S are interpolated. Stricter settings cause more invalid pixels, which often lead to longer gaps. It depends on the user and the subsequent analysis which maximum gap length is still considered appropriate to interpolate. Although the maximum gap length is mainly used as a feasibility indicator for interpolation, it also shows differences between collections. While the above-noted effect of UI5 with slightly more omitted pixels in C4 than C5 also causes a somewhat longer gap length, all other settings show remarkably longer data gaps for C5 data. The additional use of the reliability SDS makes a high

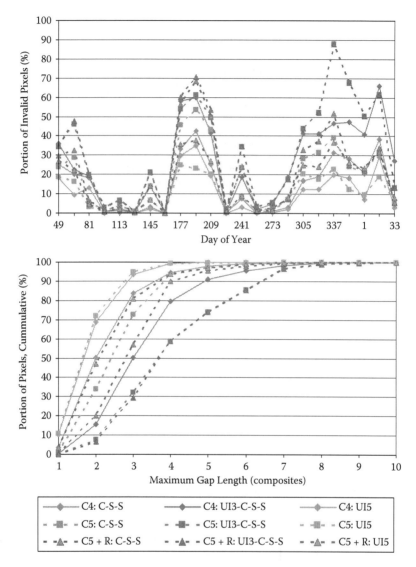

FIGURE 13.1 Temporal plot of the number of invalid pixels (left) and maximum gap length (right) for quality settings (see Table 13.1) of C4, C5, and C5+R data of Germany.

difference for lenient setting UI5 and is attributed to the additional cloud and snow/ice masking. The impact of the reliability SDS becomes lower for moderate setting C-S-S and negligible for strict setting UI3-C-S-S. While a strict QA-SDS setting already excludes most of the possible pixels due to reasons such as low angles or other failed corrections that contribute to a high score of the usefulness index, those pixels are still regarded as valid by lenient quality settings.

The spatial distribution of data availability is shown in Figure 13.2 for the number of invalid pixels. Generally, lower data quality is found in upland regions in

TABLE 13.2

Mean and Highest Differences in the Annual Course of the Number of Invalid Pixels [%] between C4 C5 and C4–C5+R

Settings	C4 – C5		C4 – C5 + R	
	Mean	Max.	Mean	Max
C-S-S	6.4	17.9	9.0	25.6
UI3-C-S-S	7.7	41.4	8.3	41.4
UI5	4.0	11.8	6.1	16.4

middle Germany and the alpine region in the South. Considerable differences are apparent among quality settings, where the strictest setting, UI3-C-S-S, also indicates less data availability in northern Germany due to frequent cloud coverage. Interestingly, the increasing continental characteristic with less cloud cover during the summer months in southern Germany is clearly revealed by this regional analysis. The comparison of C4 and C5 data visualizes no spatial differences for lenient setting UI5 but decreasing data availability for increasingly stricter settings where substantially more pixels are regarded invalid for upland areas and in the lowland of northern Germany. The difference when using the reliability SDS is also illustrated in Figure 13.2 and becomes particularly clear for lenient and moderate settings UI5 and C-S-S.

A third analysis was made for selected regions of Germany using the CORINE land cover classification (Keil et al., 2005) and natural regions of Germany (Meynen and Schmithüsen, 1953). Schleswig is located in northern Germany and dominated by grazing land for sheep and cattle. The Harz upland is Germany's northernmost upland with peaks above 1000 m. In particular, its western portion is dominated by dense coniferous forests. These coniferous stands are compared to the Alpine region in southern Germany.

Figure 13.3 (NDVI) and Figure 13.4 (EVI) show average time series plots of C4, C5, and C5+R for the land cover types and regions mentioned above. The time series correspond with the quality settings of Table 13.1 and also include the original dataset without quality analysis. The original time series shows substantial short-term temporal variability, in particular for the NDVI, which is attributed to atmospheric disturbances (clouds, aerosols) and snow/ice cover. It should be noted that the NDVI and EVI SDS themselves do not indicate cloud coverage, the major source of remarkably decreasing vegetation index values. Instead, they contain a value in the valid data range. Only the examination of the additional information, the QA-SDS and the reliability SDS for C5, reveals these influences. This is typical for many MODIS land products where only the additional data quality specifications indicate the data usability (Roy et al., 2002). Lenient setting UI5 often follows the original plot and therefore does not seem to be an improvement. On the other hand, the strict setting UI3-C-S-S eliminates most pixels and requires the interpolation of long periods. This also causes unrealistic features and does not resemble expected phenologies,

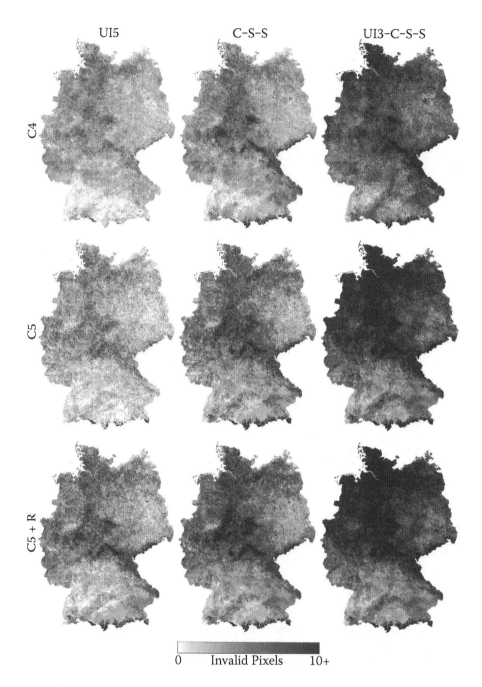

FIGURE 13.2 Spatial distribution of the number of invalid pixels for quality settings (Table 13.1) of C4, C5, and C5+R data of Germany. Note: More than 10 out of 23 invalid pixels are depicted in black.

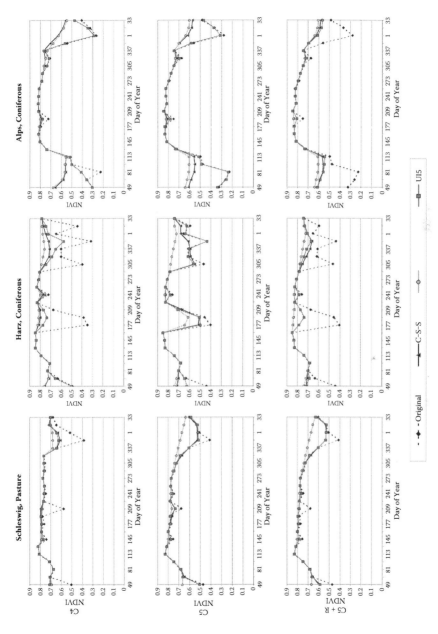

FIGURE 13.3 NDVI time series plots of natural regions and land cover types for quality settings (see Table 13.1) of C4, C5, and C5+R.

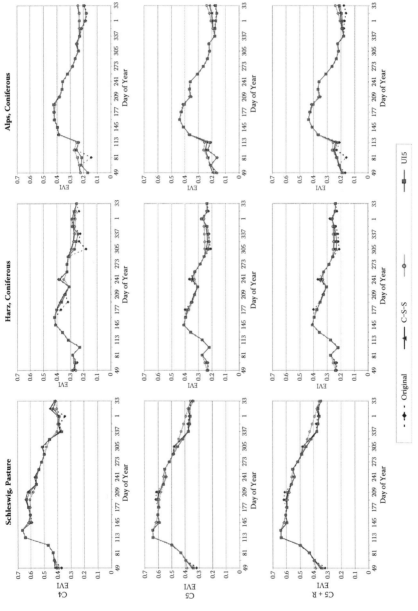

FIGURE 13.4   EVI time series plots of natural regions and land cover types for quality settings (see Table 13.1) of C4, C5, and C5+R.

e.g., during the winter season in Schleswig for the NDVI. The time series of this study are best generated with the moderate setting C-S-S, which yields expected phenological patterns. For an in-depth discussion on quality settings and time series generation, see Colditz et al. (2008).

The following discussion will focus on the differences in data collections. The NDVI data of Figure 13.3 show clear differences for interpolated time series between C4 and C5 products. The stricter characteristics of C5 quality settings are indicated for the winter period of Schleswig. Differences between C5 and C5+R data are well illustrated in Figure 13.3 for the Harz upland. While the summer cloud period from day 161 to day 225 is sufficiently well interpolated by C4, even the strictest quality setting of C5 results in a clear decrease in NDVI if only C5 QA-SDS is considered. The additional use of the reliability SDS, however, interpolates this three-composite gap of invalid data successfully and shows expected phenological plots with a pronounced plateau phase during the summer. This proves that, in contrast to C4, setting mixed pixels in the QA-SDS of C5 does not necessarily mark all cloudy pixels. In other words, in addition to the QA-SDS, the reliability SDS should be considered to exclude all invalid data. For the Harz and the Alps, snow coverage during the winter period is also successfully interpolated with the reliability SDS, resembling a phonological plot without snow cover.

Differences between C4 and C5 are less obvious for the EVI data (Figure 13.4). The temporal dynamic range of meaningfully interpolated EVI data is higher than for NDVI. It seems that the dynamic range slightly increased in C5. EVI values, however, are much less susceptible to atmospheric disturbances and temporary surface changes, as indicated by the smoother plot of the original data. This can be attributed to well-working backup algorithms if the EVI cannot be retrieved directly. The changes of the backup approach from SAVI to EVI2 seemed to improve this characteristic, indicated by the original plot for the Harz upland. While C4 data decreased during the summer cloud period and varied in wintertime, the original plot remained close to the interpolated results for C5 data. The changes in quality assessment slightly improved the resulting EVI time series of C5 data, e.g., for the spring in the Alps and during the summer period for Schleswig.

13.5 CONCLUSIONS

The QA-SDS provides meaningful information for data analysis and time series generation of MODIS data. The Time Series Generator (TiSeG) analyzes the QA-SDS and displays the data availability in time and space according to user-defined settings. Following, data gaps can be flagged or interpolated with spatial or temporal functions. This software package was adjusted to the modified MOD13 QA-SDS of C5 products and extended to incorporate the newly introduced reliability layer.

The comparison between C4 and C5 quality settings revealed a stricter specification for data quality of C5 data in particular for rigorous quality settings. The quality specifications for cloud and shadow, aerosol, and snow/ice are improved and yield a more conservative result, i.e., regard more pixels as invalid. This results in less data availability and an increase in the gap length by one composite. The higher sensitivity to atmospheric disturbances yields a higher quality of the time series, which is measured by better temporal connectivity. Spatial patterns of data availability indicate that higher elevations are frequently flagged as invalid due to both snow/ice and clouds. Lowlands in northern Germany are marked as invalid by strict settings due to frequent cloud coverage.

The regional time series analysis highlights the necessity of a critical weighting between quality and quantity. Low data quality will include invalid pixels and does not yield improved time series. Highest data quality, on the other hand, will only consider most accurate data but often results in too few data for meaningful interpolation. It can be concluded that the reliability SDS should also be analyzed when using the new C5 products. In particular, cloud coverage was better indicated by this new layer. While flag mixed clouds in C4 also indicated full cloudy pixels, the same flag in C5 only labels partly cloudy pixels. Therefore, the reliability dataset is the only means to mask cloudy pixels in C5. In comparison with the NDVI, the EVI shows better consistency and a higher temporal dynamic range.

REFERENCES

Anyamba, A., Tucker, C. J., and Mahoney, R., 2002. From El Niño to La Niña: vegetation response patterns over east and southern Africa during the 1997–2000 period. *Journal of Climate*, 15(21), pp. 3096–3103.

Asner, G. P., Townsend, A. R., and Braswell, B. H., 2000. Satellite observation of El Niño effects on Amazon forest phenology and productivity. *Geophysical Research Letters*, 27(7), pp. 981–984.

Atkinson, P. M. and Foody, G. M., 2002. Uncertainty in remote sensing and GIS: Fundamentals. In Foody, G. M. and Atkinson, P. M., Eds., *Uncertainty in Remote Sensing and GIS*. John Wiley and Sons. New York, 326 pp.

de Beurs, K. M. and Henebry, G. M., 2004. Land surface phenology, climatic variation, and institutional change: analyzing agricultural land cover change in Kazakhstan. *Remote Sensing of Environment*, 89(4), pp. 423–433.

Brockmann, C., 2004. Demonstration of the BEAM software. A tutorial for making best use of VISAT. *Proceedings of MERIS User Workshop*, ESA-ESRIN, Frascati, Italy, 2004.

Colditz, R. R., Conrad, C., Wehrmann, T., Schmidt, M., and Dech, S. W., 2006a. Generation and assessment of MODIS time series using quality information, IGARSS 2006. *IEEE International Geoscience And Remote Sensing Symposium*, Denver, CO.

Colditz, R. R., Conrad, C., Schmidt, M., Schramm, M., Schmidt, M., and Dech, S. W., 2006b. Mapping regions of high temporal variability in Africa, *ISPRS Mid-Term Symposium 2006—Remote Sensing: From Pixels to Processes*, Enschede, The Netherlands.

Colditz, R. R., Conrad, C., Wehrmann, T., Schmidt, M., and Dech, S. W., 2008. TiSeG: A flexible software tool for time series generation of MODIS data utilizing the quality assessment science data set. *IEEE Transactions on Geoscience and Remote Sensing*, accepted.

Conrad, C., Colditz, R. R., Petrocchi, A., Rücker, G. R., Dech, S. W., and Schmidt, M., 2005. Time Series Generator—Ein flexibles Softwaremodul zur Generierung und Bewertung von Zeitserien aus NASA MODIS Datenprodukten. In: J. Strobl, T. Blaschke and G. Griesebner, Eds., *AGIT. Beiträge zum 17. AGIT-Symposium Salzburg*, Salzburg, pp. 100–105.

Didan, K. and Heute, A., 2006. MODIS vegetation index product series collection 5 change summary. http://landweb.nascom.nasa.gov/QA_WWW/forPage/MOD13_VI_C5_Changes_Document_06_28_06.pdf (accessed 16 Jan. 2007).

el Saleous, N. Z., Vermote, E. F., Justice, C. O., Townshend, J. R. G., Tucker, C. J., and Goward, S. N., 2000. Improvements in the global biospheric record from the Advanced Very High Resolution Radiometer (AVHRR). *International Journal of Remote Sensing*, 21(6–7), pp. 1251–1277.

Holben, B. N., 1986. Characterization of maximum value composites from temporal AVHRR data. *International Journal of Remote Sensing*, 7(11), pp. 1417–1434.

Huete, A. R., 1988. A soil-adjusted vegetation index (SAVI). *Remote Sensing of Environment*, 25(3), pp. 295–309.

Huete, A., Justice, C. O., and van Leeuwen, W. J. D., 1999. MODIS Vegetation Index (MOD 13), Algorithm Theoretical Basis Document (ATBD) Version 3.0, 129 pp.

Huete, A. R., Didan, K., Miura, T., Rodriguez, E. P., Gao, X., and Ferreira, L. G., 2002. Overview of the radiometric and biophysical performance of the MODIS vegetation indices. *Remote Sensing of Environment*, 83(1–2), pp. 195–213.

Jönsson, P. and Eklundh, L., 2002. Seasonality extraction by function fitting to times-series of satellite sensor data. *IEEE Transactions on Geoscience and Remote Sensing*, 40(8), pp. 1824–1832.

Justice, C. O., Vermote, E. F., Townshend, J. R. G., DeFries, R. S., Roy, D. P., Hall, D. K., Salomonson, V. V., Privette, J. L., Riggs, G., Strahler, A. H., Lucht, W., Myneni, R. B., Knyazikhin, Y., Running, S. W., Nemani, R. R., Wan, Z., Huete, A. R., van Leeuwen, W. J. D., Wolfe, R. E., Giglio, L., Muller, J.-P., Lewis, P., and Barnsley, M. J., 1998. The Moderate Resolution Imaging Spectroradiometer (MODIS): land remote sensing for global change research. *IEEE Transactions on Geoscience and Remote Sensing*, 36(4), pp. 1228–1249.

Justice, C. O., Townshend, J. R. G., Vermote, E. F., Masuoka, E., Wolfe, R. E., el Saleous, N. Z., Roy, D. P., and Morisette, J. T., 2002. An overview of MODIS Land data processing and product status. *Remote Sensing of Environment*, 83(1–2), pp. 3–15.

Keil, M., Kiefl, R., and Strunz, G., 2005. CORINE land cover—Germany, Oberpfaffenhofen, German Aerospace Center.

Landmann, T., Breda, F., Di Gregorio, A., Latham, J., Sarfatti P., and Giacomo, D., 2005. Looking towards a new African Land Cover Dynamics Data Set: the medium resolution data-base for Africa (MEDA), *3rd Proceedings of AFRICA GIS*, Tshwane, South Africa, 2005.

Leptoukh, G., Berrick, S., Rui, H., Liu, Z., Zhu, T., and Shen, S., 2005. NASA GES DISC Online Visualization and Analysis System for Gridded Remote Sensing Data, *Proceedings of the 31st International Symposium of Remote Sensing of the Environment (ISRSE)*, St. Petersburg, Russia, 2005.

Lobell, D. B. and Asner, G. P., 2004. Cropland distributions from temporal unmixing of MODIS data. *Remote Sensing of Environment*, 93, pp. 412–422.

Lunetta, R. S., Knight, J. F., Ediriwickrema, J., Lyon, J., and Worthy, L. D., 2006. Land-cover change detection using multi-temporal MODIS NDVI data. *Remote Sensing of Environment*, 105, pp. 142–154.

Meynen, E. and Schmithüsen, J., 1953. Handbuch der naturräumlichen Gliederung Deutschlands. Remagen, Bundesanstalt für Landeskunde u. Raumforschung.

Morisette, J. T., Privette, J. L., and Justice, C. O., 2002. A framework for the validation of MODIS Land products. *Remote Sensing of Environment*, 83(1–2), pp. 77–96.

Roerink, G. J., Menenti, M., and Verhoef, W., 2000. Reconstructing cloudfree NDVI composites using Fourier analysis of time series. *International Journal of Remote Sensing*, 21(9), pp. 1911–1917.

Roy, D. P., Borak, J. S., Devadiga, S., Wolfe, R. E., Zheng, M., and Descloitres, J., 2002. The MODIS Land product quality assessment approach. *Remote Sensing of Environment*, 83(1–2), pp. 62–76.

Running, S. W., Thornton, P., Nemani, R. R., and Glassy, J., 2000. Global terrestrial gross and net primary productivity from the Earth Observing System. *Methods in Ecosystem Science*, pp. 44–57.

Sellers, P. J., Los, S. O., Tucker, C. J., Justice, C. O., Dazlich, D. A., Collatz, G. J., and Randall, D. A., 1996. A revised land surface parameterization (SiB2) for atmospheric GCMs. Part II: The generation of global fields of terrestrial biophysical parameters from satellite data. *Journal of Climate*, 9(4), pp. 706–737.

Tottrup, C. and Rasmussen, M. S., 2004. Mapping long-term changes in savannah crop productivity in Senegal through trend analysis of time series of remote sensing data. *Agriculture, Ecosystems and Environment*, 103(3), pp. 545–560.

Tucker, C. J., Newcomb, W. W., and Dregne, H. E., 1994. AVHRR data sets for determination of desert spatial extent. *International Journal of Remote Sensing*, 15(17), pp. 3547–3565.

Viovy, N., Arino, O., and Belward, A., 1992. The Best Index Slope Extraction (BISE): a method for reducing noise in NDVI time-series. *International Journal of Remote Sensing*, 13, pp. 1585–1590.

14 Modeling DEM Data Uncertainties for Monte Carlo Simulations of Ice Sheet Models

Felix Hebeler and Ross S. Purves

CONTENTS

14.1 INTRODUCTION

All modeling is susceptible to the introduction of uncertainties to model results throughout the modeling chain. During data acquisition, systematic error, measurement imprecision, or limited accuracy of sensors can introduce ambiguities to measured values. Preprocessing and preparation of data to meet model needs, such as reprojecting, scaling, or resampling the data, introduce uncertainty. Finally, the methods and algorithms used as well as effects such as computational precision during modeling can introduce further uncertainties to results.

As all modeling is a mere abstraction of much more complex processes that in many cases might not be fully understood, uncertainties are also an intrinsic part of the approach. Uncertainties are thus not necessarily a problem in modeling, but rather an inherent component of the process, as long as the sources and bounds of the uncertainties associated with individual models are known and understood. Where this is the case, sensitivity tests can be conducted to assess the susceptibility of model results to uncertainties in certain data, parameters, or algorithms and compare these uncertainties with the sensitivity of model runs to variations in individual parameters. Decision makers have become increasingly familiar with such methodologies, through, for example, the scenarios presented in IPCC reports (IPCC, 2001). While uncertainties inherent in spatial data have been the focus of a number of research projects in the GIScience community, many users of spatial data either completely neglect this source of uncertainty or consider it less important than, for example, parameter uncertainties. However, even if a modeler is aware of the uncertainties introduced through, for instance, a Digital Elevation Model (DEM), it is not always straightforward or even possible to assess them, e.g., when metadata from the data producers are incomplete, incorrect, or missing. If this information cannot be reconstructed, assumptions have to be made that might or might not be realistic and sensible for testing the impact of uncertainties in spatial data on a model. In this chapter, we use the term "error" when referring to the deviation of a measurement from its true value. This implies that the elevation error of a DEM can only be assessed where higher accuracy reference data are available (Fisher and Tate, 2006). Error is inherent in any DEM, but is usually not known in terms of both magnitude and spatial distribution, thus creating uncertainty. "Uncertainty" is used in this context, where a value is expected to deviate from its true measure, but the extent to which it deviates is unknown, and it can only be approximated using uncertainty models (Holmes et al., 2000).

14.1.1 MOTIVATION

Ice sheet models, which are commonly used to explore the linkage between climate and ice extent either during past glacial periods or to explore the response of the Earth's remaining large ice masses (the Greenland Ice Cap and the Antarctic Ice Sheet) to future climate change, run at relatively low resolutions of the order of 1 to 20 km, for a number of reasons. Since the models run at continental or even global scales, computational capacities as well as assumptions in model physics limit possible modeling resolutions. Furthermore, climate models used to drive such models commonly run at even lower resolutions, and until recently the highest resolution global topographic datesets had nominal resolutions of the order of 1 km. Ice sheet modelers commonly resample the highest available resolution data to model resolutions; for example, in modeling ice extents in Patagonia, a 1 km resolution DEM was resampled to 10 and 20 km, respectively (Purves and Hulton, 2000). While it is often assumed that data accuracy of 1 km source data is essentially irrelevant when resampled to 10 or 20 km, previous work has suggested that these uncertainties can have a significant impact on modeled ice sheet extents and volumes (Hebeler and Purves, 2005). Despite the recognized need (Kyriakidis et al., 1999), most DEM data are still distributed with little metadata; usually, at best, global values such as RMSE or standard deviation of error are given (Fisher and Tate, 2006). Information on the spatial distribution of uncertainties is almost always not available, and assumptions made about the distribution of uncertainties are often debatable (Fisher and Tate, 2006; Wechsler, 2006; Oksanen and Sarjakoski, 2005; Weng, 2002; Holmes et al., 2000).

Following the approach of Hagdorn (2003) in reconstructing the Fennoscandian ice sheet during the last glacial maximum (LGM), we wanted to test the sensitivity of the model results to DEM data uncertainty. Hagdorn used GLOBE DEM data as input topography, for which accuracy figures are given as global values depending on the data source, e.g., vertical accuracy of 30 m at the 90% confidence interval for data derived from DTED (Hastings and Dunbar, 1998), with no information on spatial configuration or dependencies of uncertainties or error. Thus, in order to assess the impact of uncertainty in the DEM on the ISM, a realistic model of GLOBE DEM uncertainty must also be developed that both describes dependencies of error values on the DEM and sensibly reconstructs the spatial configuration of uncertainty.

14.1.2 Aims

In this chapter we set out to address three broad aims, which can be described as follows:

To quantify the error in DEMs for a range of appropriate regions, using higher resolution data, and to assess the extent to which this error correlates with DEM characteristics

To develop a general model of DEM error for use in areas where higher resolution data are not available and simulate the spatial and numerical distribution of the remaining uncertainty stochastically

To apply the DEM uncertainty model in Monte Carlo simulations of ISM runs for Hagdorn's experiments (Hagdorn, 2003) and assess the impact of modeled topographic uncertainty on ISM results.

The third aim can thus be considered as a case study of the application of a set of general techniques aimed at modeling DEM uncertainties and allowing their impact on model results to be compared with other potential sources of uncertainty.

14.2 MATERIALS AND METHODS

The availability of SRTM data makes the evaluation of GLOBE and GTOPO30 data accuracy possible for large areas of the globe (Jarvis et al., 2004; Harding et al., 1999), and thus it is possible to retrospectively evaluate previous experiments that used GLOBE DEM as input data. However, since our study area of Fennoscandia lies outside the region covered by SRTM data (CIAT, 2006), no direct assessment of error using higher accuracy reference data is possible.

Our approach was thus as follows. First, regions with similar topography and data sources to Fennoscandia, but lying within regions covered by SRTM data, were identified. Second, error surfaces were generated by assuming the SRTM data to be a higher quality data source for these regions. A model of error, incorporating a stochastic component, which represents a generalized uncertainty model for all regions, was then developed. Using this model it is possible to perform MCS simulations with the ISM, since the stochastic component of the uncertainty model means that multiple uncertainty surfaces can be generated.

14.2.1 DEM DATA

For the analysis of typical GLOBE DEM uncertainty, three datasets were selected based on previous tests that showed that uncertainty in the GLOBE DEM data was highest in high altitude and high relief areas. Such areas are also central to ice sheet inception (Marshall, 2002; Sugden et al., 2002) and thus are likely to be particularly susceptible to uncertainty. To derive the uncertainty model for Fennoscandia, GLOBE data for the European Alps, the Pyrenees, and the eastern part of Turkey were selected. These regions have relatively similar properties in terms of hypsometry (Figure 14.1) and statistics describing elevation values (Table 14.1) and were all compiled from DTED data, with the exception of the Italian part of the Alps, where data were sourced from the Italian national mapping agency (Hastings and Dunbar,

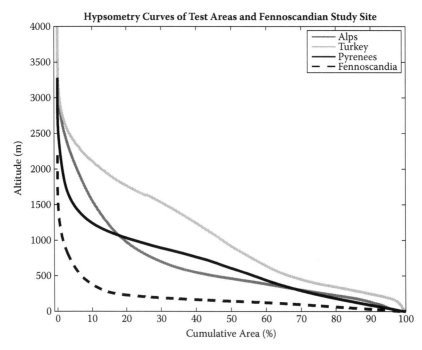

FIGURE 14.1 Hypsometry of the three selected test areas (solid lines) and the Fennoscandian study area (dashed), calculated from GLOBE DEM data at 1 km resolution. Test areas show relative large proportions of the high areas that are of interest in the study site DEM of Fennoscandia. Altitudes above 4000 m cropped for better visibility.

TABLE 14.1

**Descriptive Statistics for the Three Test Areas
and the Fennoscandian Study Site Used**

DEM	Alps	Pyrenees	Turkey	Scand
Altitude	1–4570 m	1–3276 m	1–4938 m	0–2191 m
Mean	692.8 m	651.9 m	1066.5 m	189.5 m
Std. Dev.	624.8 m	481.2 m	738.4 m	207.4 m
Skewness	1.65	0.86	0.55	3.09
Kurtosis	5.46	3.86	2.29	15.0
Source	DTED[a]	DTED	DTED	DTED
Size (cells)	1,083,108	720,000	816,837	6,094,816

[a] Italian data provided by Servizio Geologico Nazionale (SGN) of Italy.

1998). For the three selected test areas, hole-filled SRTM data at 100 m resolution (CIAT, 2006) were resampled to align with the GLOBE DEM at 1 km resolution (GLOBE Task Team & others, 1999), using the mean of all SRTM cells within the bounds of the corresponding GLOBE data cell (Jarvis et al., 2004). Waterbodies were eliminated from all datasets, and error surfaces for the respective test areas were calculated by subtracting the GLOBE data from the averaged SRTM data. SRTM data in this approach are thus used as ground truth and considered error free. Like any data source, SRTM does of course contain errors (Sun et al., 2003; Heipke et al., 2002)—however, their magnitude and spatial distribution were considered negligible for this experiment. Calculations on the datasets were conducted using the original, unprojected WGS84 spatial reference that both SRTM and GLOBE DEM data are distributed in. For calculation of the slope and related parameters, all DEMs were projected to Albers Equal Area projections (using WGS84 geoid), with the projection parameters chosen to minimize distortion for every region and minimize any further uncertainty introduced by the process (Montgomery, 2001).

14.2.2 UNCERTAINTY MODEL

Having derived error surfaces, they were first visually inspected. Descriptive statistics were calculated for each of the three areas and hypsometric curves and histograms compared. To assess spatial autocorrelation of both the DEM and the calculated error surfaces, semivariogram maps were derived for both the complete datasets as well as characteristic regions (e.g., for areas with high relief). Additionally, local Moran's I was calculated for all surfaces (Wood, 1996). Error, error magnitude, and error sign were then tested for correlation with a set of terrain attributes and parameters (Table 14.2), where all neighborhood analysis was conducted with a 3 × 3 window, which was found to give the highest correlation values in pretests. Stepwise regression analysis was used to find the best descriptive variables for modeling error in each of the three testing areas. The derived regression factors were averaged to formulate a general regression model for all three areas. Using this general regression,

TABLE 14.2

Attributes, Derivatives, and Indices Used during Correlation Analysis

Altitude	Value of GLOBE cell
Error	Deviation of GLOBE from mean SRTM value
Error magnitude	Magnitude of error
Sign	Sign of error (+1/–1)
Aspect	Direction of first derivative of elevation
Slope	Magnitude of first derivative of elevation
Plan curvature	2nd derivative orthogonal to direction of steepest slope
Profile curvature	2nd derivative in direction of steepest slope
Total curvature	Compound curvature index
Maximum- Mean-extremity* Minimum-	Deviation of center cell from max/mean/min of 3 × 3 neighborhood
Roughness (altitude)	Standard deviation of altitude in a 3 × 3 neighborhood
Roughness (slope)	Standard deviation of slope in a 3 × 3 neighborhood

* Extremity index calculated after Carlisle (2000).

the residuals for each of the areas were also analyzed to assess their dependency on the properties of the original DEM (Table 14.2). Again, a method to reproduce the characteristics common to the residuals of all three test area was sought and combined with the first regression equation. In order to reproduce the spatial autocorrelation encountered in the original error surfaces, the uncertainty surfaces modeled using the above method were then transformed to a normal distribution and filtered using a Gaussian convolution filter (Ehlschlaeger et al., 1997; Hunter and Goodchild, 1997) using kernel sizes derived from autocorrelation analysis of the original error surfaces. The modeled uncertainty surfaces were next compared with the derived true error surfaces in terms of both their spatial and statistical distributions. The developed uncertainty model was used to calculate a suite of 100 uncertainty surfaces for Fennoscandia that were superimposed on the original GLOBE DEM and used as input topographies for an MCS using the ISM.

14.2.3 Ice Sheet Model Runs

The ISM used in these experiments is the GLIMMER model (Hagdorn et al., 2006), which was developed as part of GENIE (Grid Enabled Integrated Earth system model) and is freely available. For our experiments, we followed the approach of Hagdorn (2003) and ran simulations at 10 km resolution for the 40,000 years from approximately 120 ka to 80 ka BP. Climate forcing (essentially describing temperature and input mass) is based on an equilibrium line altitude (ELA) parameterization (the ELA is the altitude at which net accumulation is zero—above the ELA mass accumulates, and below it ablates) derived from the Greenland ice core project (GRIP) data. Model runs have a time step of 1 year, and simulated ice thickness (and thus extent) is output to file every 500 years. Input topographies for the GLIMMER simulations consist of the GLOBE DEM data with added uncertainty derived from 1 km uncertainty surfaces created by the uncertainty model, projected to Albers Equal Area projection and resampled to 10 km resolution using bilinear interpolation. This method was chosen because it is a standard resampling technique applied by ice sheet modelers, and therefore it is more representative for the study than the method of averaging of all contribution cells used in resampling SRTM to 1 km (compare Section 14.2.1).

14.3 RESULTS

14.3.1 Uncertainty Model

14.3.1.1 Error Properties

Initial visual inspection of the derived error surfaces shows the high spatial correlation of error along prominent terrain features within the dataset (compare Figure 14.2 and Figure 14.3), with reduced autocorrelation in areas of low relief. The distribution of error magnitude and sign also suggests some error dependencies on data sources, most visible through the lower overall error in the Italian part of the Alps seen in Figure 14.3. Global autocorrelation analysis using semivariogram maps showed the range of autocorrelation to lie between 2 and 4 km for each dataset, with directional

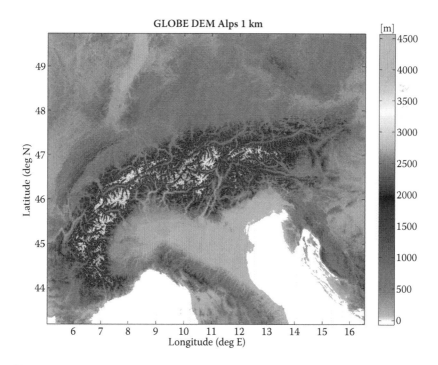

FIGURE 14.2 GLOBE DEM of the Alps test area at 1 km resolution (WGS84).

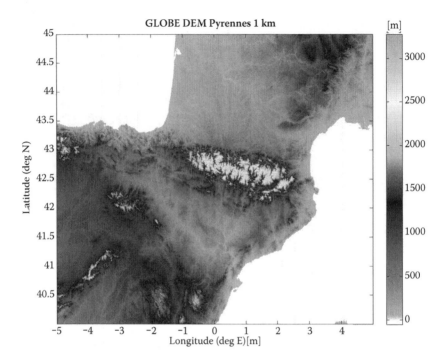

FIGURE 14.3 GLOBE DEM of the Pyreness test area at 1 km resolution (WGS84).

TABLE 14.3
Descriptive Statistics of Derived GLOBE DEM Error from Three Test Areas

	Range	Mean	Std. Dev.	Skewness	Kurtosis
Alps	−1140 to 1169 m	3.3 m	82.2 m	0.05	11.61
Pyrenees	−920 to 797 m	4.2 m	68.8 m	−0.14	14.2
Turkey	−817 to 964 m	3.0 m	70.7 m	−0.04	11.29

trends following the orientation of prominent terrain features in the original DEMs. These semivariogram maps are strongly influenced by the semivariogram properties of high relief areas, since areas of low relief show little to no spatial autocorrelation at these resolutions. Calculated values of local Moran's I reinforce these findings. The statistic distribution of error (Table 14.3) shows comparable distributions for all three areas.

14.3.1.2 Error Correlation

Correlation analysis of error with the parameters presented in Table 14.2 showed relatively weak correlations with coefficients of between 0.2 and 0.5 for mean extremity, curvature, and aspect for all datasets. Testing the magnitude of error for correlation resulted in higher correlation coefficients for minimum extremity, roughness of

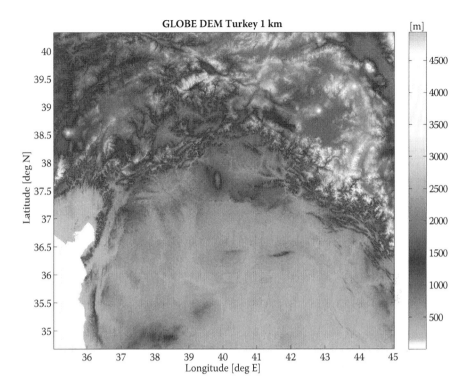

FIGURE 14.4 GLOBE DEM of the Turkey test area at 1 km resolution (WGS84).

altitude, slope, and altitude with values of up to 0.66. In a third analysis using binary logistic regression, the sign of error showed some correlation with aspect and minimum extremity, with 55 to 65% of the original error sign modeled correctly, depending on the test area. All parameters that exhibited a significant correlation with either error or error magnitude were included in a stepwise regression analysis. The best fit for modeling error was achieved with three parameters (mean extremity, curvature, and aspect) yielding an r^2 of around 0.23. Regression of the magnitude of error gave an average r^2 of 0.42 (Table 14.4) using only two variables, roughness (altitude) and minimum extremity. Taking the mean of the corresponding factors from all three test areas gave the following regression equation for modeling the amount of error:

$$abs(\varepsilon') = 0.53 \times roughness + 0.031 \times extremity_{min} + 7.6 \qquad (14.1)$$

TABLE 14.4
r^2 Values of the Regression Modeling the Amount of Error for the Three Test Areas

Error Magnitude	Alps	Pyrenees	Turkey
Local model	0.441	0.406	0.423
Global model	0.430	0.393	0.422

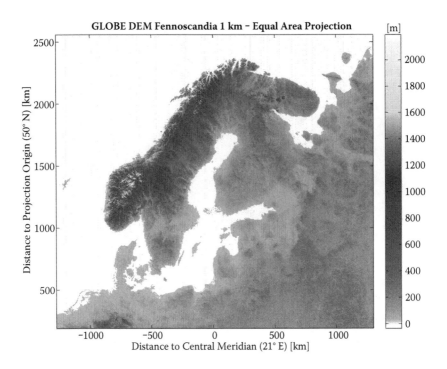

FIGURE 14.5   GLOBE DEM of the Fennoscandia study site at 1 km resolution (AEA).

This regression was found to capture 50 to 70% of the measured error magnitude for the three test areas. As results of regression on error were considerably weaker, only the regression on error magnitude was used in the uncertainty model. Slope and its derivatives are therefore not used in the model and the analysis was continued on the unprojected WGS84 datasets. Using Equation. 14.1, residuals were calculated for the three test areas and analyzed. Residuals showed to be centered around a mean of 0 with a standard deviation of 43 to 50 m, minimum values of around –300 m, and their maxima at 600 to 900 m. This resulted in mildly skewed (skewness 1.7–2.4) distributions with high kurtosis of 10 to 18. The residuals were found to be well approximated using a modified random normal distribution ($N[0, 45]$). Squaring the residuals and randomly reassigning the signs to center the distribution around 0 again, then downscaling through a division by 100, proved to be a simple and satisfactory way to simulate regression residuals, while introducing a stochastic component to the uncertainty model. Since only the magnitude of error showed a useful correlation, the sign of the modeled uncertainty was modeled separately for the uncertainty model. Although Equation 14.2, derived from binary logistic regression, showed agreement of only 55 to 65% of modeled against true error sign, the regression proved to capture the spatial correlation of the error sign well, at the cost of an overestimation of positive error of the order of 10 to 20%:

$$S = -0.0012 \times extremity_{\text{mean}} + 0.002 \times aspect - 0.2 \qquad (14.2)$$

where $-1 \leq S \leq 1$. Further analysis confirmed that the closer the modeled values were to either +1 or −1, the higher the probability that the error's sign was modeled correctly. For the three test areas, almost all values higher than 0.6 or lower than −0.6, respectively, modeled the error sign correctly. Thus, a stochastic element was introduced for modeling error sign, where a random number r was drawn from a standard normal distribution for every value of S. Where $r \leq abs(S) + f$, with the correction factor $f = 0.35$, the modeled sign was kept, otherwise the sign was assigned randomly. This resulted in a ratio of positive to negative modeled error close to the measured error, while retaining most of the spatial characteristics of the sign distribution. Combining the three steps, that is, modeling the dependence of the error, residuals (resid), and error sign, resulted in the following uncertainty model:

$$U_{tot} = (abs(\varepsilon) + resid) \times S \qquad (14.3)$$

Finally, the modeled surfaces, though correctly representing the statistical distribution of error, did not yet take full account of the spatial autocorrelation of error. A Gaussian convolution filter (Oksanen and Sarjakoski, 2005) was thus applied to the modeled uncertainty raster by transforming the distribution of modeled uncertainty to a normal distribution and applying a convolution filter with a kernel range of 3 km (3 cells). After the filtering, the uncertainty raster was transferred back to its original distribution. QQ-plots show the distribution to be altered only minimally, with the added advantage that unrealistically noisy parts of the surface were effectively smoothed.

14.3.1.3 Modeled Uncertainty Surfaces

Modeled uncertainty surfaces show a good correspondence in spatial configuration with the derived error surfaces. The general dependencies visible in the derived error surfaces (Figure 14.6) are generally preserved in the modeled uncertainty (Figure 14.7), due to the regression component of the model. The small-scale distribution of modeled uncertainty is generally noisier than that of the error, with the autocorrelation introduced through convolution filtering clearly visible (Figure 14.7, inset). Comparing the histograms of the derived error with the modeled uncertainty (Figure 14.8) shows good accordance, with an underestimation of values close to zero and an overestimation of values around the standard deviation of the distribution. Extreme error values are not reproduced by the uncertainty model, and the overall sum of the modeled uncertainty for any of the test areas is within 10% of the range of derived error. Modeling a suite of 100 uncertainty surfaces for Fennoscandia (2366 × 2576 cells), the descriptive statistics proved to vary little (Table 14.5). Calculating the mean, range, and standard deviation of the modeled uncertainty for every raster cell across all 100 runs (Figure 14.9) illustrates the influence of the deterministic and the stochastic parts of the uncertainty model. For areas with mean positive or negative error, the strong influence of the sign regression results in predominately positive or negative errors. Likewise, areas of high uncertainty are likely to be the result of the regression modeling the magnitude of error following dominant landscape features. However, the two stochastic elements in the determination of error

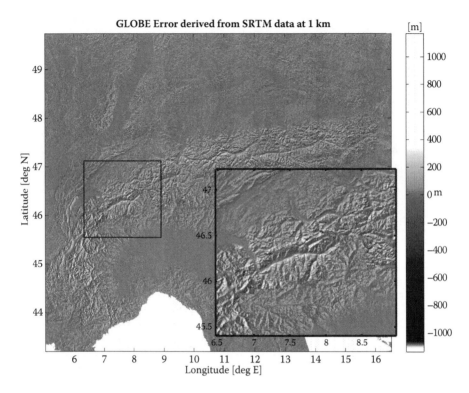

FIGURE 14.6   GLOBE error surfaces for the Alps derived using SRTM reference data.

sign and modeling of the residuals introduce a stochastic component that results in the imposition of noise across the raster, shown through the standard deviation and range of modeled uncertainty (Figures 14.11 and 14.12).

14.3.1.4   Sensitivity Study

Figure 14.13 and Figure 14.14 show a suite of representations of the influence of the modeled uncertainty in ISM results as a result of the driving temperature (Figure 14.13) imposed together with the parameterization of mass balance. Figure 14.13 shows the development through time of ice sheet extent and volume and the uncertainty induced in these values as a function of the DEM uncertainty, while Figure 14.14 illustrates the variation in ice sheet extent for a variety of snapshots in time. These results clearly show that, first, uncertainty is greatest during ice sheet inception (standard deviation [STD] in extent ~12%), where uncertainties in elevation can raise or lower individual ice nucleation centers above or below the ELA. As ice centers grow and coalesce, the effects of uncertainty in topography decrease (STD in extent ~3%), as the ice mass itself becomes the predominant topography. However, during periods of retreat (e.g., around 20 ka model years), uncertainty again increases. Figure 14.14 clearly shows how, with a mature ice sheet (e.g., after around 37 ka model years), most uncertainty in ice sheet extent is found at the edges of the ice sheet. Once the ice sheet has reached a certain size, e.g., after approx 10 ka model years, the range

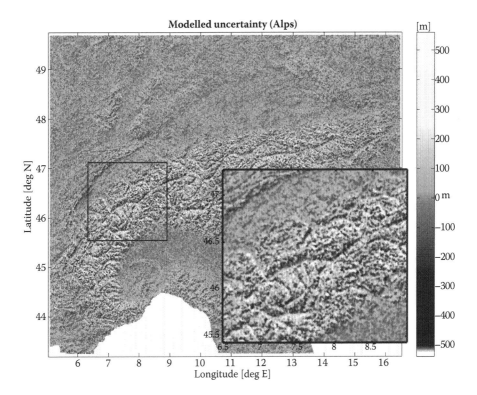

FIGURE 14.7   Modeled GLOBE DEM uncertainty surface for the Alps, with detail inset.

of uncertainty in the position of the ice front for these simulations varies between 40 and 100 km for all later model stages. The variation is less at the NW ice front, as the bathometry rapidly lowers off the Norwegian coast and the ISM ablates all ice at altitudes lower than –500 m. Variation of ice extent across the MCS runs is thus much higher toward Finland and the Baltic Sea.

14.4   DISCUSSION

In Section 14.1.2 we set out three broad aims for this work, namely, to quantify DEM error for a variety of regions where higher-quality data were available, to develop a general model of uncertainty based on these findings, and to apply this model to assess the uncertainty introduced into the results of ISM runs as a result of uncertainty in DEMs.

14.4.1   QUANTIFYING DEM ERROR

In assessing DEM error, we sought to identify areas that had broadly similar characteristics, based on the assumption that dependencies and characteristics of DEM error based on a DEM might be expected to be broadly similar for similar regions. Table 14.5 gives the descriptive statistics for error surfaces calculated for the three regions, which are broadly similar, suggesting that this assumption is reasonable.

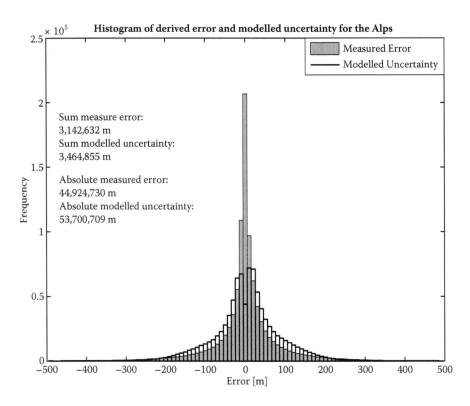

FIGURE 14.8 Histogram of the derived error for the Alps test area, compared to that of an example of a stochastically generated uncertainty surface for the same area.

TABLE 14.5

Mean and Standard Deviation of the Distribution Statistics of 100 Modeled Uncertainty Surfaces for Fennoscandia

	Mean	Max	Min	Std. Dev.	Skewness	Kurtosis	Sum
Mean	0.64 m	560 m	−561 m	40.5 m	0.0 m	8.7 m	3.8×10^6 m
Standard deviation	0.02 m	54.8 m	53.7 m	0.02 m	0.0 m	0.03 m	8.9×10^4 m

However, a further inherent assumption is that the variation in error is mainly described by terrain parameters within each region. In fact, this was found not to be the case in the Alpine region, where error values notably decreased at the Swiss/Italian border in the Italian region of the Alps, where the original GLOBE data have a different source. The error surfaces themselves (e.g., Figure 14.16) show strong correlations of error with terrain features and, most strikingly, that error increases and is more spatially autocorrelated in areas of high relief. Initial attempts to correlate error with a range of parameters were relatively unsuccessful with low correlations, however, the absolute error was found to be relatively strongly correlated with roughness and minimum extremity. Roughness in particular increases with

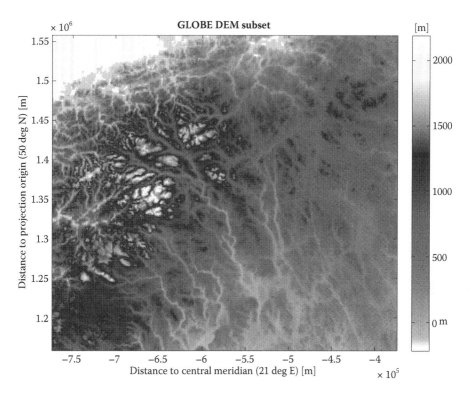

FIGURE 14.9 Subset of the Fennoscandian DEM (Inset in Figure 14.5).

relief, thus suggesting that the use of such a parameter is sensible. Local models with different coefficients were averaged for the three regions to create a global model (Equation 14.1) and the differences between the r^2 values generated by the local and global models found to be small, thus justifying the application of this global model in areas with similar terrain characteristics. Examination of the residuals for the error model showed no correlations with terrain parameters and no spatial autocorrelation. Thus this component of the error model was treated as uncertainty, along with the sign of the magnitude of error, and is discussed further below. The sign of the magnitude of the error was also examined for correlation with terrain parameters, and weak dependencies were found (around 55–65% of the signs were correctly modeled by a binary logistic regression) based on aspect and mean extremity. These parameters, in particular aspect, introduce spatial autocorrelations to the error model similar to those seen running along terrain features. However, as discussed in Section 14.3.1, a purely deterministic approach to modeling error sign significantly overestimates positive errors, and thus a further stochastic term was introduced.

14.4.2 Developing an Uncertainty Model

The uncertainty model given in Equation 14.3 has three terms: absolute error, a residual, and an error sign. Of these three terms, the first is purely deterministic,

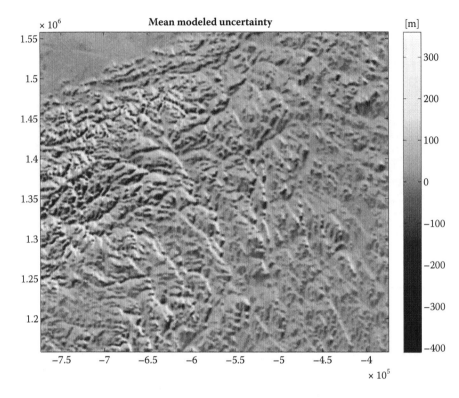

FIGURE 14.10 Mean of the modeled uncertainty for the subset DEM (Figure 14.5) averaged over 100 surfaces.

while the latter both contain stochastic elements, resulting in the generation of an uncertainty model. Importantly for our application, the uncertainty model can be generated purely from a single DEM, thus allowing us to model uncertainty in regions where high-quality data are not available. Figure 14.6–14.8 show a comparison between one uncertainty surface for the Alps and the calculated error for the same region. The influence of the stochastic elements is immediately clear, with considerably more noise in areas of lower relief and overall, and overall greater total error (i.e., the area under the curve in Figure 14.8). However, the range of error for the uncertainty surface is lower than that for the calculated error and the sum of positive and negative values (see Figure 14.8) similar. Figures 14.9–14.12 show how the uncertainty surfaces for Fennoscandia are themselves related to terrain features. For example, the mean modeled uncertainty is greatest in regions of high relief. The range of uncertainty illustrates clearly that areas where ice sheet inception is likely have the highest uncertainty in elevation (of the order of 800 m). Application of the convolution filter effectively smoothes extreme outliers and reduces the range of uncertainty within a given distance. This is important in many modeling applications, since outliers, in particular, can lead to model instabilities (e.g., through unphysical steep slopes for a given resolution). One important limitation of the model

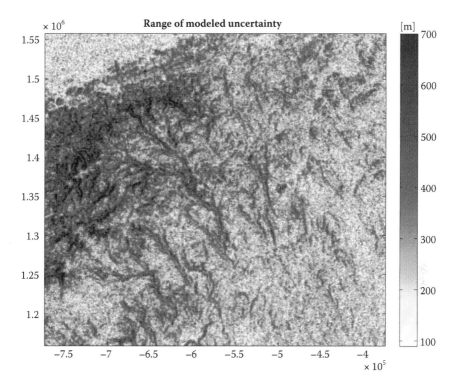

FIGURE 14.11   Range of modeled uncertainty for the subset DEM (Figure 14.5) averaged over 100 surfaces.

as it stands lies in the similarity between the three test regions and Fennoscandia. Overall, Fennoscandia has less and lower areas of high relief by comparison to our three test regions, and therefore uncertainty may be overestimated. However, as long as this assumption is clearly stated, we believe that the application of the model is valid. Since for Fennoscandia no higher accuracy reference data are available, other approaches of modeling DEM uncertainty including autocorrelation, such as stochastic conditional simulation (Kyriakidis et al., 1999), would be difficult to implement. However, if a measure of spatial autocorrelation of the error could be correlated to DEM attributes or compound indices, local information on spatial correlation could be used for improving the uncertainty surfaces produced, e.g., by using automated variogram analysis with stochastic conditional simulation (Liu and Jezek, 1999).

14.4.3   Case Study: ISM in Fennoscandia

The developed uncertainty model proved to deliver surfaces that are both suitable for Monte Carlo simulations through the inherent stochastic elements as well as fit to run an ISM at a considerably low resolution of 10 km. Earlier experiments (Hebeler and Purves, 2004) have shown that uncertainty modeled using random error in excess of 100 m STD can destabilize the ISM at resolutions as low as 20 km. This effect is

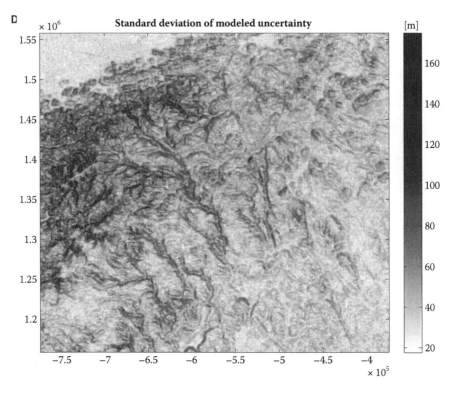

FIGURE 14.12 Standard deviation of modeled uncertainty for the subset DEM (Figure 14.5) averaged over 100 surfaces.

mainly due to unreasonably high slope gradients introduced by the added uncertainty. By contrast, the uncertainty model presented in this chapter produces topographically sound surfaces by both incorporating information on the underlying topography as well as convolution filtering, thus avoiding unrealistic terrain configurations. With a mean of zero and standard deviation of 40 m, the introduced uncertainties for Fennoscandia are effectively smaller than those with standard deviations of up to 150 m of previous experiments (Hebeler and Purves, 2005), but nevertheless prove to result in significantly different model results, especially during the inception and retreat phases of the ISM. This implies that care has to be taken when interpreting results during these phases (Sugden et al., 2002). DEM uncertainties can influence model results in both ice sheet size and configuration during susceptible stages that may otherwise be attributed to climate or mass balance changes. On the other hand, even though the relative variation of large ice sheets, e.g., the reconstructed Fennoscandian ice sheet after 15,000 and 31,000 model years, are relatively small in the order of 2 to 5% (Figure 14.13), the absolute difference in modeled extent is on the order of 50 to 100 km. Differences of modeled and empirically derived ice extents of ice sheets during the LGM of this order of magnitude have fueled debate over the years (Hulton et al., 2002; Wenzens, 2003). In order to relate the impact of these DEM uncertainties to the effect other parameters have on ISM results, further

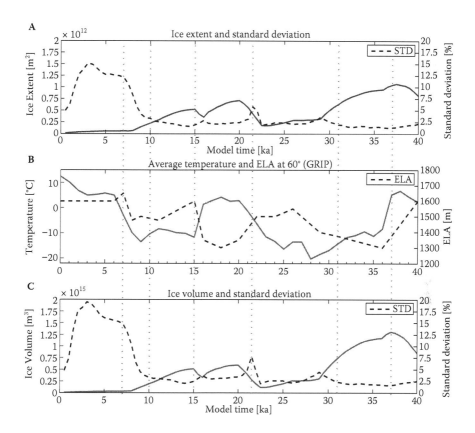

FIGURE 14.13 Mean ice extent (A) and volume (C) with their respective relative standard deviation (dashed lines) across 100 MCS runs plotted against modeling time. Climate forcing (temperature and ELA) shown in B, with vertical gray lines marking snapshot times shown in Figure 14.14.

sensitivity studies are necessary. For example, stepwise variation of climate forcing, e.g., temperature and mass balance, could be applied and compared to the range of modeled ice sheet configurations this chapter delivered.

14.5 CONCLUSIONS

In this chapter, we have successfully captured the dependency of GLOBE DEM error for mountainous terrain with the underlying topography and integrated this relationship into an uncertainty model. By applying this uncertainty model, we produced spatially correlated, realistic uncertainty surfaces that are suitable for the use in Monte Carlo simulations. Even though the amount of DEM uncertainty derived from GLOBE data was shown to have significant impact on ISM results for the Fennoscandian ice sheet during the LGM, sensitivity studies of ISM parameters and climate forcing are needed to relate the impact of DEM uncertainty, e.g., to that of temperature change. Future experiments will explore whether the developed

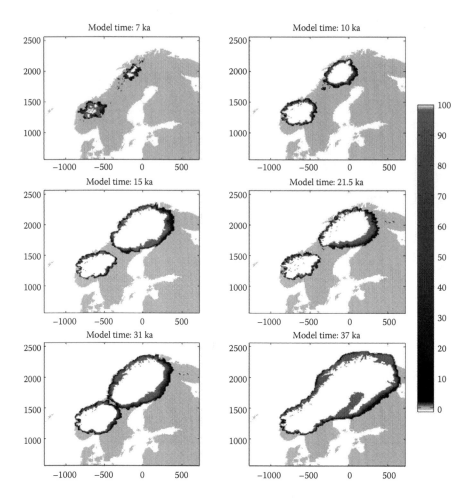

FIGURE 14.14 Frequency of DEM cells glaciated across 100 MCS runs after 7, 10, 15, 21.5, 31, and 37 ka model time. Present-time Fennoscandian coastline plotted for comparison.

uncertainty model could be improved by refining the selection of test areas or through a better reproduction of local spatial autocorrelation. Porting the uncertainty model to other topographies and source data, and testing it on different resolutions, for example, using SRTM and LIDAR data, will allow us to explore the sensitivity of other process models to DEM uncertainty.

ACKNOWLEDGMENTS

Felix Hebeler would like to thank Dr. Phaedon Kyriakidis (UC Santa Barbara), Prof. Peter Fisher (University of Leicester), and Dr. Jo Wood (City University London) as well as Dr. Juha Oksanen (Finnish Geodetic Institute) for their advice and encouragement. This research is funded by the Swiss National Science Foundation (SNF Project Number 200021-100054).

REFERENCES

Carlisle, B. H., 2000. The highs and lows of DEM error—developing a spatially distributed DEM error model. In: *Proceedings of the 5th International Conference on GeoComputation*, University of Greenwich, United Kingdom, pp. 23–25.

CIAT, 2006. International Centre for Tropical Agriculture: void filled seamless SRTM data V3, available from the CGIAR-CSI SRTM 90m Database: http://srtm.csi.cgiar.org.

Ehlschlaeger, C. R., Shortridge, A. M., and Goodchild, M. F., 1997. Visualizing spatial data uncertainty using animation. *Computers & Geoscience* 23(4), pp. 387–395.

Fisher, P. F. and Tate, N. J., 2006. Causes and consequences of error in digital elevation models. *Progress in Physical Geography* 30(4), pp. 467–489.

GLOBE Task Team & others, 1999. The Global Land One-kilometer Base Elevation (GLOBE) Digital Elevation Model, Version 1.0. Digital database on the World Wide Web (URL: http://www.ngdc.noaa.gov/mgg/topo/globe.html) and CD-ROMs. National Oceanic and Atmospheric Administration, National Geophysical Data Center, 325 Broadway, Boulder, CO 80303, USA.

Hagdorn, M.,2003. Reconstruction of the past and forecast of the future European and British ice sheets and associated sea-level change. Unpublished PhD thesis, University of Edinburgh.

Hagdorn, M., Rutt, I., Payne, T., and Hebeler, F., 2006. GLIMMER—The GENIE Land Ice Model with Multiply Enabled Regions—Documentation. http://glimmer.forge.nesc. ac.uk/. Universities of Bristol, Edinburgh and Zurich.

Harding, D. J., Gesch, D. B., Carabajal, C. C., and Luthcke, S. B., 1999. Application of the shuttle laser altimeter in an accuracy assessment of GTOPO30, a global 1-kilometer digital elevation model. *International Archives of Photogrammetry and Remote Sensing* 17-3/W14, pp. 81–85.

Hastings, D. A. and Dunbar, P. K., 1998. Development & assessment of the global land one-km base elevation digital elevation model (GLOBE). *ISPRS Archives* 32(4), pp. 218–221.

Hebeler, F. and Purves, R. S., 2004. Representation of topography and its role in uncertainty: a case study in ice sheet modeling. In: *GIScience 2004: Proceedings of the Third International Conference on Geographic Information Science*, pp. 118–121.

Hebeler, F. and Purves, R. S., 2005. A comparison of the influence of topographic and mass balance uncertainties on modeled ice sheet extents and volumes. In: *EosTrans. AGU, Fall Meet. Suppl. Vol.* 86, Abstract C23A-1154. No 52.

Heipke, C., Koch, A., and Lohmann, P., 2002. Analysis of SRTM DTM—methodology and practical results. *Journal of the Swedish Society for Photogrammetry and Remote Sensing: Photogrammetry Meets Geoinformatics*, 1, pp. 69–80.

Holmes, K. W., Chadwick, O., and Kyriakidis, P., 2000. Error in a USGS 30-meter digital elevation model and its impact on terrain modeling. *Journal of Hydrology* 233, pp. 154–173.

Hulton, N. R., Purves, R. S., McCulloch, R., Sugden, D. E., and Bentley, M., 2002. The last glacial maximum and deglacation in southern South America. *Quaternary Science Reviews* 21, pp. 233–241.

Hunter, G. J. and Goodchild, M. F., 1997. Modelling the uncertainty of slope and aspect estimates derived from spatial databases. *Geographical Analysis* 19(1), pp. 35–49.

IPCC, 2001. *Climate Change 2001: Synthesis Report. A Contribution of Working Groups I, II, and III to the Third Assessment Report of the Intergovernmental Panel on Climate Change*. Cambridge University Press, Cambridge, United Kingdom, and New York, NY.

Jarvis, A., Rubiano, J., Nelson, A., Farrow, A., and Mulligan, M., 2004. Practical use of SRTM data in the tropics: Comparisons with digital elevation models generated from cartographic data. Working Document 198, 32 p. International Centre for Tropical Agriculture (CIAT), Cali, Colombia.

Kyriakidis, P. C., Shortridge, A., and Goodchild, M., 1999. Geostatistics for conflation and accuracy assessment of digital elevation models. *International Journal of Geographical Information Science* 13(7), pp. 677–707.

Liu, H. and Jezek, K. C., 1999. Investigating DEM error patterns by directional variograms and Fourier analysis. *Geographical Analysis* 31, pp. 249–266.

Marshall, S. J., 2002. Modelled nucleation centres of the Pleistocene ice sheets from an ice sheet model with subgrid topographic and glaciologic parameterizations. *Quaternary International* 95–96, pp. 125–137.

Montgomery, D. R., 2001. Slope distributions, threshold hill-slopes, and steady-state topography. *American Journal of Science* 301(4–5), p. 432.

Oksanen, J. and Sarjakoski, T., 2005. Error propagation of DEM-based surface derivatives. *Computers & Geoscience* 31(8), pp. 1015–1027.

Purves, R. S. and Hulton, N. R., 2000. Experiments in linking regional climate, ice-sheet models and topography. *Journal of Quaternary Science* 15, pp. 369–375.

Sugden, D. E., Hulton, N. R., and Purves, R. S., 2002. Modelling the inception of the Patagonian icesheet. *Quaternary International* 95–96, pp. 55–64.

Sun, G., Ranson, K. J., Kharuk, V. I., and Kovacs, K., 2003.Validation of surface height from shuttle radar topography mission using shuttle laser altimeter. *Remote Sensing of Environment* 88(4), pp. 401–411.

Wechsler, S. P., 2006. Uncertainties associated with digital elevation models for hydrologic applications: a review. *Hydrology and Earth System Sciences Discussions* 3, pp. 2343–2384.

Weng, Q., 2002. Quantifying uncertainty of digital elevation models derived from topographic maps. In: D. Richardson and P. van Oosterom (eds.), *Advances in Spatial Data Handling*, chapter 30, pp. 403–418, Springer, London.

Wenzens, G., 2003. Comment on: "The Last Glacial Maximum and deglaciation in southern South America" by N. R. J. Hulton, R. S. Purves, R. D. McCulloch, D. E. Sugden, M. J. Bentley [*Quaternary Science Reviews* 21 (2002), 233–241]. *Quaternary Science Reviews* 22(5–7), pp. 751–754.

Wood, J. D., 1996. The geomorphological characterisation of digital elevation models. PhD thesis, University of Leicester, UK.

Section IV

Applications

INTRODUCTION

This section is treated with a variety of applications where quality of spatial data was made explicit. In several applications, scale is an important issue in quality assessment, which we see in the approaches to map generalization, flood modeling, and translations between habitat classifications.

In Chapter 15, Wijaya, Marpu, and Gloaguen apply a variety of geostatistical texture classifiers to Landsat imagery of tropical rainforests in Indonesia and show how the assumption that neighboring pixels are not independent can improve the classification of a tropical forest.

In Chapter 16, Podolskaya, Anders, Haunert, and Sester tackle the issue of quality in map generalizations. During map generalization, as they say, two conflicting objectives have to be met: reducing the amount of data while keeping the map similar to the input map. The authors approach quality as a compromise between the extent to which these opposite goals are reached and demonstrate their approach of measuring this quality for polygons of buildings in a cadastral dataset. The application demonstrates the strength of the approach and is very relevant and applicable in the many situations where the detailed cadastral data have to be generalized.

Rientjes and Alemseged, in Chapter 17, show the effect of uncertainty in hydro-dynamic modeling of floods in an urban area. In their study, flooding of an urban area is modeled with digital terrain models of varying spatial resolution as input. The authors state they cannot be conclusive about the effectiveness of high-resolution DSMs for this application, since it could not yet be separated from other aspects of setup and parameterization of the model. Especially the way buildings were represented had an important effect on the modeling results.

Data scale and grain of process are also crucial in the work of Comber, Fisher, and Brown in Chapter 18, where they move from crisp mappings to bounded belief in a case of translation between various habitat classifications to answer different landscape questions. Their approach with context-sensitive Boolean maps also indicates the role of the user: Conservation managers have to able to define the "best" decision.

In Chapter 19, the quest of a user to find appropriate spatial data for environmental studies is modeled using similarities with impedance mismatch. With this metaphor, Guemeida, Jeansoulin, and Salzano provide a refreshing insight and terminology to describe the matching of metadata.

Chapter 20, by Dias, Edwardes, and Purves, shows an analysis of visitor tracks in a nature reserve, handling the uncertainty in location during spatio-temporal clustering of individual tracks. The visitors were provided with information in different ways, and the study reveals that the behavior of visitors in terms of where they walk and how much time is spent at certain locations is influenced by the way the information was provided to them.

In these last three sections of this Applications chapter, there is a clear focus on the users of spatial data, with their needs and definitions of quality, which also points our view to the next section, where the communication with the user of the data is the central issue.

15 Geostatistical Texture Classification of Tropical Rainforest in Indonesia

Arief Wijaya, Prashanth R. Marpu, and Richard Gloaguen

CONTENTS

15.1 INTRODUCTION

Mapping of forest cover is an ultimate way to assess forest cover changes and to study forest resources within a period of time. On the other hand, forest encroachment has hardly stopped recently due to excessive human exploitation on forest resources. Forest encroachment is even worse in the tropical forest, which is mostly located in developing countries, where forest timber is a very valuable resource. The need for an updated and accurate mapping of forest cover is an urgent requirement in order to monitor and to properly manage the forest area.

Remote sensing is a promising tool for mapping and classification of forest cover. A huge area can be monitored efficiently at a very high speed and relatively low cost using remote sensing data. Interpretation of satellite image data mostly applies per pixel classification rather than the correlation with neighboring pixels. Geostatistics is a method that may be used for image classification, as we can consider spatial variability among neighboring pixels (Jakomulska and Clarke, 2001). Geostatistics and

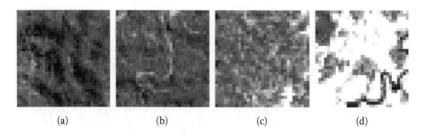

(a)   (b)   (c)   (d)

FIGURE 15.1   Different texture of land cover classes represented on the study area: (a) dense forest, (b) logged over forest, (c) burnt area/open forest, and (d) clear-cut forest/bare land.

the theory of regionalized variables have already been introduced to remote sensing (Woodcock et al., 1988).

This chapter attempts to carry out image classification by incorporating texture information. Texture represents the variation of gray values in an image, which provides important information about the structural arrangements of the image objects and their relationship to the environment (Chica-Olmo and Abarca-Hernandez, 2000). The chapter aims to explore the potential of pixel classification by measuring texture spatial variability using geostatistics, fractal dimension, and conventional gray level co-occurrence matric (GLCM) methods. This is encouraged by several factors, like: (1) texture features can improve image classification results, as we include extra information; (2) image classification on forest area, where visually there are no apparent distinct objects to be discriminated (e.g., shape, boundary), can take into benefit the use of texture variation to carry out the classification; and (3) texture features of land cover classes in forest area, as depicted on Figure 15.1, are quite different visually even if the spectral values are similar; therefore, the use of texture features may improve the classification accuracy.

15.2 STUDY AREA

The study focuses on a forest area located in the Labanan concession forest, Berau municipality, East Kalimantan Province, Indonesia, as described in Figure 15.2. This area geographically lies between 1° 45′ to 2° 10′ N, and 116° 55′ and 117° 20′ E.

The forest area belongs to a state-owned timber concession-holder company where timber harvesting activity is carried out, and the area is mainly situated inland of coastal swamps and formed by undulating to rolling plains with isolated masses of high hills and mountains. The variation in topography is a consequence of the folding and uplifting of rocks, resulting from tension in the earth crust. The landscape of Labanan is classified into flat land, sloping land, steep land, and complex landforms, while the forest type is often called as lowland mixed dipterocarp forest.

15.3 DATA AND METHOD

15.3.1 Data

Landsat 7 ETM of path 117 and row 59 acquired on May 31, 2003, with 30 m resolution was used in this study. The data were geometrically corrected using WGS 84

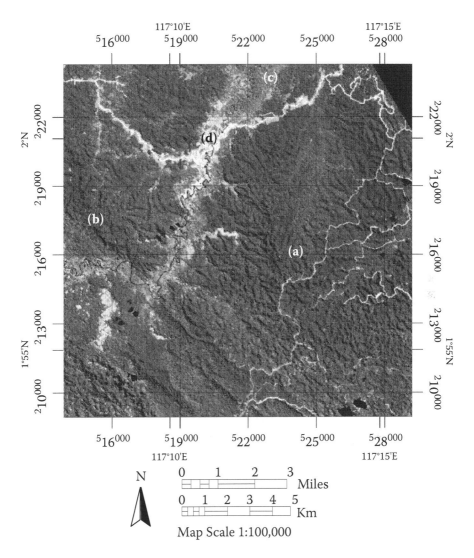

FIGURE 15.2 Study area represented using a combination of Bands 4, 5, and 3 in the RGB channel. Important land cover classes are marked here, namely, logged over forest (a), dense forest (b), burnt area/open forest (c), and clear-cut forest/bare land (d).

datum and UTM projection with an root mean square (RMS) error of less than 1.0 pixel. Subsequently, atmospheric corrections on the satellite data were conducted using an ATCOR module (Richter, 1996). A subset of the Labanan concession area (512 × 512 pixels) was used for the classification in order to optimize effort and time for forest cover classification and validation. During the dry season, 531 sampling units were collected on September 2004; 364 units were used to train the classification and 167 units were used as a test dataset. Five forest cover classes were identified, namely, logged over forest, clear-cut forest/bare land, dense forest, and burnt areas/open forest.

15.3.2   METHOD

15.3.2.1   Gray-Level Co-occurrence Matrix

The gray-level co-occurrence matrix (GLCM) is a spatial dependence matrix of relative frequencies in which two neighboring pixels that have certain gray tones and are separated by a given distance and a given angle occur within a moving window (Haralick et al., 1973). The GLCM texture layers could be computed from each band of Landsat data. To provide the largest amount of texture information, the following strategy was adapted in selecting the satellite band for computing the GLCM texture layers. A covariance matrix showing the variance of each land cover class for each band was computed and the band corresponding to the highest mean variance of forest classes was selected. Compared to other spectral bands, Band 5 of Landsat ETM has the highest mean variance value, as summarized on Table 15.1. Using a window size of 5 × 5 at every pixel and a grayscale quantization level of 64, four GLCM layers were derived from the Landsat image, using variance, homogeneity, contrast, and dissimilarity as defined by Haralick et al. (1973).

15.3.2.2   Geostatistics Features

To incorporate geostatistical texture features in the classification, a semivariogram was computed in the neighborhood of every pixel. Generally, spatial variability increases gradually with the distance separating the observations up to a maximum value (the sill) representing the maximum spatial variance. The distance at which the sill is reached represents the range of variation, i.e., the distance within which observations are spatially dependent. Respectively, the size of the moving window used to extract texture information of spectral data has an important role to provide an accurate estimation of the semivariance, which eventually affects the classification accuracy. This chapter uses 5 × 5 and 7 × 7 moving windows to derive geostatistics texture layers.

   A semivariogram is an univariate estimator, which describes the relationship between similarity and distance in the pixel neighborhood. $Z(x)$ and $Z(x + h)$ are two values of the variable Z located at points x and $x + h$. The two locations are separated by the lag of h. The semivariogram values are calculated as the mean sum of squares

---

TABLE 15.1
Variance Matrix of Forest Cover Classes Training Data

Land Cover Class	Band 1	Band 2	Band 3	Band 4	Band 5	Band 6	Band 7
Logged over forest	3.49	3.03	8.46	23.56	50.98	0.70	15.46
Burnt areas/open forest	2.59	1.76	2.59	24.19	18.46	0.46	5.04
Road network	65.68	165.39	332.63	74.12	386.45	1.91	294.98
Clear-cut forest/bare land	2.70	5.20	2.90	33.52	32.49	0.83	8.70
Dense forest	2.92	1.56	1.49	1.82	11.08	0.49	4.73
Hill shadow	2.62	2.78	2.58	40.28	30.32	0.57	6.63
Mean variance of total classes	**13.33**	**29.95**	**58.44**	**32.91**	**88.30**	**0.83**	**55.92**

of all differences between pairs of values with a given distance divided by 2, as described in the following equation (Carr, 1995):

$$\gamma(h) = \frac{1}{2n} \sum_{i=1}^{n} \left(Z(x_i) - Z(x_i + h) \right)^2 \tag{15.1}$$

where n is number of pairs of data.

Another spatial variability measure is the madogram, which instead of measuring squares of all differences takes the absolute values (Deutsch and Journel, 1998; Chica-Olmo and Abarca-Hernandez, 2000):

$$\gamma(h) = \frac{1}{2n} \sum_{i=1}^{n} \left| Z(x_i) - Z(x_i + h) \right| \tag{15.2}$$

By calculating the square root of the absolute differences, we can derive a spatial variability measure called a rodogram, as shown in the following formula (Lloyd et al., 2004):

$$\gamma(h) = \frac{1}{2n} \sum_{i=1}^{n} \left| Z(x_i) - Z(x_i + h) \right|^{\frac{1}{2}} \tag{15.3}$$

Alternatively, three multivariate estimators to quantify the joint spatial variability (cross-correlation) between two bands, namely, pseudo-cross variogram and pseudo-cross madogram, were also computed. The pseudo-cross variogram represents the semivariance of the cross increments and is calculated as follows:

$$\gamma(h) = \frac{1}{2n} \sum_{i=1}^{n} \left(Y(x_i) - Z(x_i + h) \right)^2 \tag{15.4}$$

The pseudo-cross madogram is similar to the pseudo-cross variogram, but, again, instead of squaring the differences, the absolute values of the differences area taken, which leads to a more generous behavior toward outliers (Buddenbaum et al., 2005):

$$\gamma(h) = \frac{1}{2n} \sum_{i=1}^{n} \left| Y(x_i) - Z(x_i + h) \right| \tag{15.5}$$

Using Band 5 of satellite data, the spatial variability measures were computed and the median values of semivariance at each computed lag distance were taken, resulting in full texture layers for each calculated spatial variability measure. These texture layers were then put as additional input for the classification.

15.3.2.3 Fractal Dimension

Fractals are defined as objects that are self-similar and show scale invariance (Carr, 1995). Fractal distribution requires that the number of objects larger than a specified size has a power law dependence on the size. Every fractal is characterized by a fractal dimension (Carr, 1995). Given the semivariogram of any spatial distribution, the fractal dimension (D) is commonly estimated using the relationship between the fractal dimension of a series and the slope of the corresponding log-log semivariogram (m) plot (Burrough, 1983; Carr, 1995):

$$D = 2 - \frac{H}{2} \qquad\qquad (15.6)$$

15.4 RESULTS AND DISCUSSION

15.4.1 Results

Before geostatistics texture layers were derived, we observed whether there was textural variation among the different classes. Using training data, a semivariogram of land cover classes on the study area was sequentially computed for a lag distance (range) of 8 pixels.

As shown in Figure 15.3, semivariance computed for every lag distance may provide useful information for data classification as those values of each forest class reveal the spatial correlation for lag distances of less than 8 pixels. However, there is an exceptional case for road network and clear-cut forest/bare land classes, which show spatial variability on a larger lag. This may be a problem for computing semivariance for this particular class as the calculation of per-pixel semivariance on large lag distance is computationally expensive. Compromising with other forest classes, texture layers were computed using 5×5 and 7×7 moving windows. Using the different spatial variability measures explained before, semivariance values for each pixel were calculated and the median of these values was used, resulting in texture information of the study area. The results of the geostatistics texture layers are described in Figure 15.4.

Classification of the satellite image was done using the following data combinations: (1) ETM data; (2) ETM data and GLCM texture; and (3) ETM data and geostatistics texture. Two classification methods, using a minimum distance algorithm and the Support Vector Machine (SVM) method, were applied for the purpose of the study. The SVM method is originally a binary classifier, which is based on statistical learning theory (Vapnik, 1999). Multiclass image classification using the SVM method is conducted by combining several binary classifications with segmenting data with the support of an optimum hyperplane. The optimum performance of this method was mainly affected by a proper setup of some parameters involved in the algorithm. This study, however, was not trying to optimize the SVM classification; therefore, those parameters were arbitrarily determined. For the classification, a radial basis function kernel was used, where γ in kernel and classification probability threshold were, respectively, 0.143 and 0.0, while the penalty parameter was 100. The motivation of using the SVM and minimum distance was to study the

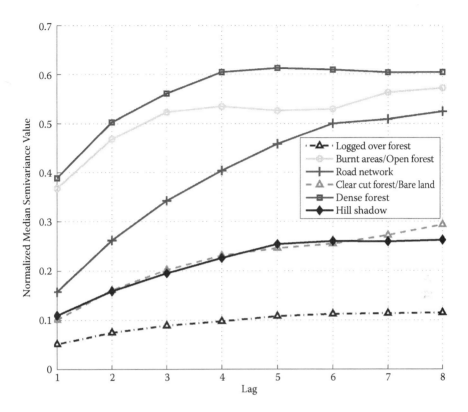

FIGURE 15.3 Variogram plot of training data shows the spatial variability of land cover classes over the study area.

performance of texture data given two completely different algorithms in the classification, and the results are summarized in Table 15.2. Applying these classifiers, the results showed that 74% and 76% of the accuracies were achieved when Bands 3, 4, and 5 of the Landsat image and multispectral Landsat data (i.e., Bands 1–5, 7) were used in the classification, respectively. As multispectral bands give higher accuracies, these bands were used together with texture data for further classification. The GLCM texture layers slightly improved the classification accuracies, when variance, contrast, and dissimilarity were used in the classification. The GLCM texture classification performed by the SVM resulted in 81% accuracy when a combination of the ETM data and all the GLCM texture layers was applied.

The geostatistics texture layers, on the other hand, performed quite satisfactorily, resulting in more than 80% of accuracies when the fractal dimension, madogram, rodogram, and a combination of those texture layers were used in the classification. The classification resulted in 81.44% of accuracy and a kappa of 0.78 when the image data, fractal dimension, madogram, and rodogram were classified by the SVM method; the results are depicted in Figure 15.5.

Indeed, the SVM performed better than the minimum distance, when texture data were used. It has already been proven that the SVM performed well when

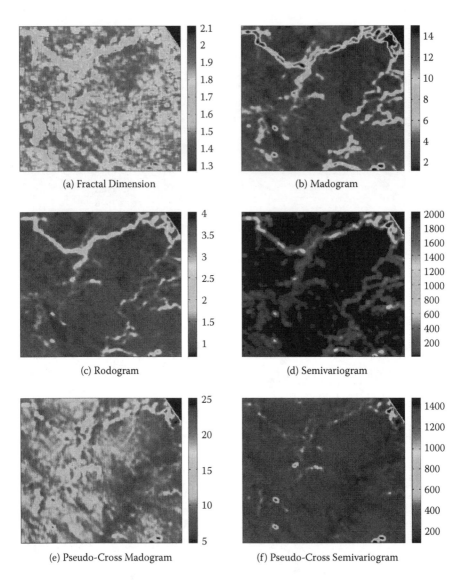

FIGURE 15.4 Different texture layers derived from spatial variability measures of geo-statistics methods: (a) fractal dimension, (b) madogram, (c) rodogram, (d) semivariogram, (e) pseudo-cross madogram, and (f) pseudo-cross semivariogram.

dealing with large spectral data resolution, such as hyperspectral, as reported by several recent studies (Gualtieri and Cromp, 1999; Pal and Mather, 2004, 2005).

15.4.2 DISCUSSION

The geostatistics texture layers performed quite well in the classification. However, semivariogram and pseudo-cross semivariogram texture layers were not giving sat-isfactory classification results when those layers were classified by the minimum

TABLE 15.2

Overall Accuracy Assessment (OAA) of the Classification

	Min. Distance		SVM	
	OAA	Kappa	OAA	Kappa
ETM Data				
ETM 6 Bands	76%	0.71	76%	0.71
ETM Band 3, 4 ,5	74%	0.69	74%	0.71
ETM 6 Bands, Geo-texture Windows 5×5				
ETM 6 Bands, Fractal	76%	0.71	81%	0.77
ETM 6 Bands, Madogram	77%	0.72	78%	0.74
ETM 6 Bands, Rodogram	76%	0.71	80%	0.76
ETM 6 Bands, Semivariogram	57%	0.48	77%	0.72
ETM 6 Bands, Pseudo-cross Semivariogram	47%	0.36	77%	0.72
ETM 6 Bands, Pseudo-cross Madogram	76%	0.71	75%	0.71
ETM 6 Bands, Fractal, Madogram, Rodogram	77%	0.72	81%	0.77
ETM 6 Bands, Geo-texture Windows 7×7				
ETM 6 Bands, Fractal	76%	0.71	79%	0.75
ETM 6 Bands, Madogram	78%	0.73	80%	0.76
ETM 6 Bands, Rodogram	76%	0.71	81%	0.77
ETM 6 Bands, Semivariogram	50%	0.39	77%	0.73
ETM 6 Bands, Pseudo-cross Semivariogram	47%	0.37	76%	0.71
ETM 6 Bands, Pseudo-cross Madogram	76%	0.71	76%	0.71
ETM 6 Bands, Fractal, Madogram, Rodogram	78%	0.73	81%	0.78
ETM 6 Bands, GLCM				
ETM 6 Bands, Variance	77%	0.72	77%	0.72
ETM 6 Bands, Contrast	77%	0.72	75%	0.70
ETM 6 Bands, Dissimilarity	72%	0.67	77%	0.73
ETM 6 Bands, Homogeneity	62%	0.54	77%	0.72
ETM 6 Bands, Variance, Contrast, Dissimilarity, Homogeneity	63%	0.55	81%	0.77

distance method. This is due to the nature of the semivariogram and pseudo-cross semi-variogram, which calculate the mean square of the semivariance for all observed lag distance, either using monovariate or multivariate estimators. This eventually may reduce the classification accuracy because of the presence of data outliers. Combined with the madogram and rodogram, the classification resulted in higher accuracies with the SVM and minimum distance method. This is obvious as the madogram, which calculates the sum of the absolute value of the semivariance for all observed lag distance, and the rodogram, which computes the sum of the square roots of those semivariances, have "softer" effects on the presence of outliers compared to those of the semivariogram. This study observed that by changing the size of the moving window from 5 × 5 to 7 × 7 slightly improved the classification accuracy. This is because the scale of land cover texture is similar to the 7 × 7

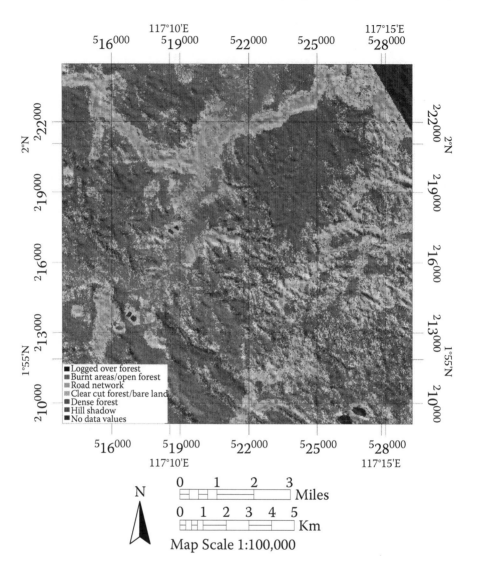

FIGURE 15.5   The final classification result image

window size; therefore, this window size provides more texture information than the other. However, computation of texture layer using a larger sized moving window is absolutely not efficient in terms of time; thus, to initially find the optimum size of the moving window may be an alternative to reducing efforts and time for the computation of the geostatistics texture layers. Selection of the proper-sized moving window will provide better texture information from the spectral image data. In general, additional texture layers for image classification, derived from either the GLCM or geostatistics, have effectively improved the classification accuracy. Although this study found that by applying different GLCM texture layers as well as geostatistics layers in a single classification process considerably improved classification accuracy, one should be very careful to apply the same method for different types of data. The selection of classification algorithm depends on the data distribution.

15.5   CONCLUSIONS AND FUTURE WORK

This study found that texture layers derived from the GLCM and geostatistics methods have improved classification of spatial data of Landsat images. Texture layers are computed using the moving window method. The selection of the moving window size is very important since the extraction of texture information from the spectral data is more useful when the texture characteristics corresponding to the observed land cover classes are already known.

The Support Vector Machine as well as the minimum distance algorithm performed well in classifying elaborating texture data as additional input of Landsat ETM data. Moreover, the SVM resulted in average higher accuracies compared to those of the minimum distance method.

The authors observed that, for future work, it is also possible to compute geostatistics texture layers with adjustable moving window sizes, depending on the size of the texture polygon for particular land cover class being observed. This may be an alternative to extracting better texture information from the spectral data.

ACKNOWLEDGMENTS

The Landsat ETM and ground truth data were collected when the first author joined the MONCER Project during his master's studies at the International Institute for Geo-Information Science and Earth Observation, The Netherlands. Therefore, the first author would like to thank Dr. Ali Sharifi and Dr. Yousif Ali Hussein, who made possible the data collection used for the purpose of this study.

REFERENCES

Buddenbaum, H., Schlerf, M., and Hill, J., 2005. Classification of coniferous tree species and age classes using hyperspectral data and Geostatistical methods. *International Journal of Remote Sensing* 26(24), pp. 5453–5465.

Burrough, P. A., 1983. Multiscale sources of spatial variation in soil. II. A non-brownian fractal model and its application in soil survey. *Journal of Soil Science* 34(3), pp. 599–620.

Carr, J., 1995. *Numerical Analysis for the Geological Sciences.* Prentice-Hall, Inc., Upper Saddle River, NJ.

Chica-Olmo, M. and Abarca-Hernandez, F., 2000. Computing geostatistical image texture for remotely sensed data classification. *Computers and Geosciences* 26(4), pp. 373–383.

Deutsch, C. and Journel, A., 1998. *GSLIB: Geostatistical Software Library and User's Guide.* Second edition, Oxford University Press, New York.

Gualtieri, J. and Cromp, R., 1999. Support vector machines for hyperspectral remote sensing classification. *Proceedings of SPIE—The International Society for Optical Engineering* 3584, pp. 221–232.

Haralick, R., Shanmugam, K., and Dinstein, I., 1973. Textural features for image classification. *IEEE Transactions on Systems, Man and Cybernetics* 3(6), pp. 610–621.

Jakomulska, A. and Clarke, K., 2001. Variogram-derived measures of textural image classification. In: P. Monesties, Ed., *geoENV III—Geostatistics for Environment Applications*, Kluwer Academic Publisher, Dordrecht, The Netherlands, pp. 345–355.

Lloyd, C., Berberoglu, S., Curran, P., and Atkinson, P., 2004. A comparison of texture measures for the per-field classification of mediterranean land cover. *International Journal of Remote Sensing* 25(19), pp. 3943–3965.

Pal, M. and Mather, P., 2004. Assessment of the effectiveness of support vector machines for hyperspectral data. *Future Generation Computer Systems* 20(7), pp. 1215–1225.

Pal, M. and Mather, P., 2005. Support vector machines for classification in remote sensing. *International Journal of Remote Sensing* 26(5), pp. 1007–1011.

Richter, R., 1996. Atmospheric correction of satellite data with haze removal including a haze/clear transition region. *Computers and Geosciences* 22(6), pp. 675–681.

Vapnik, V., 1999. An overview of statistical learning theory. *IEEE Transactions on Neural Networks* 10(5), pp. 988–999.

Woodcock, C. E., Strahler, A. H., and Jupp, D. L. B., 1988. The use of variograms in remote sensing: II. real digital images. *Remote Sensing of Environment* 25(3), pp. 349–379.

16 Quality Assessment for Polygon Generalization

Ekatarina S. Podolskaya, Karl-Heinrich Anders, Jan-Henrik Haunert, and Monika Sester

CONTENTS

16.1 INTRODUCTION

The quality of a map can be understood as its ability to satisfy the needs of users (TSNIIGAiK, 2003). Evaluating and ensuring the quality is one of the primary goals of map generalization (Cheng and Li, 2006). Experts suggest several directions; for example, Müller et al. (1995) proposed to clarify the expectations in terms of data quality and to analyze the potential errors introduced by using digitized maps in a GIS. The problem of quality assessment in general is not new; it has been tackled in several approaches (e.g., van Smaalen, 2003; Galanda, 2003; Bard, 2004; Skopeliti and Tsoulos, 2001).

The aims of our chapter are as follows: (1) to propose a method for data quality assessment of polygon generalization by adapting the approaches of different authors and (2) to apply the method to buildings in cadastral datasets and to areas of different land cover in a topographic database.

The topographic database contains digital landscape models (DLM) of four different scales that were taken from the German "Authoritative Topographic-Cartographic Information System" (ATKIS) (www.atkis.de). These DLM are called

211

Basis-DLM (1:25.000), DLM50 (1:50.000), DLM250 (1:250.000), and DLM1000 (1:1.000.000). We used DLM50 and DLM250 for our investigations. Buildings at scale 1:10.000 were taken from the German cadastre. This dataset was generalized to scale 1:20.000 with the software CHANGE, developed at the IKG-Institute of Cartography and Geoinformatics Leibniz University of Hannover (http://www.ikg .uni-hannover.de).

The chapter is organized according to the following structure. After this introduction, we describe related work, i.e., elements of quality assessment, existing ideas on quality assessment of polygon generalization, and measures for polygon generalization. The third section is devoted to two aims of generalization. Our method of integrating different objectives into a single quality measure is considered in the fourth section. The application of our method to a German cadastral dataset and the official German topographic database ATKIS is presented and discussed in the next section. The chapter ends with some conclusions and directions for future research.

16.2   RELATED WORK

Data quality assessment requires three steps: specification of requirements, definition of data quality measures, and evaluation of data quality (Joao, 1998). Elements of data quality assessment have been discussed in different papers and books, and are also standardized in national and international standards. Mayberry (2002) proposed the following components: accuracy, integrity, consistency, completeness, validity, timeliness, and accessibility. The factors affecting the quality of spatial data are given in Burrough and McDonnell (1998): Currency, completeness, consistency, accessibility, accuracy and precision, sources of errors in data, and sources of errors in derived data and in the results of modeling and analysis. Guptill and Morrison (1995) described the elements of data quality: lineage, accuracy (positional, attribute, and semantic accuracy), completeness, logical consistency, and temporal information. Quantitative (completeness, accuracy, correctness of identification of objects, logic coordination of structure, and representation of objects), and qualitative (purpose, lineage, or source of data) indicators are used for quality assessment in Kolokolova (2005). Thus, researchers suggest and use identical elements for the characteristic of data quality assessment. The data quality concept in map generalization has been described in the following components: object completeness of target scale to the initial scale as well as details of the qualitative characteristic of the phenomenon (Garaevskaya and Malusova, 1990).

In recent years, some investigations for developing the evaluation model with quantitative parameters have been undertaken. Bard and Ruas (2004) define the quality using the deviation from a given ideal. In this way, specifications for ideals are used (e.g., minimum size for legibility) and compared with the generalized situation. The ideal is defined using scale-dependent functions.

A paper by Frank and Ester (2006) describes the method of quality assessment of a whole map. For a comparison of two maps they use values for shape, location, and information. The approach takes into account changes in individual objects in the form of shape similarity, groups of objects using the location similarity, and changes across the entire map using semantic content similarity.

But, despite of this research, we do not have a comprehensive investigation of quality assessment in polygon generalization. First of all, we should conclude that in the majority of the suggested methods various levels of data quality assessment from one separate object up to a whole map have been proposed: macro (for the map), meso (for groups of objects), and micro (for individual objects). Such a concept is, e.g., used by Peter (2001). Secondly, the evaluation of generalization quality depends on the choice of an optimal set of these measures. There is a large number of measures for polygonal maps that can be used for map quality evaluation. A very detailed description of such measures is presented in Peter (2001). We can give here only a very brief classification of these measures into seven classes with their relation to the map levels:

Size (*micro, meso, macro*): Absolute and relative geometric properties of a polygon, e.g., area or perimeter

Shape (*micro, meso, macro*): For instance, shape descriptors could be *compactness, convexity, principal components* (Peura and Iivarinen, 1997), or *Fourier descriptors* (Zahn and Roskies, 1977)

Distance (*micro, meso*): Geometric proximity of polygons, e.g., Hausdorff distance

Topology (*micro, meso*): Occurrence of self-intersections, orientation changes, aggregation, or separation

Density (*meso, macro*): Preservation of the distribution of polygons, number of polygons in a certain area, or covered area by polygons in a certain region

Pattern (*meso, macro*): Preservation of patterns, e.g., alignments, grid-, ring-, or star-structures (Anders, 2006; Heinzle et al., 2006)

Semantic/Information (*meso, macro*): Based on hierarchical ontologies or concept hierarchies, it is possible to include semantics into similarity measures (Anders, 2004; Rodriguez and Egenhofer, 2004)

Obviously, there is a large variety of measures to quantify the quality of a polygon generalization. Some of these measures are difficult to assess and implement (e.g., patterns). In this chapter, we define some new polygon measures on the micro level, but with the focus on an integrated quality measure.

16.3 TWO AIMS OF GENERALIZATION

In general, there are two conflicting aims in generalization: on the one hand, the amount of data has to be reduced; on the other hand, the resulting map has to be similar to the original one. We try to use measures for these two goals and integrate them using a simple weighted addition.

16.3.1 REDUCING THE AMOUNT OF DATA

The amount of map information decreases when its scale is reduced. We have considered two types of reduction: reducing the amount of objects (polygons) and reducing the amount of detail (vertices) of individual objects. Reducing the amount of polygons can be achieved with the following generalization operations: A polygon is not represented in another scale according to rules for this scale (*elimination*); a polygon is merged with another polygon (*aggregation*).

16.3.2 KEEPING THE MAP SIMILAR TO THE INPUT MAP

Keeping the map similar to the input map is the second main goal of map generalization. Similarity can be defined in terms of object size before and after generalization, or the respective perimeter values. Another measure for the analysis of shape similarity based on *the stepping turning function* is described by Frank and Ester (2006). The similarity value can then be computed using the two turning functions of the polygon before and after generalization:

$$V_{TF} = 1 - \frac{\text{Area}\left(TF_1 \Delta TF_2\right)}{\text{Max}\left[\text{Area}\left(TF_1\right), \text{Area}\left(TF_2\right)\right]} \tag{16.1}$$

where TF_1 is the shape between the x-axis and the turning function of the polygon in map M1, TF_2 is the shape between the x-axis and the turning function of the polygon in map M2, and $TF_1 \Delta TF_2$ is the symmetric difference of TF_1 and TF_2, i.e., the shape between both turning functions.

16.4 INTEGRATION OF DIFFERENT MEASURES

In order to combine two opposite goals of generalization, we integrate the measures described above:

1. Reduction of polygon vertices (V_N)
2. Keeping the map similar to the input map is based on area of polygon (V_A), perimeter of polygon (V_P), and turning function (V_{TF})

For the quality assessment, we use the values of parameters V_{TF}, V_N, V_A, and V_P from Equation 16.1 through Equation 16.4. Values equal or close to 1 indicate good quality, whereas bad quality is denoted with values equal or close to 0:

$$V_N = \frac{\left|N_2 - N_1\right|}{\text{Max}(N_1, N_2)} \tag{16.2}$$

$$V_A = 1 - \frac{\left|\text{Area}\left(p_2\right) - \text{Area}\left(p_1\right)\right|}{\text{Max}\left[\text{Area}\left(p_2\right), \text{Area}\left(p_1\right)\right]} \tag{16.3}$$

$$V_P = 1 - \frac{\left|\text{Perimeter}\left(p_2\right) - \text{Perimeter}\left(p_1\right)\right|}{\text{Max}\left[\text{Perimeter}\left(p_2\right), \text{Perimeter}\left(p_1\right)\right]} \tag{16.4}$$

where p_1 and p_2 are two corresponding polygons having N_1 and N_2 vertices, respectively.

The overall quality measure of polygons is calculated as a weighted sum of these measures:

$$V = c_{TF} \cdot V_{TF} + c_N \cdot V_N + c_A \cdot V_A + c_P \cdot V_P \tag{16.5}$$

with $c_{TF} + c_N + c_A + c_P = 1$, where c_{TF}, c_N, c_A, and c_P are the weights of the different quality measures.

There are two approaches to using the weights in the quality assessment of generalization. First, the biggest weight can be given to the parameter that is the most important for the user, and results in good quality of this parameter (Frank and Ester, 2006). Second, we can assign arbitrary weights to all parameters. Then we receive results for different weight combinations and can make a choice as to what is the most preferable variant with respect to the visual quality of the result. Table 16.1 shows the possible sets of weights. The rational behind Variant 1 is the fact that the two opposing goals, reduction (parameter V_N) and preservation (parameters V_A, V_P, V_{TF}), are weighted equally. Obviously, the number of variants is not limited to the presented variants.

TABLE 16.1
Variants of Weights

	c_N	c_A	c_P	c_{TF}
Variant 1	0.5	0.167	0.167	0.167
Variant 2	0.167	0.5	0.167	0.167
Variant 3	0.167	0.167	0.5	0.167
Variant 4	0.167	0.167	0.167	0.5

16.5 IMPLEMENTATION OF OUR APPROACH

16.5.1 BUILDINGS IN SCALES 1:10.000 AND 1:20.000

The buildings from the cadastral dataset (original scale approx. 1:10.000) have been generalized using the software package CHANGE. The measures used in this chapter are all based on a comparison of properties of a polygon before and after generalization. For aggregated buildings we calculated V_N, V_A and V_P by defining N_1, Area (p_1), and Perimeter (p_1) in Equation 16.2 through Equation 16.4 to be the sums of these values for the individual components in the original scale. We analyzed three samples of buildings in scales 1:10.000 and 1:20.000 with different structures (Figure 16.1).

Visual control is one of the most important components of quality assessment. To visualize the obtained results, we display the obtained quality values with Variant 1 by different gray values for individual buildings (Figure 16.2 through Figure 16.4). Dark gray values represent low quality, to draw the attention to problematic cases. The legend in Figure 16.2 applies to all three samples. Intuitively, one would assume that the generalization of more complex buildings is more difficult. In our opinion, the results of Variant 1 reflect the quality of the map best. However, more tests need to be done to come to an assured conclusion about the appropriate weights settings.

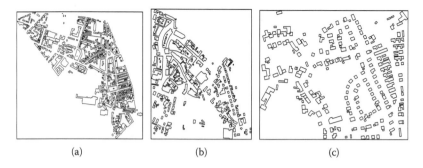

(a)   (b)   (c)

FIGURE 16.1   Source dataset from cadastre (scale 1:10.000).

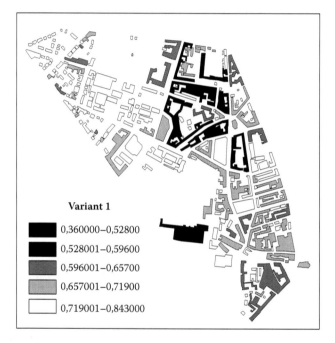

FIGURE 16.2   Sample (a).

16.5.2   LAND COVER POLYGONS FROM ATKIS: AGGREGATED DLM50 AND DLM250

Land cover polygons are different from buildings in several aspects of geometry and semantics. Figure 16.5 shows polygons of the DLM50 after application of an aggregation method based on global optimization techniques (Haunert and Wolff, 2006). The optimization criteria were compactness and semantic similarity of feature classes. We refer to this dataset as "aggregated DLM50." In order to create an appropriate representation for the target scale 1:250.000, a line simplification algorithm by de Berg

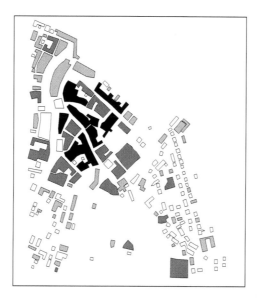

FIGURE 16.3 Sample (b).

et al. (1995) was applied after this aggregation, leading to the result in Figure 16.6. The line simplification results in simple "one-to-one" relations between features of the input dataset (aggregated DLM50) and the output dataset (DLM250).

We summarize these results as follows: In comparison to buildings, the average quality values are significantly higher and the variation is smaller. It is important to note that we cannot conclude from this that the applied generalization procedure for land cover polygons is better than the method for building simplification. As mentioned earlier, important differences for both problems exist. Thus, in order to classify the results into categories such as "good" or "bad," different classification schemes need to be applied.

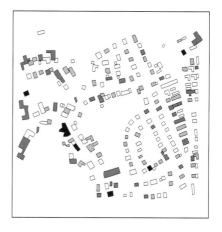

FIGURE 16.4 Sample (c).

Future research on map quality will be along the four directions: developing methods for quality assessment of generalization n-polygons into m-polygons and developing measures on the meso and macro levels.

16.6 CONCLUSIONS

A method of quality assessment for polygon generalization has been suggested and explored. The proposed procedure offers a possibility to calculate measures

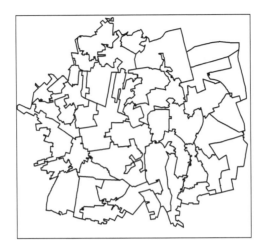

FIGURE 16.5   Source dataset DLM50.

FIGURE 16.6   Quality assessment of land cover polygons (DLM250).

for quality assessment and to visually inspect them. This allows selecting different weights for the parameters in order to highlight different preferences. The ideal is a situation when the accumulated quality measure exactly fits the expectation of a human cartographer. This parameter setting then, in turn, can be used to quickly inspect new datasets. Although there is a correlation between the visual quality assessment and the quality value calculated with our measures, there is still room for improvement. Obviously, the measures fit more to man-made objects and less to natural ones.

ACKNOWLEDGMENTS

All colleagues at the Institute of Cartography and Geoinformatics, Leibniz University of Hanover for providing a friendly and positive working atmosphere with useful discussions are thanked.

The funding of the Mikhail Lomonosov program from the German Academic Exchange Service (DAAD) and the Russian Ministry of Education and Science is gratefully acknowledged.

REFERENCES

Anders, K.-H., 2004. Parameterfreies hierarchisches Graph-Clustering-Verfahren zur Interpretation raumbezogener Daten. Dissertation, Universität Stuttgart. Persistent Identifier: urn:nbn:de:bsz:93-opus-20249.

Anders, K.-H., 2006. Grid Typification. In: Riedl, A., Kainz, W. & Elmes, G. A., Eds., *Progress in Spatial Data Handling, 12th International Symposium on Spatial Data Handling.* Springer-Verlag, pp. 633–642.

Bard, S., 2004. Quality assessment of cartographic generalization. *Transactions in GIS,* 8(1), pp. 63–81.

Bard, S., and Ruas A., 2004. Why and how evaluating generalized data? In: *Proceedings of 12th International Symposium. Progress in Spatial Data Handling.* Springer-Verlag, pp. 327–342.

Burrough, P., and McDonnell, R., 1998. *Principles of Geographical Information Systems.* Oxford University Press, p. 223.

Cheng, T., and Li, Zh., 2006. Toward quantitative measures for the semantic quality of polygon generalization. *Cartographica,* 41(2), pp. 135–147.

de Berg, M., van Kreveld, M., and Schirra., S., 1995. A new approach to subdivision simplification. In: *Twelfth International Symposium on Computer-Assisted Cartography,* Volume 4, pp. 79–88, Charlotte, NC.

Frank, R., and Ester, M., 2006. A quantitative similarity measure for maps. In: *Proceedings of 12th International Symposium. Progress in Spatial Data Handling.* Springer-Verlag, pp. 435–450.

Galanda, M., 2003. Automated polygon generalization in a multi agent system, PhD thesis, Department of Geography, University of Zurich, Switzerland.

Garaevskaya, L. S., and Malusova, N. V., 1990. *The Practical Manual on Cartography.* Nedra, pp. 46–47.

Guptill, C., and Morrison J. L., 1995. *Elements of Spatial Data Quality.* Elsevier Science Ltd., p. 202.

Haunert, J.-H., and Wolff, A., 2006. Generalization of land cover maps by mixed integer programming. In: *Proceedings of 14th International Symposium on Advances in Geographic Information Systems,* 10–11 November 2006, Arlington, VA, pp. 75–82.

Heinzle, F., Anders, K.-H., and Sester, M., 2006. Pattern recognition in road networks on the example of circular road detection. In: Raubal, M., H. J. Miller, Frank, A. U., and Goodchild, M. F., Eds., *Geographic Information Science, GIScience 2006, Münster, Germany.* LNCS 4197, Springer-Verlag, pp. 253–267.

Joao E. M., 1998. *Causes and Consequences of Map Generalization.* Taylor & Francis, p.220.

Kolokolova, I., V. 2005. Automatic and interactive quality assessment in technologies of automated generalization. In: *Proceedings of Scientific Congress "GEO-Sibir 2005,"* 25–29 April 2005, Novosibirsk, Russia. pp. 287–290.

Mayberry, M., 2002. Paper "Data quality: Before the map is produced" http://www.directionsmag.com/article.php?article_id=250.

Müller, J. C., Weibel, R., Lagrange J. P., and Salge F., 1995. Generalization: state of the art and issues. In: *GIS and Generalization. Methodology and Practice. GIS Data I.* Taylor & Francis. J. C. Muller, J. P. Lagrange, and R. Weibel, Eds., pp. 3–7.

Peter, B., 2001. Measures for the generalization of polygonal maps with categorical data. In: *Fourth ICA Workshop on Progress in Automated Map Generalization,* 2–4 August 2001, Beijing, China.

Peura, M., and Iivarinen, J., 1997. Efficiency of simple shape descriptors. In: *3rd International Workshop on Visual Form*, Capri, Italy.

Rodriguez, A., and Egenhofer, M., 2004. Comparing geospatial entity classes: an asymmetric and context dependent similarity measure. *Geographical Information Science*, 18(3): 229–256.

Skopeliti, A., and Tsoulos, L., 2001. The accuracy aspect of cartographic generalization. In: *Proceedings of the GIS Research UK 9th Annual Conference GISRUK 2001.* Wales, UK, pp. 22–25.

van Smaalen, J. W. N., 2003. Automated aggregation of geographic objects. A new approach to the conceptual generalization of geographic databases. Doctoral diss. Wageningen University, The Netherlands.

TSNIIGAiK, 2003. OST 68-3.4.2-2003: The standard of branch. Digital maps. Methods of estimation of data quality. The general requirements. Moscow, Russia. http://gis-lab .info/docs.html

Zahn, C. T., and Roskies, R. Z., 1977. Fourier descriptors for plane closed curves. *IEEE Transactions on Computers*, C-21(3):269–281.

17 Effectiveness of High-Resolution LIDAR DSM for Two-Dimensional Hydrodynamic Flood Modeling in an Urban Area

Tom H.M. Rientjes and Tamiru H. Alemseged

CONTENTS

17.1 INTRODUCTION

Over the past decade, floodings of river floodplains and urbanized areas have become a common feature in many parts of the world due to reasons such as climate and land use changes. It appears that the frequency of flood events has increased but also that the extremeness of the events in terms of magnitude of flow discharges and flood inundation extent increased. By intensified agricultural practices in the relative flat river plains, damages in rural areas have increased while in urban areas damages have grown dramatically by increased building activities but also, as observed in many less developed countries, by illegal settlements at river banks and river beds.

For developing proper flood management strategies, engineers and policy makers require information on flood characteristics such as flood levels, flow velocities, discharge volumes, and inundation extent that serve for designs of flood mitigation and prevention measures. Knowledge of flood characteristics and detailed land use and property inventories allow for flood vulnerability and risk assessments. As such, reliable flood forecasting and simulation tools must be available, and the effectiveness of new data sources and their integration in simulation tools must be explored and assessed. In distributed flood modeling it is common practice to utilize hydrodynamic model approaches to simulate spatially distributed flood characteristics such as inundation extent as well as flood depth and flow velocity that change over space and time. Model algorithms of such approaches are based on the shallow water equations commonly known as the St. Venant's equations (Saint-Venant, 1871). In many flood studies, these equations are reported to yield information that is satisfactorily accurate for practical applications, although model approaches are very different with respect to the procedure for the discretization of selected model domains, the dimensionality of the flow algorithms (i.e., one-dimensional, two-dimensional), the conceptualization of flow processes and their interactions such as river-floodplain mass exchanges, as well as the specific flow conditions simulated such as kinematic, diffusion, and hydrodynamic flow conditions (see, e.g., Bates and de Roo, 2000; Horrit and Bates, 2001b, 2002; Werner, 2001; Dutta et al., 2007; Alho and Aaltonen, 2008). In this literature discussions and debate on generic aspects and effectiveness of the applied model approaches are presented and consensus has been reached that trade-offs in model performance exist between the various approaches. For instance, a one-dimensional model approach could be too simple to accurately simulate flood wave propagation in complex terrain while for the same terrain a two-dimensional model may not be effective by the high data demand for input data. Such becomes even more manifest when model simulation results are to be seen as time- and space-averaged outputs for sequential model calculation time steps for the spatial units the model equations are solved. A lengthy discussion on the relations between model complexity, effectiveness, and the reliability of model approaches is ignored here since such is not within the scope of this chapter. For reasons of brevity, here we only mention that, for the selection of an effective model approach, the various generic modeling issues must be considered. These issues relate, for instance, to the representation of small- and large-scale heterogeneity of the system, the applicability of selected flow equations, the selection of model boundary conditions for model forcing, as well as to the availability of real-world data on system properties such as topography, land use, and observed flow data. Moreover, advanced calibration tools must be available to analyze model performance, which changes over space and time in response to model forcing. In this respect reference is made to the works of Gupta et al. (1998), Khu and Madson (2005), Hogue et al. 2000), Vrugt et al. (2003), and de Vos and Rientjes (2007), all of which have a specific focus on the use of advanced multiobjective model evaluation procedures in the field of river flow and runoff modeling. Application of such procedures in distributed flood modeling, however, is unprecedented and research still has a focus on assessing the effectiveness of new data sources to improve model performance and on developing more efficient numeric solvers.

Research on numeric solvers has a focus on developing novel finite difference, finite element, or finite volume approaches as well as on developing multidimensional flow algorithms to allow for simulations in complex terrain (see Horritt and Bates, 2001a, 2002; Horritt et al., 2007; Alho and Aaltonen, 2007). Other developments are in the field of integration and assimilation of earth observation and remote sensing data to allow for the use of high-resolution topographic and digital terrain data in modeling (see Marks and Bates, 2000; Mason et al., 2003; Bates, 2004; Mignot et al., 2006; Wilson and Atkinson, 2007; Horritt et al., 2007). These developments are triggered by developments in geographic information system (GIS) technology for processing and visualization of model input and model output in a manner society, planners, and policy makers easily understand.

Despite many progresses, there are relevant problems that make flood modeling a topic of ongoing research. Examples are the use of input data of poor quality as, for instance, due to inappropriate low spatial and temporal resolution; the simulating of relevant small-scale processes such as turbulence as well as the accurate forecasting of extreme events. Also, model calibration and validation of simulated flood events is very challenging when input data are not adequate or of poor quality, or when scale issues cause a deterioration of results and poor model performance. Generally, the dimensions or scale of the grid elements of selected model domains are much larger as compared to the scale at which field observations are available, and thus simulated flood characteristics only represent averages at the scale of the grid elements. In flood modeling, these elements serve as model calculation units or building blocks that constitute the model domain. In this respect, Horritt and Bates (2001a, b) mentioned that model process representation is a subject of ongoing research and debate, and the representation required is likely to be a function of the type and required accuracy of predictions, the quality of the model parameterization, and the scale at which the model operates. Following this reasoning, it is obvious that selected spatial and temporal model resolutions will have significant impacts on simulation results. A clear description, however, on how selected grid size and calculation time steps propagate into simulation results and how the selected grid size affects flood model behavior and outcomes is not commonly shown, and assessing such effects is at the core of our work, with a specific focus on the use of high-resolution LIDAR data in an urban area. By rescaling the LIDAR data that are of $1^2 \, m^2$ resolution, flood simulations at various grid sizes are executed and results are assessed for typical flood characteristics such as extent of inundation area, flow velocities, and inundation depth. Assessments are also made for applied land surface parameterizations and on the propagation effects of mathematical boundary conditions.

In this chapter, an extreme flood event in the urbanized area of the city of Tegucigalpa Honduras is simulated. Airborne light detection and ranging (LIDAR) data are used as the main source for creating a digital surface model (DSM) while the SOBEK one-dimensional/two-dimensional model approach (see www.sobek.nl; Dhondia and Stelling, 2004; Verwey, 2001) is selected for flood simulation. This approach applies a finite difference computational scheme and requires a spatially distributed model domain with rectangular grid elements that all are of equal size.

17.2   LIDAR DIGITAL SURFACE MODELS

LIDAR technology provides DSMs that are suitable for a range of applications (see, e.g., Priestnall et al., 2000). Specific to LIDAR data is the accurate topographic representation that results from combining global positioning systems (GPS) and laser distance measuring and inertial measuring unit (IMU) technologies. Maps of surface heights can be produced with a height precision of about 15 cm (depending on the nature of the ground cover) and at spatial resolutions of 1^2 m^2 or lower. Baltsavias (1999) describes some advantages of LIDAR technology; LIDAR allows for complete area coverage, it is an indirect measuring technique that does not require encoding of three-dimensional coordinates, and objects much smaller than the footprint area of the object itself can be identified. Other advantages include its rapid collection and the possibility to repeat flights.

The use of LIDAR data in flood modeling in urbanized areas is particularly attractive since LIDAR data provide elevation values of the bare ground surface while objects of relatively small scale such as roads, buildings, and possibly dykes remain visible. The availability of a DSM allows for easy land surface model parameterization where land utilization is represented and parameterized by surface roughness values to denote land surface friction and obstructions to flow. Cobby et al. (2003) prove the effectiveness of LIDAR in large-scale floodplain modeling, but proving such effectiveness in urban areas only has gained little attention, although the effectiveness of urban flood modeling already was suggested by Priestnall et al. (2000). In the same work it was noted that, for many applications, relatively simple filtering procedures could be applied to extract discrete features such as buildings from the LIDAR DSM, but the use of more complex methods including artificial Neural Network (ANN) was also suggested.

In our research we selected simple slope and minimum filters to obtain footprints of buildings. The identification of an accurate building footprint map involved a trial-and-error procedure where the output of the procedure was compared to an ortho-photo of the study area, as shown in Figure 17.1. A widely accepted filtering method for defining bare ground surface elevations and feature detection is not yet

FIGURE 17.1   Building footprints overlain on an ortho-photo.

available and this, to the knowledge of the authors, remains a research topic with many challenges.

The elevation data as obtained from LIDAR are in the form of mass points that need to be transformed to a DEM structure as required by the hydrodynamic model approach. Grid-based DEM structures are still the most commonly used in distributed flood modeling despite the fact that such structures fail to represent the various shapes of slopes such as topographic convergence, divergence, convexity, and concavity (see, e.g., Tachikawa et al., 1996).

The effect of DEM resolution on topographic attributes and hydrologic model outputs is explored and discussed in several studies. Zhang and Montgomery (1994), Hutchinson and Dowling (1994), Hutchinson and Dowling, Jenson (1991), and Callow et al. (2007) indicate that the DEM grid cell size significantly affects both the representation of the land surface and its topographic attributes, but they also show that the results of selected conceptual rainfall-runoff model approaches are largely affected. In hydrodynamic flood modeling, Cobby et al. (2003) and Bates et al. (2003) report on the effect of the DEM structure that was based on the triangular finite element discretization. For our study, however, a raster DEM with uniform and equally sized grid elements is required since the SOBEK model approach that is selected for hydrodynamic flood modeling requires such DEM as input.

17.3 HYDRAULIC MODELING

In this study, the SOBEK flood model approach is adopted that is described by Dhondia et al. (2006). In the SOBEK approach, one- and two-dimensional algorithms are combined that allow for the simulation of water flow in river reaches (one-dimensional) as well as the overflow of river banks and flowover floodplains (two-dimensional). In this study the two-dimensional approach is used by the availability of very high resolution LIDAR data. The use of the one-dimensional approach is ignored since grid elements of the LIDAR data are much smaller than the actual widths of the river sections, but also because in an urban environment, river sections preferably must be simulated at a spatial resolution similar to the floodplain grid resolution. This aspect of the model setup is described in Alemseged and Rientjes (2005) and is not further described here. A second rationale for only using the two-dimensional approach is that such an approach allows for the change of flow characteristics in two flow directions by solving the momentum balance equations in two directions. This is of particular importance when flow patterns in heterogeneous and complex terrains such as in urban areas have to be simulated. Water movement in SOBEK is described by a finite difference approximation with continuity equations of mass and momentum at the core of the flow algorithm. For such modeling, the land surface requires parameterization and mathematical boundary conditions that govern the flow of water at the upstream and downstream ends of the model domain must be defined.

17.3.1 LAND SURFACE PARAMETERIZATION

In the real world, topography is a critical factor that affects the propagation of a flood wave in a channel and its surrounding floodplain. Clearly, topographic properties

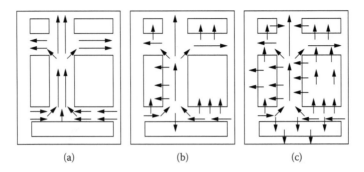

FIGURE 17.2 Representations of buildings (a) solid objects, (b) partially solid objects, and (c) hollow objects. Assumed flow vectors are added.

such as roads, dykes, trees, buildings, or ditches may obstruct the flow but also could conduct or accelerate the flow of water.

In flood modeling, parameterization of river reaches and floodplains is through assigning roughness coefficients to grid elements to reflect on land surface friction as it relates to land surface characteristics. Commonly, these values are available and presented in tabulated form (see, e.g., Engman, 1986) and are generally obtained through laboratory experiments under controlled conditions. In flood modeling, the typical approach is to define a fixed and unique roughness value for each floodplain grid element. In such procedures all real-world properties within the scale of the grid element are lumped and averaged, and it is implicitly assumed that all objects have similar effects on the simulated flood characteristics. In the case of urban flood modeling, it thus is assumed that obstructing objects such as buildings are assumed to act like a hollow object that conducts the flow of water, although such cannot be considered a realistic representation.

In this study, buildings are represented as solid objects, hollow objects, or partially solid objects, as illustrated in Figure 17.2. For representation of a building as a solid object, a roughness value (i.e., Manning's value [in $s\ m^{-1/3}$]) of 1 is applied while a value of 0.025 is applied for areas without buildings.

When buildings are considered a hollow object, a roughness value similar to the other grid elements of the floodplain is selected. For further analysis of the effect of DSM resolution, surface roughness values of 0.07 for the floodplain and 0.04 for the channel were specified. These values were adopted based on a review of tabulated roughness coefficients and other relevant studies.

In flood model calibration, it is common practice to update and fine-tune roughness values until the model performance is considered satisfactory (see, e.g., Werner et al., 2005). In this study model calibration and performance assessments through-fine tuning is not performed by a lack of reliable flood observation data, but model performance assessments are through performance comparisons and sensitivity analysis.

17.3.2 Mathematical Boundary Conditions

In flood modeling, mathematical boundary conditions are commonly specified at boundary elements at the upstream and downstream ends of the model domain to

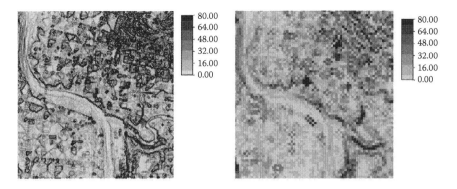

FIGURE 17.3 Slope maps of re-sampled DSMs of 5 and 15 m resolution.

govern the inflow and outflow of water in the model. For this study, an accurate and reliable time series of inflow discharges at the inflow elements was not available, and therefore synthetic discharge hydrographs and related peak flows for a 50-year flood event have been constructed through regional flood frequency analysis. For assessing the effect of the hydrograph shape, three different shapes are constructed that follow a normal distribution (bell shape), a triangular distribution, and a lognormal (skewed) distribution. At outflow elements at the downstream model boundary, a head-dependent flow condition has been specified. In this study, analyses on downstream boundary effects were performed by varying the hydraulic head values and the type of boundary condition (fixed head, free flow, and normal depth). Additionally, grid resolutions at the downstream end of the model domain were altered and the effects were analyzed as well.

17.4 RESULTS

17.4.1 LIDAR DSM RESOLUTIONS

In this work, DSMs of various grid resolutions are prepared from the LIDAR data and serve to assess the effects of surface elevations and related attributes such as slope gradient and slope aspect on model simulations. Figure 17.3 shows the effect of re-sampling on local slope gradients and indicates that a significant loss of small-scale information is observed when grid resolution decreases from 5 to 15 m. Here, re-sampling stands for the aggregation or disaggregation of grid elements of the LIDAR DSM and implies that new DSMs of different grid resolution are constructed based on available attribute data of the reference* map. By re-sampling of the LIDAR DSM to DSMs of lower resolution, the size of the grid elements becomes larger and new attribute values depend on the re-sampling method and the number of grid elements used for re-sampling.

In this study the LIDAR DSM is re-sampled by the nearest neighbor, bi-linear, and bi-cubic methods and DSMs are constructed for resolutions of 4.5, 7.5, and 10 m. Results of this re-sampling are presented in Table 17.1.

---

* This is the LIDAR map with grid resolution of $1^2 m^2$.

TABLE 17.1

Results from Re-sampling

Method	Resolution (m)	Mean (m)	Min. Error (m)	Max. Error (m)	Std. Dev. (m)	RMSE (m)
Nearest neighbor	4.5	0.56	−3.01	24.3	3.10	3.13
	7.5	0.13	−16.75	22.9	3.56	3.54
	10.0	−0.18	−13.58	13.5	2.82	2.81
Bi-linear	4.5	0.56	−3.01	24.3	3.10	3.13
	7.5	0.13	−14.07	22.1	3.27	3.25
	10.0	−0.14	−12.47	13.2	2.72	2.71
Bi-cubic	4.5	0.53	−3.01	24.3	3.11	3.13
	7.5	0.19	−15.42	23.5	3.45	3.44
	10.0	−0.45	−29.96	13.1	14.30	4.32

For the lower resolution DSMs (7.5 and 10 m), the results reveal significant differences between the re-sampling methods in terms of magnitude of errors that are generated. Such is somewhat obscured for the 4.5 m grid element size since all methods resulted in a comparative magnitude of minimum and maximum errors. The bi-linear method resulted in smaller errors as compared to the bi-cubic method in two of the three cases. This method also constructs a smoother surface model as compared to the nearest-neighbor method because it allows for averaging across multiple values.

For the nearest-neighbor and bi-linear methods, the smallest error is observed for the largest grid element size (i.e., 10 m). For the bi-cubic method, the results show that an increase of grid element size results in an increase of error, possibly due to both sharpening and smoothing of the input map.

17.4.2 Effects of DSM Resolution

In hydraulic modeling, the geometry of river reaches and floodplains has to be represented by the land surface model. Obviously, the size of the grid elements of such models to a large extent defines the detail at which properties can be represented, since surface grid element values only represent averaged values. Values are lumped values and imply that the heterogeneity of real-world properties within the grid elements scale is ignored. An example of such a lumping effect is illustrated in Figure 17.4, which shows the cross sections below the junction of the river and the tributary as extracted from DSMs of different resolutions. It is illustrated that the lumping effect at 25 m resolution is the largest and that smaller scale properties that obstruct the exchange of water between the river and floodplain in the real world become obscured or even unseen in the DSMs. Thus effects of re-sampling propagate into simulations since, on the one hand, relevant small-scale flow conveyance properties are ignored and, on the other hand, model performance in general is affected by "fixed" and parameterized hydraulic flow properties that actually change as a function of grid resolution.

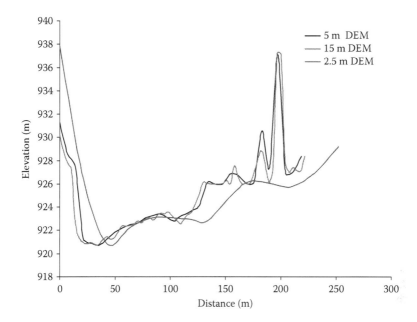

FIGURE 17.4 Channel cross sections below the junction of the river with its tributary.

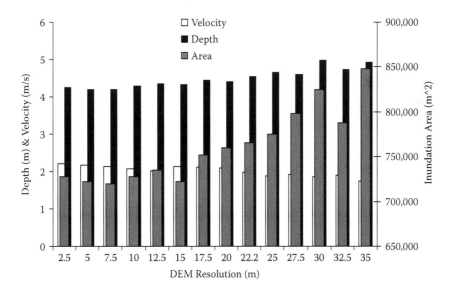

FIGURE 17.5 Maximum of average flow characteristics as a function of DSM resolution.

For analyses of simulation results, the averages of maximum flow depths and maximum flow velocities are stored. Averages represent maxima of averaged values as calculated over the model domain for single time steps. In Figure 17.5, for instance, it is shown that the average of the maximum flow velocity decreases when DSM

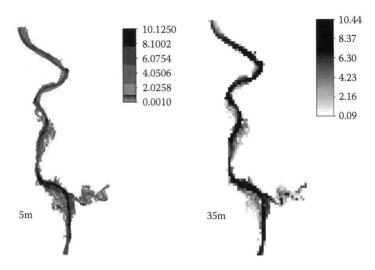

FIGURE 17.6 Average of maximum flood depth for 5 and 35 m DSM resolutions.

resolutions increase from 2.5 to 35 m, although some variation on this pattern can be observed. Also, the maxima of averaged inundation depth and inundation area change, and these changes can be directly related to changes in hydraulic gradients across the model domain.

Obviously, re-sampling has a direct effect on simulated flood characteristics and thus has a direct effect on simulated discharges, flood behavior, and model performance in general. In addition, based on simple hydraulic reasoning, it can be argued that the effects of re-sampling on hydraulic flow behavior are manifold but also that quantifying each of these effects is far from trivial.

This is illustrated in Figure 17.6, in which the maximum inundation area and related water depths for DSMs with resolutions of 5 and 35 m are shown. By visual comparison, the illustrations show a similar flood behavior, but they also show that the differences can be observed locally in the model domain. These differences are also observed in flow velocities (see Figure 17.5) and are caused by re-sampling, which results in local differences in the river and floodplain representations.

An aspect in modeling that possibly could cause a similar effect relates to the applied mathematical boundary condition. For these simulations, however, similar discharge hydrographs are selected for the upstream boundary condition and equal volumes of water are expected to be entered and stored in the model domain. Although cross-sectional areas increase with increased element size, it is assumed that such effects can be ignored since both inflow and outflow boundary elements are of equal size. These assumptions are to satisfy the law of mass conservation for the simulation period.

17.4.3 EFFECTS OF LAND SURFACE PARAMETERIZATION

Effects of land surface parameterization are analyzed through a base line study where the available LIDAR data are re-sampled to a DSM of 5 m resolution. Flood

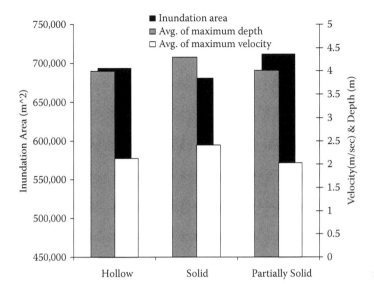

FIGURE 17.7 Flow characteristics for distinct building representations.

simulations are performed where buildings are represented as solid objects, hollow objects, or partially solid objects. The flow characteristics of the results are show in Figure 17.7, where averages of maximum inundation depth and maximum flow velocity and inundation extent are shown. The diagram clearly shows that building representation has a direct effect on flood characteristics. Variations are in all three characteristics, but plausibility to explain and to understand results is not trivial. For instance, when surface roughness is set largest as for solid objects, flow velocities on average become highest while the inundation area becomes smallest. For this situation, the average inundation depth is largest. Based on simple reasoning on hydraulic flow behavior, this result can be questioned by the fact that flow velocities generally decrease when surface roughness increases. Moreover, it is also surprising that for these conditions the inundation extent increases.

In order to interpret the result in Figure 17.7, it must be realized that the presented flow characteristics do not necessarily coincide over time but that observations can be for different calculation time instants. It must also be realized that buildings are geometrically not parameterized, and thus a change of conveyance is not considered or accounted for. As described earlier, in the real world solid, objects obstruct and possibly block flow discharges. A detailed description and an in-depth discussion on the effects of building representation are ignored here since such should be supported by more advanced model simulations and calibration procedures where model performance can be evaluated over model space and over time.

17.4.4 Effects of Boundary Conditions

Assessments after applied boundary conditions indicate that both upstream and downstream conditions only have a relatively small effect on model simulations.

TABLE 17.2

Simulated Flow Characteristics for Different Types of Inflow Hydrograph Shapes

Hydrograph Shape	Inundation Area (m²)	Maximum Depth (m)		Maximum Velocity (m/s)	
		Max	Average	Max	Average
Triangular	693100	9.58	3.99	17.8	2.1
Bell shape	698575	9.64	4.03	13.1	2.0
Skewed	695100	9.61	4.03	12.9	2.18

For the synthetic discharge hydrographs at the upstream model boundary, Table 17.2 shows the maximum inundation area; the maximum and the average of the maximum simulated depths and flow velocities as observed during the simulation period. The results show that, in general, changes in the flow characteristics are small except for the maximum flow velocity, which appears to be extremely and unrealistically high for all three simulations.

Assessments after applied downstream boundary conditions primarily focussed on propagation effects that commonly are referred to as boundary effects. Simulations for a DSM with a resolution of 20 m have been performed for fixed head, free flow, and normal depth conditions. To assess the effect of the grid resolution, simulations have peen performed for a DSM with downstream elements of 2 m length. For these assessments, the one-dimensional HEC-RAS approach has been used since such an approach allows for local grid densification. For illustration purposes, Figure 17.8 shows diagrams of applied fixed and free flow conditions.

Results on these simulations are extensively described in Alemseged and Rientjes (2007) and indicate that the selected boundary conditions propagate up to distances of 1500 m but commonly only propagate up to a few hundred meters. Also, effects are largest at the boundary element itself and quickly dampened out when distance increases. Considering the size of the model domain of 4 × 1.5 km², these distances are considered small and thus the effects are considered minor. The results also show that propagation and dampening are at smaller distances when the grid elements are of smaller size. For grid elements of 2 m length, propagation distances reduce to a maximum of 500 m as subjected to the specified boundary condition.

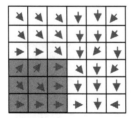 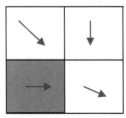

FIGURE 17.8 Example of averaging effect on flow vector distribution when re-sampling from 5 to 15 m resolution.

17.5   DISCUSSION

In this research, much focus has been on assessing the effects of selected grid reso-lutions and on assessing the effects of land surface parameterization on simulated flood characteristics such as flow velocities, extent of inundation area, and inunda-tion depth. Principle to the modeling is that our real world requires schematization and conceptualization that are to be handled by a model approach, but also that model input data need to be prepared for model calculation units at which flow equa-tions are solved. In flood modeling these units are the grid elements that, when com-bined, make up a DSM. In this research, a very high resolution LIDAR DSM of 1^2 m^2 was available and served for constructing DSMs of coarser resolutions. The LIDAR DSM had a rectangular grid structure with grid elements of equal size; re-sampled DSMs are based on such a grid structure as well.

In this research, DSMs are constructed through re-sampling by the nearest-neighbor, bi-linear, and bi-cubic methods. Statistics on errors of re-sampling to 4.5, 7.5, and 10.0 m resolutions are shown in Table 17.1. For all three re-sampling meth-ods, the statistics indicate a considerable loss of information by transforming the high resolution DSMs to lower ones. The largest errors are observed for the bi-cubic method, while it appears that the bi-linear method produces the best results. A clear trend in error statistics, however, is not observed and does not reveal superiority. By these results, we argue that it is important to understand the principle of re-sampling in a sense that any newly calculated grid element value represents an averaged or lumped property over the size of the grid element. By increasing the grid element size, averaging takes place over increasingly larger spatial domains and implies that small-scale heterogeneity becomes obscured. Figure 17.8 indicates that averaging may have an effect on flow vectors.

In hydrodynamic flood modeling, re-sampling and lumping result in the gener-alization of features such as dykes and other flow obstacles that obstruct or conduct water flows, but lumping also causes local storage areas with sizes smaller than the selected grid element to become obscured. Therefore, in modeling it is of great inter-est to know to what extent the resolution of a DSM and its associated generalization affect the model outputs. In Figure 17.5, relations between DSM grid resolution and maxima of inundation area, averaged flow velocities, and averaged water depth are presented. The figure shows that grids of coarser resolution result in an increase of maximum inundation area but also that the maximum averaged flow velocities decrease. A weak relation is observed between the increased grid resolution and the maximum of the averaged water depth. By these results, it can be argued that re-sampling and lumping over larger grid elements in general cause hydraulic gradients to decrease. Since flow velocities can directly be related to hydraulic gradients, and since lower velocities cause lower flood wave propagation, it is obvious that addi-tional storage of water in the model is observed. This happens through the increase of inundation extent as well as through the small, but gradual, increase of the maxi-mum averaged flood inundation depth. These simple hydraulic flow principles are reflected in the model simulation results, but the results do not reveal the most opti-mal and effective grid resolution. Moreover, the authors argue that all the results of

Figure 17.5 are acceptable and none of the simulations can be rejected because of unrealistic model behavior.

Important in urban flood modeling is the parameterization of buildings and how to quantify the effects of buildings on flow dynamics. In this research simulations are performed where buildings, as extracted from the LIDAR DSM, are represented as solid objects, partially solid objects, or hollow objects. Representation is only through modifying the surface roughness coefficient and, as such, a reduction of conveyance is not explicitly considered. The results for a DSM of 5 m grid resolution indicate that building representation affects the simulation results. Crucial for acceptance of these results, however, is the arbitrary procedure of surface roughness parameterization. In our research, fixed surface roughness coefficients have been applied by a lack of a sound scientifically based procedure to uniquely define the optimal values. In our work, roads, ditches, and other small-scale properties have not been modeled explicitly, although such properties are visible in the high-resolution LIDAR DSM. The effectiveness and necessity of using high-resolution data in urban flood modeling therefore have not been proven. In this respect it must be questioned if all urban properties that affect flood propagation can realistically be simulated and also if current flood model approaches such as SOBEK are capable of doing so. Much work, however, on these aspects is not presented in literature and still remains a major challenge to the authors.

17.6 CONCLUSIONS

In this study, it is shown that the resolution of a DSM has a significant effect on the flow characteristics and flood patterns across the model domain. DSMs of different resolutions have been constructed through the re-sampling of a LIDAR DSM of $1^2 m^2$ and indicate that averaging of small-scale topographic features across grid elements of larger scale causes relevant losses of information. Averaging in particular affects the distribution of grid elevation and thus also the local slope gradients in the land surface model. Besides this aspect, averaging also affects the land surface parameterization that reflects on surface roughness as required by the hydrodynamic flow models. Based on the simulation results, it can be concluded that DSM resolutions directly affect inundation extent as well as water depths and flow velocities across the model domain. The results indicate that a coarser resolution may be associated with, on average, larger flood depths, but also with lower flow velocities. It is, however, concluded that all simulation results are acceptable simply by the fact that none of the results can be considered unrealistic. In order to be more conclusive, advanced model performance evaluation through model calibration is required. For this study, reliable calibration data were not available and are considered constrained.

In this study, assessments are also made after the effects of applied upstream and downstream boundary conditions. The results show that applied boundary conditions only have a minor effect on simulation results. The effects of boundary conditions dampened out at relatively short distances of a few hundred meters with a maximum of 1500 m from the model boundary. Tests for smaller sized downstream grid elements indicate that propagation distances reduce even further.

By this study it can be concluded that the method and procedure to represent buildings affect the simulation results. Buildings are represented as solid objects, partially solid objects, or as hollow object, and the effects of these representations are assessed by varying the surface roughness values. The results for a DSM of 5^2 m^2 resolution indicate that such a representation also has a direct effect on the flow characteristics. The results show that flow velocities on average increase when surface roughness is set high, such as when buildings are represented as solid blocks. Based on hydraulic reasoning, this behavior may be questioned and we conclude that building representation through only setting roughness coefficients is not sufficient to simulate real-world hydrodynamic flow effects. In principle, effects of reduced conveyance must be considered as well, but the effects of small-scale properties such as roads and ditches that accelerate or obstruct the flow of water also must be considered.

In the discussion on the effectiveness of high-resolution DSMs in hydrodynamic flood modeling in an urban area, we cannot be conclusive. This research shows that many aspects that relate to model setup and model parameterization have a direct effect on model performance but also that any model simulations must be associated with error and uncertainty. In this research, it is shown that much uncertainty is introduced by the grid resolution of the DSM.

ACKNOWLEDGMENT

The authors would like to thank Dr. Cees Van Westen for providing all the necessary data for this study.

REFERENCES

Alemseged, T. H., and Rientjes, T. H. M., 2007. Uncertainty issues in hydrodynamic flood modeling. In: *Proceedings of the 5th Int. Symp.on Spatial Data Quality SDQ 2007, Modeling Qualities in Space and Time*, ITC, Enschede, The Netherlands, p. 6.

Alemseged, T. H., and Rientjes, T. H. M., 2005. Effects of LIDAR DEM resolution in flood modeling: A model sensitivity study for the city of Tegucigalpa, Honduras. ISPRS WG III/3, III/4, V/3 Workshop "Laser scanning 2005," Enschede, The Netherlands.

Alho, P., and Aaltonen, J., 2008. Comparing a 1D hydraulic model with a 2D hydraulic model for the simulation of extreme glacial outburst floods. *Hydrol. Process.*, 22(10), pp. 1537–1547.

Baltsavias, E. P., 1999. A comparison between photogrammetry and laser scanning. *ISPRS Journal of Photogrammetry & Remote Sensing*, Vol. 54, No. 2-3, pp. 83–94.

Bates, P. D., 2004. Remote sensing and flood inundation modeling. *Hydrol. Process.*, 18, pp. 2593–2597.

Bates, P. D., Marks, K. J., and Horritt, M. S., 2003. Optimal use of high resolution topographic data in flood inundation models. *Hydrol. Process.*, 17, pp. 537–557.

Bates, P. D., and De Roo, A. P. J., 2000. A simple raster-based model for flood inundation simulation. *J. Hydrology*, 236, pp. 54–77.

Callow, J. N., Van Niel, K. P., and Boggs, G. S., 2007. How does modifying a DEM to reflect known hydrology affect subsequent terrain analysis? *Journal of Hydrology*, 332(1-2), pp. 30–39.

Cobby, D. M., Mason, D. C., Horritt, M. S., and Bates, P. D., 2003. Two-dimensional hydraulic flood modeling using a finite-element mesh decomposed according to vegetation and topographic features derived from airborne scanning laser altimetry. *Hydrol. Process.*, 17, pp. 1979–2000.

Dhondia, J. F., and Stelling, G. S., 2004. Applications of one dimensional-two dimensional integrated hydraulic model for flood simulation and damage assessment. URL: www .sobek.nl. Access date: July 22, 2004.

Dutta, D., Alam., J., Umeda, K., and Hayashi, M., 2007. A two-dimensional hydrodynamic model for flood inundation simulation: a case study in the lower Mekong river basin. *Hydrol. Process.*, 21, pp. 1223–1237.

Engman, E. T., 1986. Roughness coefficients for routing surface runoff. *J. Irrig. Drainage Eng.*, 112(1), pp. 39–53

Gupta, H. V., Sorooshian, S., and Yapo, P. O., 1998. Toward improved calibration of hydrologic models: multiple and noncommensurable measures of information. *Wat. Resour. Res.* 34(4), pp. 751–763.

Hogue, T. S., Sorooshian, S., Gupta, H., Holz, A., and Braatz, D., 2000. A multi-step automatic calibration scheme for river forecasting models. *J. Hydrometeorol.* 1, pp. 524–542.

Horritt, M. S., and Bates P. D., 2001a. Effects of spatial resolution on a raster based model of flood flow. *J. Hydrology*, 253, pp. 239–249.

Horritt, M. S., and Bates, P. D., 2001b. Prediction floodplain inundation: raster-based modeling versus the finite-element approach. *Hydrol. Process.*, 15, pp. 825–842.

Horritt, M. S., and Bates, P. D., 2002. Evaluation of 1D and 2D numerical models for predicting river flood inundation. *J. Hydrology*, 268, pp. 87–99.

Horritt, M. S., Di Baldassarre, G., Bates, P. D., and Brath, A., 2007. Comparing the performance of a 2-D finite element and a 2-D finite volume model of floodplain inundation using airborne SAR imagery. *Hydrol. Process.*, 21(20), pp. 2745–2759.

Hutchinson, M. F. and Dowling, T. I., 1994. A continental hydrological assessment of a new grid-based digital elevation model of Australia, *Hydrol. Process.*, 5(1), p. 45–58.

Jenson, S. K., 1991. Applications of hydrologic information automatically extracted from digital elevation models. *Hydrol. Process.*, 5(1), pp. 31–34.

Khu, S. T., and Madsen, H., 2005. Multiobjective calibration with Pareto preference ordering: an application to rainfall–runoff model calibration. *Water Resour. Res.*, 41, W03004.

Marks, K., and Bates, P., 2000. Integration of high-resolution topographic data with floodplain models. *Hydrol. Process.*, 14, pp. 2109–2122.

Mason, D. C., Cobby, D. M., Horritt, M. S., and Bates, P. D., 2003. Floodplain friction parameterization in two-dimensional flood models using vegetation heights derived from airborne scanning laser altimetry. *Hydro. Process.*, 17, pp. 1711–1732.

Mignot, E., Paquier, A., and Haider, S., 2006. Modeling floods in a dense urban area using 2D shallow water equations. *J. Hydrology*, 327, pp. 186–199.

Priestnall, G., Jaafar, J., and Duncan, A., 2000. Extracting urban features from LIDAR digital surface models. *Comput. Environ. Urban Syst.,* 24, pp. 65–78.

Saint-Venant, Barré de, 1871. Theory of unsteady water flow, with application to river floods and to propagation of tides in river channels. *Competus Rendus*, 73, pp. 148–154 and 237–240.

Tachikawa, Y., Takasao, T., and Shiiba, M., 1996. TIN-based topographic modeling and runoff prediction using a basin geometric information system. In: *HydroGIS 96: Applications of Geographic Information Systems in Hydrology and Water Resources Management (Proceedings of the Vienna Conference, 1996)*, IAHS Publ. no. 235.

Verwey, A., 2001. Latest developments in floodplain modeling – 1D/2D integration. *Conference on Hydraulics in Civil Engineering*, The Institute of Engineers, Australia.

Vrugt, J. A., Gupta, H. V., Bastidas, L. A., and Bouten, W., 2003. Effective and efficient algorithm for multiobjective optimization of hydrologic models. *Wat. Resour. Res.*, 39 (8), p. 1214.

Vos, de, N. J., and Rientjes, T. H. M., 2007. Multi-objective performance comparison of an artificial neural network and a conceptual rainfall-runoff model. *Hydrological Sciences Journal (Journal des sciences hydrologiques)*, 52(3), pp. 397–413

Werner, M., 2001. Impacts of grid size in GIS based flood extent mapping using a 1D flow model. *Phy. Chem. Earth (B)*, 26(7–8), pp. 517–522.

Werner, M. G. F., Hunter, H. M., and Bates, P. D., 2005. Identifiability of distributed floodplain roughness values in flood extent estimation. *J. Hydrology*, 314, pp. 139–157.

Wilson, M. D., and Atkinson, P. M., 2007. The use of remotely sensed land cover to derive floodplain friction coefficients for flood inundation modeling. *Hydrol. Process.*, 21(25), pp. 3576–3586.

Zhang, W., and Montgomery, D. R., 1994. Digital elevation model grid size, landscape representation and hydrologic simulations. *Water Resour. Res.*, 30(4), pp. 1019–1028.

18 Uncertainty, Vagueness, and Indiscernibility

The Impact of Spatial Scale in Relation to the Landscape Elements

Alexis J. Comber, Pete F. Fisher, and Alan Brown

CONTENTS

18.1 INTRODUCTION

Countryside agencies such as the Countryside Council for Wales (CCW) are responsible for reporting on and monitoring the rural environment. The CCW is increasingly being asked to monitor the landscape pressures and effects relating to a series of drivers such as agri-environmental impacts, climate change, and changes to structural support for farmers. Countryside agencies would like to be able to describe the landscape under a range of different policy initiatives. These include the traditional environmental roles relating to land cover habitats (e.g., Annex 1, Priority Habitats), but increasingly relate to new questions. Each of these has their own set of constructs within which the landscape is viewed.

The problem addressed in this chapter is how to translate different habitat classifications from existing ones, given some additional information (e.g., field survey, other

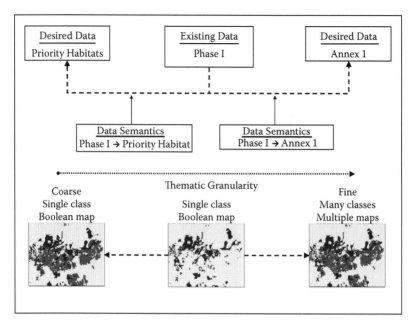

FIGURE 18.1   Issues in translating between different habitat classifications based on data semantics.

data, remote sensing information). The CCW has the national Phase I habitat data (JNCC, 2003), but would like to be able to describe the landscape in terms of other habitats with different grains as a result of EU and national biodiversity legislation:

- Priority habitats as described in the UK Biodiversity Action Plan (UK Government, 1994)
- Annex I habitats from the EC Habitats Directive, which lists important high-quality conservation habitat types and species in its annexes (Commission of the European Communities, 1992)
- Phase II or National Vegetation Classification (NVC) habitats as specified by Rodwell (2006).

Traditionally, conservation agencies use their understanding of habitat semantics to integrate data: habitat A is translated into habitat B by considering the range of attributes or vegetation sub-classes classes within A and in B. However, this process has occurred more or less covertly. Examination of data semantics allows sets of relations between classes to be constructed in light of the grain of process. The translation from Phase I to Annex 1 habitats, for example, represents a refinement in grain. A scheme of this integration process is shown in Figure 18.1.

18.2   BACKGROUND

Countryside agencies are faced with two problems: The first is how to translate information from their existing data holdings to answer new landscape questions. The data may be thematically or spatially coarser than would be ideal to answer these.

Second, it is difficult to incorporate the uncertainties associated with data translations. This is essential as any uncertainty involved will necessarily depend on the question being asked of the data (Comber et al., 2004a, 2006).

The multiplicity of questions that may be asked of any dataset raises a number of issues: 1) how to generate a range of possible maps that manipulate the data (e.g., fusions and aggregations) in different ways; 2) how to choose the most appropriate map for the task in hand (i.e., what to display); and 3) how to understand and quantify the uncertainty that relates to this specific application.

A recent mapping initiative sought to evaluate how updated maps of Phase I habitats in Wales could be re-worked to answer other questions at different scales and granularities. This chapter considers the uncertainty of feature representation where a number of habitats could be identified at any particular point, and how beliefs and preferences can be incorporated in a consistent way into the final map. The resulting maps are called *context-sensitive*, because they have been produced to meet a particular need.

Bayes and Dempster-Shafer are applied to two example questions relating to different habitat granularities and scales for which practical management (specifically the monitoring of burning activity) requires decisions that relate to patch size and landscape context. For example, there may be patches of bog within the upper bound of a potential bog that are too small to be managed independently of the surrounding heath. Similarly, the potential upper bounds may lie beyond those patches assigned to the habitat class so that patches of heath are treated as a bog because their mosaic with the bog is too intimate, in which case bog management takes priority over heath management.

Both examples can be considered from a legal or a conservation perspective. The legal one relates to the legitimacy of burning activity and the conservation one relates to the monitoring of important (Annex I) habitats. In legal situations, Bayesian or probabilistic approaches are more appropriate where there is less uncertainty about the evidence and there is a need to identify the *probable* outcomes, as they may have implications (e.g., prosecutions). Where the evidence has more uncertainty, approaches that identify upper and lower bounds are more appropriate as they explicitly incorporate the uncertainty by showing the *possible* extent of different habitats.

18.3   UNCERTAINTY

All land cover maps incorporate some uncertainty, even if this is not obvious: Error and uncertainty arise at every stage in the production of maps from remotely sensed imagery (Fisher, 1997; Comber et al., 2005a, 2005b). Remote sensing of land cover is predicated on the assumption that the land cover features of interest can be statistically separated and discerned from remotely sensed imagery. Most land cover datasets are Boolean classifications that allocate each data object (pixel, parcel) into one class and the membership of any class is binary. There are uncertainties associated with the process of mapping land cover from remotely sensed imagery, relating to:

- The discerning power of the image may not be able to identify land cover at the required level of grain (spatial resolution).
- The target land covers may themselves not be spectrally homogenous (spectral resolution).

The end result is that statistical clusters in N image bands may not relate to the desired or target land covers resulting in class to class confusions. These issues are well described in the literature (e.g., Freidl et al. [2001] describe issues relating to spatial resolution, Comber et al. [2004b] to spectral resolution), but are rarely accommodated operationally where the end result is that a Boolean allocation decision is made for each object and any uncertainty is often conveniently ignored.

There are a number of issues with this land cover mapping model:

1. Land covers may be composed of heterogeneous mixtures of vegetation that may be beyond the spectral and spatial resolutions of the remotely sensed data. This is often the case in upland seminatural landscapes.
2. Many land cover initiatives seek to augment analyses of remotely sensed data with other information.
3. Land cover maps are used for many other purposes besides that for which they were originally constructed and are used to answer multiple landscape questions, not just the extent and distribution of habitats, such as Phase I.

Therefore, there is a growing interest in being able to re-allocate data objects into different classes for different landscape questions: context-sensitive maps. The re-allocation may be based on the uncertainty associated with the original Boolean allocation and/or due to different weights being given to the supporting evidence, for instance, from ancillary data.

Most approaches to managing uncertainty in the GIS and the nature conservation communities adopt a probabilistic approach under the assumption that the various pieces of data and evidence are independent (i.e., they are not correlated with other data or evidence). This is problematic for a number of reasons. First, the much of the environmental data are spatially autocorrelated. Second, the classic error assessment method, tabulating predicted against observed in a correspondence matrix, assumes that like is being compared with like. This is not the case. Field surveys relate to land cover to plant communities, while remotely sensed classes exist in spectral or image band feature space. These are fundamentally different mental constructs of land cover (see Comber et al., 2005b for a full description). Third, the landscape objects themselves are assumed to be well defined (i.e., not vague, indeterminate, or ambiguous—see Fisher et al., 2006) and can therefore be assessed using crisp probabilistic measures to give measures of error.

Two examples illustrate these problems with independence in the mapping land cover. First, any time series of satellite imagery will contain a mixture of correlated and noncorrelated information, which cannot be treated as independent (though they are often treated as conditionally independent). Second, consider how plant presence and plant cover are modeled in sample stands. It is often acceptable to consider the presence of plant species in a large sample stand of several square meters to be independent. However, when the size of the sample is reduced, the presence of plants becomes positively correlated at a scale that picks out habitat patches (e.g., blanket bog with pools and dry areas). At a smaller scale still, the same species might be negatively correlated because they start to exclude one another.

18.4 METHODS

18.4.1 FORMALISMS

18.4.1.1 Bayes and Dempster-Shafer

Bayes' theorem computes the probability of a hypothesis or event h given the evidence e in support of that event, $P(h|e)$:

$$P\left(h|e\right) = \frac{P(h) * P\left(e|h\right)}{P(e)} \tag{18.1}$$

Dempster-Shafer can be considered as an extension to Bayesian statistics that contains an explicit description of uncertainty and plausibility. It assigns a numerical measure of the weight of evidence (mass assignment, m) to *sets of hypotheses* as well as individual hypotheses. It does not consider the evidence hypothesis by hypothesis as Bayes' theorem does; rather, the evidence is considered in light of the hypotheses. A second piece of evidence is introduced by combining the mass assignments (m and m') using Dempster's rule of combination, to create a new mass assignment m'. Dempster's rule of combination is defined by

$$m''(C) = \sum_{\substack{i,j \\ A_i \cap B_j = C}} m(A_i) m'(B_j) \tag{18.2}$$

That is, the combined mass assignment of C (m'' (C)) is equal to the sum of $m(A_i)$ m' (B_j) for the sets of hypotheses supported by the two pieces of evidence, i and j, such that set $A_i B_j$ equals C. The result of combining two assignments is that for any intersecting sets A and B, where A has mass M from assignment m and B has mass M from assignment m', the belief at their intersection is the product of M and M' (i.e., sum for each combination of A and B that overlap with C).

18.4.1.2 Bayes vs. Dempster-Shafer

The question that the Bayesian approach is answering is, "What is the belief in A, as expressed by the unconditional probability that A is true given evidence, e?" It has at its crux the notion that the evidence can be used to vary the prior probabilities, $P(h)$, and the evidence either supports or refutes the hypothesis. In principal, this approach can be applied to any problem involving uncertainty, assuming that precise probabilities can be assessed for all events. But, this is rarely the case. Dempster-Shafer (DS) accommodates explicit representations of uncertainty and plausibility, which equates to belief plus uncertainty. Therefore, a weak belief in a hypothesis does not imply a strong belief in its negation. One of the weaknesses of Dempster's rule is that it can favor a class that has low mass in two datasets over any class that has a high mass in only one dataset. The classic example is that of the two doctors, one of which is 90% certain the patient has disease *A* and 10% disease *B*; the other is 90% convinced over disease *C* and 10% disease *B*. DS will give 100% support for disease *B*, even though neither doctor thought it likely (although this can be overcome by the use of alternative fusion rules). The point is that it may be problematic to interpret the outcomes of Dempster-Shafer relative to evidence and the hypotheses. Descriptions of the arguments and counterarguments put forward by both sides of the Bayes/Dempster-Shafer dichotomy can be found in a text edited by the main protagonists from either side, Shafer and Pearl (1990), and Parsons (1994) also provides a good introduction to Dempster-Shafer.

18.5   RESULTS

The objective was to identify the potential extent of bog Priority and Annex I habitats within Upland Heathland Phase I habitats, using some additional information and the existing Phase I survey, that is, to identify the potential extent of Bog habitats at higher and lower grains than Phase I. The two analyses are as follows:

- To determine whether any given patch of Upland Heathland is one of the Annex 1 Blanket Bogs (7130)
- To identify the extent of Upland Heathland (priority habitat) that can legitimately be burned, i.e., is not a Bog

18.5.1   EXTENT OF BOG ANNEX I HABITATS

In the first example, identifying Annex I Blanket Bog habitats, there are different pieces of evidence that support a number of competing hypotheses. The evidence is the presence of NCV classes M15 (*Scirpus cespitosus—Erica tetralix*) and M16 (*E. tetralix—Sphagnum compactum*). M15 is characteristic of Annex 1 habitats Active Raised Bogs (7110) and Blanket Bogs (7130), and M16 of Northern Atlantic wet heaths (4010) and European dry heaths (4030). Other information relating to the Phase I habitat present, that is, soil wetness and peat depth, was used to identify the likely Annex 1 habitat based on the additional evidence shown in Table 18.1.

From the various evidence beliefs were generated different sets of hypotheses using Dempster-Shafer (Table 18.2). From the same data, the Bayesian probability of

TABLE 18.1

Evidence in Support of Hypotheses (H)

| Evidence | Hypotheses (Annex 1 Habitats) | | | | |
	4010 H1	4030 H2	7110 H3	7130 H4	Uncertainty
Heathland	0.167	0.167	0.167		0.5
Peat depth	0.233		0.233	0.233	0.3
Dry soil	0.25	0.25			0.5
Acid soil			0.25	0.25	0.5

TABLE 18.2

The Belief in Hypotheses from Dempster-Shafer

Hypotheses	Belief	Plausibility
H1	0.132	0.698
H1, H2	0.377	0.566
H3	0.057	0.170
H4	0.132	0.321
H3, H4	0.057	0.057
H1, H2, H3	0.057	0.057
H1, H2, H4	0.132	0.132
Theta	0.057	0.057

TABLE 18.3

The Belief in Hypotheses from Bayes

Hypotheses	Belief
H1	0.062
H2	0.265
H3	0.062
H4	0.372
H5	0.239

singleton hypotheses was calculated (Table 18.3) through the combined probability that each hypothesis will pass each evidence "test."

Bayes and Dempster-Shafer provide different answers to the question of whether this patch of land is Blanket Bog (Annex I habitat, 7130). The Dempster-Shafer results have two characteristics. First, the evidence is combined over sets of hypotheses, and, second, it generates an upper bound of belief (Plausibility) from the uncertainty inherent in the evidence. The results of applying Dempster-Shafer belief functions to the problem show that the set {H1, H2} has the most support, but when plausibility is considered, the set {H1} has the most supporting evidence. The Bayesian approach only generates support singleton hypotheses and indicates support for {H4}.

Dempster-Shafer combines evidence over a range of hypotheses and does not allocate any remaining support (i.e., the uncertainty) to ¬Belief as in Bayes. Rather, uncertainty is allocated to all hypotheses or the "frame of discernment," Theta. Dempster-Shafer shows how the various pieces of evidence support different sets of hypotheses. Bayes, by contrast, partitions the evidence between Belief and ¬Belief. The hypotheses with only two pieces of evidence are the most supported, as none of the evidence supports any one hypothesis with a belief of more than 0.5 (therefore, in this context, more evidence equates to lower belief).

18.5.2 EXTENT OF BOG PRIORITY HABITAT

In the second example, identifying Priority Habitats, some remote sensing information indicates that some landscape object (e.g., a parcel or a pixel) is a bog. However, there are uncertainties associated with remote sensing information. Ancillary data are used to support the allocation of the object into a particular priority habitat class. Upland Heathland is a priority habitat and has a *one to many* relationship with the following Phase I habitats:

Dry acid heath
Wet heath
Dry heath/acid grassland mosaic
Wet heath/marshy grassland mosaic

The lower bound of the priority habitat that can be legitimately burned is given by the extent of the union of these single feature (i.e., nonmosaic) Phase I parcels. If a suspected area of burning fell within this area, then there is confidence that any burned area is not on one of the ecologically important Blanket Bog vegetation communities. If the suspected area fell within the upper bound of the Upland Heathland priority habitat, then more evidence is needed to determine the belief in legitimacy.

The object is to calculate overall belief in Bog and in Heath priority habitat hypotheses using evidence weighted by ecological knowledge, in order to determine whether any burning is legitimate. Note that in this case disbelief in Bog equates to belief in Heath. Each outcome is initially believed to be equally likely:

$$P(\{bog\}) = P(\{heath\}) = P(\{not\_sure\}) = 1/3$$

Remote sensing information indicates a 90% probability of Bog, 10% Heath, and 30% something else. This could be based on field validation, and the probabilities do not have to sum to unity and will be normalized. The three possible worlds must be considered in light of the remote sensing evidence:

$P(\{bog\}, pass_{rs}) = 0.9/3 = 0.3 \ (0.692) \text{ normalized}$
$P(\{heath\}, pass_{rs}) = 0.1/3 = 0.033 \ (0.077)$
$P(\{not\_sure\}, pass_{rs}) = 0.3/3 = 0.1 \ (0.231)$

In this example, the area of suspected burning has the following hypothetical characteristics as evidence (Table 18.4):

- Within the upper bound of the Upland Heathland priority habitat
- Within the upper bound of the Blanket Bog priority habitat
- It is within a conservation area (e.g., SSSI)
- Most of the area is not on steep slopes (i.e., <25°)
- Most of the suspected area is above the treeline
- NVC/Phase II survey data for the area indicate that the suspected area contains areas of grass, heather, and mire communities (M15, M16)

TABLE 18.4

Evidence Supporting Bog and Heath, with Ecological Weighting

Evidence	Belief (in Bog)	Belief (in Heath)	Uncertainty
Remote sensing and priors	0.692	0.077	0.231
1. Within Upland Heath mosaic	0.1	0.5	0.4
2. Within Blanket Bog mosaic	0.5	0.1	0.4
3. Not on wet soil	0.1	0.4	0.5
4. Not on slopes	0.25	0.1	0.65
5. Below 600m	0.25	0.1	0.65
6. NVC classes M15/M16	0.5	0.1	0.4

The normalizing factor is used to update the conditional probabilities of the three classes using Bayes theorem applied to the evidence from the six "tests" in Table 18.4 using Equation 18.1:

$P(\{bog\} = P(pass_{rs}).(pass_1, pass_2, pass_3, pass_4, pass_5, pass_6) = 0.692 \times (0.1 \times 0.5 \times 0.1 \times 0.25 \times 0.25 \times 0.5) = 0.00010817$
$P(\{heath\}) = 0.077 \times (0.5 \times 0.1 \times 0.4 \times 0.1 \times 0.1 \times 0.1) = 0.00001538$
$P(\{not\_sure\}) = 0.231 \times (0.4 \times 0.4 \times 0.5 \times 0.65 \times 0.65 \times 0.4) = 0.00312000$

These are normalized by the total probability of all worlds, given all passes (0.00324356):

$P(\{bog\}|\{all\ pass\}) = 0.033493108$
$P(\{heath\}|\{all\ pass\}) = 0.000476346$
$P(\{not\_sure\}|\{all\ pass\}) = 0.966030546$

In Dempster-Shafer, each piece of evidence may be combined to determine belief in Bog (*Bel*) and disbelief in Bog, which equates to belief in Heath (*Dis*) and uncertainty (*Unc*), according to the formulation from Tangestani and Moore (2002):

$$Belief = (Bel_1.Bel_2 + Unc_1.Bel_2 + Unc_2.Bel_1)/\beta \qquad (18.3)$$

$$\beta = (1 - Bel_1.Dis_2 - Bel_2.Dis_1) \qquad (18.4)$$

Applying Dempster-Shafer and Bayesian approaches to combine the evidence generates different overall weightings:

	Dempster-Shafer	Bayes
Belief (Bog)	0.836	0.033
Belief (Heath)	0.149	0.000
Uncertainty	0.015	0.966

These two approaches for combining information and evidence generate very different results in this instance: Dempster-Shafer has partitioned the uncertainty in the evidence into belief in Bog and belief in Heath (i.e., disbelief in Bog) under the assumption of conjunctive evidence. The Bayesian approach assumes independence of evidence; using a multinomial probability approach effectively calculates the probabilities of a set of events that are believed to be possible, based on passing a series of tests.

Using data relating to soil wetness, elevation, slope, and the presence of NVC classes, maps of different degrees of legitimacy can be constructed as in Figure 18.2 using the two approaches to combining evidence, Dempster-Shafer and Bayesian inference, in the assessment of a single hypothesis. These maps in this figure were generated without using any remote sensing information (i.e., they combine evidence 1–6).

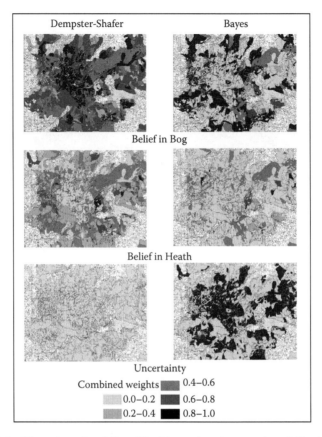

FIGURE 18.2 Upper bounds of Bog (illegitimate burning areas) and Heath (legitimate burning areas) in a test area using Dempster-Shafer and Bayesian probability (© Crown Copyright, Ordnance Survey, an EDINA Digimap/JISC supplied service).

18.6 DISCUSSION AND CONCLUSION

Remote sensing of land cover is an inherently uncertain exercise due to the spectral and spatial limitations of remotely sensed imagery. Because of this, a number of applications incorporate other information into the classification process, including ancillary data (soils, geology, elevation) and rules relating to plant and vegetation phenological cycles. The worked examples highlight a number of issues relating to fusing different information:

1. The method by which information and evidence are combined results in different mapped and modeled outcomes.
2. Different landscape questions require different weightings strategies.
3. The method by which data at different scales and grains are combined needs to be considered in relation to weights.

This indicates that decisions relating to what features to display or map depend on the intended use of that information. Decision theory is implicit for the creation of any such map through the concept of "expected value," where the value of a decision taken on the basis of the mapped data relates the probability of each possible outcome and its value. Decision theory can inform such decision making (Choquet, 1953; Chu and Halpern, 2003a, 2003b) where the evaluation of different decisions is ordinal (i.e., not cardinal), as is the case when mapping decisions are taken by conservationists and ecologists, who evaluate different elements of the landscape qualitatively rather than quantitatively: "Habitat A has a greater priority than habitat B" rather than "habitat A has three times the value of habitat B."

Conservation agencies would like to be able to develop and produce alternative, context-dependent maps relating to a series of different questions and granularities. Managers need to be able to identify the "best" decision in the face of uncertainty. Different methods for handling uncertainty in information integration activities such as Bayesian probability and Dempster-Shafer produce different outcomes, and there is also uncertainty due to the scale or grain of the object, process, or question. This situation describes an interdependence between grain, scale, uncertainty, and the integration method. Management decisions have to be made in light of the set of possible outcomes, indicating the need for a formal evaluation of the expected value of that decision using decision theory.

REFERENCES

Choquet, G., 1953. Theory of capacities. *Annales de l'institut Fourier*, 5, pp. 131–295.

Chu, F. C., and Halpern, J. Y., 2003a. Great expectations. Part I: On the customizability of generalized expected utility, *Proc. 18th International Joint Conference on AI (IJCAI)*, pp. 291–296.

Chu, F. C., and Halpern, J. Y., 2003b. Great expectations. Part II: Generalized expected utility as a universal decision rule. *Proc. 18th International Joint Conference on AI (IJCAI)*, pp. 297–302.

Comber, A., Wadsworth, R., and Fisher, P., 2006. Reasoning methods for handling uncertain information in land cover mapping. In *Fundamentals of Spatial Data Quality*, edited by R. Devillers and R. Jeansoulin, ISTE, London, pp. 123–139.

Comber, A. J., Fisher, P. F., and Wadsworth, R. A., 2005a. You know what land cover is but does anyone else?…an investigation into semantic and ontological confusion. *International Journal of Remote Sensing*, 26(1), pp. 223–228.

Comber, A. J., Fisher, P. F., and Wadsworth, R. A., 2005b. What is land cover? *Environment and Planning B: Planning and Design*, 32, pp. 199–209.

Comber, A. J., Law, A. N. R., and Lishman, J. R., 2004a. A comparison of Bayes', Dempster-Shafter and endorsement theories for managing knowledge uncertainty in the context of land cover monitoring, *Computers, Environment and Urban Systems*, 28(4), pp. 311–327.

Comber, A. J., Law, A. N. R., and Lishman, J. R., 2004b. Application of knowledge for automated land cover change monitoring. *International Journal of Remote Sensing*, 25(16), pp. 3177–3192.

Commission of the European Communities, 1992. Council Directive 92/43/EEC of 21 May 1992 on the conservation of natural habitats and of wild fauna and flora. *Official Journal of the European Communities*, L206.

Fisher, P. F.,1997. The pixel: a snare and a delusion. *International Journal of Remote Sensing* 18(3), pp. 679–685.

Fisher, P., Comber, A., and Wadsworth, R., 2006. Approaches to uncertainty in spatial data. In *Fundamentals of Spatial Data Quality*, edited by R. Devillers and R. Jeansoulin, ISTE, London, pp. 43–59.

Friedl, M. A., McGwire, K. C., and McIver, D. K., 2001. An overview of uncertainty in optical remotely sensed data for ecological applications. In *Spatial Uncertainty in Ecology,* edited by C. T. Hunsaker, M. F. Goodchild, M. A. Friedl, and T. J. Case, Springer-Verlag, New York, pp. 258–283.

JNCC, 2003. *Handbook for Phase 1 Habitat Survey – A Technique for Environmental Audit, Field Manual,* JNCC, Peterborough, UK, p. 62.

Parsons, S., 1994. Some qualitative approaches to applying the Dempster–Shafer theory. *Information and Decision Technologies*, 19(4), pp. 321–337.

Rodwell, J. S., 2006. *National Vegetation Classification: Users' Handbook*, Joint Nature Conservation Committee, Peterborough, UK.

Shafer, G., Pearl, J. 1990. *Readings in Uncertain Reasoning,* Morgan Kaufmann, San Mateo, CA.

Tangestani, M. H., and Moore, F., 2002. The use of Dempster-Shafer model and GIS in integration of geoscientific data for porphyry copper potential mapping, north of Shahr-e-Babak, Iran. *International Journal of Applied Earth Observation and Geoinformation*, 4, pp. 65–74.

UK Government, 1994. *Biodiversity: the UK Action Plan*, HMSO, London.

19 A Quality-Aware Approach for the Early Steps of the Integration of Environmental Systems

Abdelbasset Guemeida, Robert Jeansoulin, and Gabriella Salzano

CONTENTS

19.1   INTRODUCTION

19.1.1   MATCHING MATTERS!

Users discover geographical data on the Internet, as out of a large and unstructured information repository, through an opaque query system. Users are forced to rely more and more on automatic support for deciding what to choose or reject. Hence, it is more appropriate to talk about system matching rather than data merging: This insists on the fact that each component influences the whole process, and not only the data.

19.1.2   QUALITY MATTERS!

The necessity to rely on web robots for geographic information, as we do already with text queries, implies taking data quality into account. Thus, the service must compromise the declared quality of the data with the quality level that the user is able to accept. The translation between these two visions of quality has been an issue for years, and it can be considered as an "impedance mismatch" problem.

In this chapter, we develop on the "impedance" metaphor (Section 19.2) as a help to better design the various operators to insert in the data flow. This design is broken down into three levels (Section 19.3). An architecture is proposed for implementing this mediation (Section 19.4). This is illustrated in a simple example, taken from the domain of public health (Section 19.5).

19.2   IMPEDANCE MISMATCH IN GEOGRAPHIC INFORMATION FUSION

19.2.1   THE IMPEDANCE MISMATCH METAPHOR

By analogy with electrical, acoustic, or hydraulic systems, we can talk about the "user-provider impedance mismatch" in geographic information (GI). Impedance makes sense if the flow is bidirectional, e.g., alternating current, and if at least two independent magnitudes can be considered, e.g., intensity and voltage. Let's try to develop in the case of GI:

Intensity: I, the amount of data (flow)
Voltage: V, a measure of information potential, e.g., the depth of the data structure

Transformers: can modify the voltage, but only for alternating current

Impedance: complex number made of a resistance R and a reactance X: $Z = R + \iota X$

Resistance: R, a bias between the referent and its representation in the information system

Reactance: X, if positive, is an inductive reactance, measuring the tendency to preserve information. This inertia can yield an external energy, e.g., warnings and annotations, not for direct use, but which can have a later or external impact. If negative, X is a capacitive reactance, which acts as an information cache, or an aggregate operator, or a filter, possibly transferring the potential part of the information (intentional), rather than its extensional part. When systems are connected, it is important to control the difference of their impedances Z_{in} and Z_{out}, and to take the appropriate answer depending on the objective

Impedance matching, for a maximal power transmission, we need $Z_{in} = Z_{out}{}^*$, the conjugate to the impedance of the source. Then, we need to equalize the resistances $R_{out} = R_{in}$, and to oppose the reactances $X_{out} = -X_{in}$ (inductor against a capacitor, or vice versa)

Impedance bridging, for a better control of the use of a certain part of the information: it can be obtained with $Z_{in} \gg Z_{out}$

Such answers always entail some overhead, due to the introduction of intermediate devices. When using complex information systems, mismatch can arise at each level and task: early or late requirements, global or detailed architecture.

19.2.2 IMPEDANCE MISMATCH BETWEEN GEOGRAPHIC DATASETS

When considering geographical information, we have to cope with several issues.

Vector vs. Raster (object vs. field): Objects do not commensurate with field parts, nor do pixels aggregate with objects. It is similar to what has been named the "object-relational impedance mismatch" (Ambler, 2001): access by pixel sets ≠ object behavior (boundaries, topology). A vector-to-raster transformer acts as an inductor and regularizes the data flow.

Geometry vs. topology: We can theoretically derive the topology from a perfect geometry, but in real situations, the topology can depend on data quality, and we must build a "capacitor" that filters the flow of geometric coordinates and derives only a consistent topology.

Space scale: A gradual change in space resolution may not match with a homothetic zoom. In signal processing, the Shannon theorem teaches us how objects can be distinguished with respect to a channel bandwidth. We must adapt our requirements to a range compatible with the input, and if several sources mix several scales, an aggregation-disaggregation process is necessary.

Time scale: similar to space scale issue.

Fitness for use: "external quality" (user side), as opposed to internal quality (producer side). It can't be reduced to a signal-to-noise ratio, because several quality components are involved, but, as the metaphor suggests, it will become harder as the signal power lowers to the noise level.

Granularity (specialization hierarchy): The number of detail levels doesn't increase necessarily with the size of a vocabulary, and discrepancies may exist between words and their referents.

19.3   A THREE-STOREY STORY

19.3.1   Existence, Quality, and Contents Aspects

Data collection and selection precede decision making. In small applications, it is reasonable to group them into a unique process, whose impedance can be adapted to a variety of sources. In large applications, we should rather consider the data selection as one task. The impedance of this task must fit sources and the user model as well. The question is, how many user models can we manage with only one data selection system? (See Figure 19.1.)

Public health bodies, for instance, collect significant amounts of data, from independent systems, at different spatial levels (international, national, communal), and for different goals: epidemic monitoring, control of health expenses, etc., and three aspects must be considered:

Existence of relevant data
Quality, sufficient for fitting the intended use
Contents, for a consistent and effective use

19.3.2   Catalogs, Metadata, and Data Storeys

Questioning relevance, fitness, and consistency contributes to the overall impedance. But do we need to analyze the whole system, to assess the impedance that should oppose it?

Let's examine what can be gradually learned. First, we query Catalogs, to identify datasets and to get their location, coverage, format, etc. Second, the metadata of each dataset give a richer description of the contents, of the aspects of quality, and of the vocabulary and its granularity. Finally come the data. This suggests that we use these three levels to design, step by step, the components of Z_{out}.

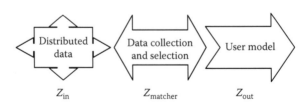

FIGURE 19.1   Processes and impedance matching.

19.3.3   A Three-Step Impedance Builder

19.3.3.1   Step 1: Existence

1. **Geometry**: To ask a catalog for intersection with a zone. It simulates resistance equalization.
2. **Time:** To match time requirements by interval equalization (easy), or by accepting a much larger time interval, with an additional "inductive" processing, if regularization is necessary.
3. **Theme:** Theme relevance is always approximate. For a good choice at the catalog level, we need to match terms (same resistance) from titles or descriptions, to establish similarities (inductor), and to build structures between terms (capacitor). It can help to combine direct or reverse geocoding and to use smart text processors. Rapid browsing can select too few, and a cautious approach too much, but we try to avoid most irrelevant sources while preserving the most crucial ones.

19.3.3.2   Step 2: Quality

Fitness for use is not easy to convert into the standard quality elements (ISO 19115), but it makes sense to use them in the description of the impedance Z_{in}.

Positional accuracy: If it undershoots requirements, we must adapt the output resistance, or anticipate downsizing the data (for step 3), if overshoot. Relative accuracy and topology preservation require a capacitor for computing constraints that will be checked in the next step.

Attribute accuracy: To include a resistance on metadata, or to prepare an inductor for the next step.

Completeness: To combine resistances (undershoot case), inductors (overshoot), and capacitors (if some inference should be derived), and, again, to prepare the operators of the next step.

Time accuracy, time completeness, time validity: Similar operations are expected there.

Lineage, logical consistency: This information must be collected (capacitor) for further processing with constraints created by other impedance elements, e.g., a topology capacitor.

19.3.3.3   Step 3: Contents

Once a reduced list of datasets has been selected, we must confront the data with the integrity constraints of the global schema. This task is cost effective, and the probable detection of conflicts can make the whole process intractable. Prior reduction of the size of the exploration space is mandatory: let's use an appropriate preference order. Let's also use an order on the confidence levels. Such partial orders can be based on statistics (e.g., Bayesian) or qualitative ranking (e.g., formal concept analysis). Data are either accepted, possibly with warnings, or rejected, in which case a new query must be issued back. Hence the information flow becomes bi-directional, and will loop until a final decision is reached.

19.4 MEDIATION ACHITECTURE

The first two steps can be achieved, mostly at an early stage, by browsing catalogs and metadata. Hence we can anticipate a reliability level for the outcome, and we can proceed, with an a priori best selection of data, completing the fitness for use assessment and accessing the actual data.

19.4.1 A REQUIREMENTS-DRIVEN VIEW OF THE THREE STEPS

We model an integrated view of these steps, as in Figure 19.2: Top and middle levels concern requirements for existence and quality, and the bottom level concerns queries on contents.

Step 1. For a given target T, let **rids(T)** = $\{S_1, S_2, \ldots, S_m\}$ be the set of ideal datasets required by T. Step 1 must determine the usable datasets **uds(T)** = $\{S'_1, S'_2, \ldots, S'_k\}$ such that

$$\forall\, i = 1, \ldots m\ \exists\, j = 1, \ldots k\ \exists\, c_{ij}\, (S_i, S'_j) > tc \qquad (19.1)$$

where $S_i \in$ ideal dataset, $S'_j \in$ real dataset, and c_{ij} is a correspondence between S_i and S'_j, better than a threshold tc, e.g., a minimal number of constraints to satisfy.

The c_{ij} are defined in the sense of Parent and Spacca-Pietra (2000). To find it, we must (i) explore theme, space, and time dimensions, and (ii) use semantic, geometric, and topological relations between the metadata of the catalogs and sources (usability study).

Capacitive action: When computing **uds(T)** for tc, some near-to-tc sets can be memorized into **uds′(T)** (capacitor). For instance, if we have the required data for a neighboring region, or at a more global scale, we just keep track of that, for saving time during a possible next call.

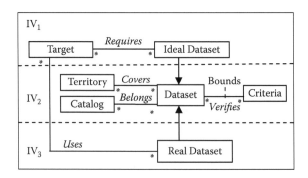

FIGURE 19.2   Integrated view of the existence, quality, and contents aspects.

Step 2. We note $\Delta(\mathbf{T})$ the gap between the required and usable datasets for a target T, with respect to some distance function $\Delta(\mathbf{T}) = \mathbf{d}(\text{rids}(T), \text{uds}(T))$. Step 2 consists of:

> Evaluating the quality of uds(T) with respect to some criteria and bounds derived from those related to the corresponding data sets rids(T)
>
> Choosing one optimal uds(T), denoted **ouds(T),** that minimizes $\Delta(T)$ (w.r.t. organizational, conceptual, or technical aspects ≈ impedance matcher)
>
> If no **ouds(T)** can be found, release threshold tc and go back to step 1, emptying the capacitor **uds'(T)** and possibly recharging it. Some extra processing (inductor) must be activated to enhance the new, but less, fitting **uds'(T):** for instance, a similarity model or an aggregation/de-aggregation model to compute approximate data from **uds'(T).**
>
> Repeat until $\Delta(\mathbf{T})$ become acceptable, with respect to a quality balance, or abort and report failure reasons

Step 3. This step integrates datasets of ouds(T), if they exist.

The "best" situation arises when rids(T) ≡ ouds(T) (i.e., (1) is satisfied with c_{ij} = identity, $\forall\, i, j = 1, \ldots m$), while the "worst" situation arises when ouds(T) = $\varnothing$.

19.4.1.1   Role of Quality Issues

Steps 1 and 2 reduce the volume of data sources addressed to step 3, compromising between (i) queries and quality needs expressed by the target system, and (ii) existing data and their quality. If necessary, they mutually adapt (i) and (ii) to obtain an acceptable transformation of (i) and/or acceptable costs for inductor and capacitors (filters, caches …) in (ii).

19.4.1.2   Classification of Datasets and Targets

Steps 1 and 2 also classify datasets and targets. For instance, "required and available" or "qualified" datasets (resp.: **rads(T)**, **qds(T)**), and "described" and "well described" targets (resp.: **DT**, **WDT**) are defined by

$$\text{rads}(T) = \text{rids}(T) \cap \text{uds}(T)$$

$$\text{qds}(T) = \{ds | ds \in \text{rads}(T) \text{ and ds satisfies some criteria}\} \qquad (19.2)$$

$$\text{DT} = \{T \mid \text{rids}(T) \subseteq \text{rads (T)}\}$$

$$\text{WDT} = \{T \mid \text{rids}(T) \subseteq \text{qds}(T)\}$$

These concepts introduce order relations at each level (existence, quality) for datasets and related targets.

19.4.2   ARCHITECTURE: AN LAV APPROACH FOR THE MEDIATOR

The proposed architecture couples a reasoning system and a mediator system. The first operates on application ontology; the second is based on (i) a global schema, (ii) a set of sources with real data, and (iii) a set of relations between the global schema and the local sources.

We follow a *local_as_views* (LAV) approach (Lenzerini, 2002), which characterizes the local sources as views over the global schema and gives priority to global requirements, extensibility of the system and quality of the sources, for a reliable decision. (See Figure 19.3.)

19.4.2.1   Level A_1: Application Ontology and Virtual Schema

Application ontology can be interpreted as a specialization of the approach ontology and the domain ontology (Van Heijst et al., 1997). It starts with a few concepts, properties, and roles, extracted by the class model in Figure 19.2. To represent the quality of the decision process (steps 1 and 2), we derive new classes based on formulas 19.2.

Classes and relations are specified in a Description Logics (DL) (Calvanese et al., 2004) and operated by a reasoning system. The DL belongs to a family of knowledge representation formalisms based on first-order logic and the notion of class (concept). DL expressivity is related to the set of supported constructors (Guemeida et al., 2007).

Global schema is the domain ontology, independently developed from data sources, to formulate user global queries (Visser, 2004). In our approach, it is a virtual object-oriented schema: the concepts, with typed attributes, are connected by binary relations.

19.4.2.2   Level A_2: Mediation Schema

The mediation level is described as an LAV integration technique (Amann et al., 2002). Correspondences between global concepts and data sources are expressed by a set of mapping rules. Queries on the global concepts are formulated in an OQL variant. A global query is broken into a set of local queries, executed by the local systems, and the partial results are merged.

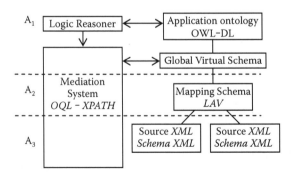

FIGURE 19.3   Three-level architecture.

19.4.2.3 Level A$_3$: Local Sources

The local level contains the existing data sources. Their schemas are completed by some metadata, corresponding to the quality criteria (step 2). Sources corresponding to global requirements are marked. The other data sources, which are definitely out of scope, are ignored. The next section describes this process.

19.4.3 TECHNICAL IMPLEMENTATION CHOICES

19.4.3.1 Knowledge Representation Formalisms

To perform step 1 and step 2, a knowledge base (KB) corresponds to the application ontology. This KB consists of a set of terminological axioms (TBox) and a set of assertional axioms (ABox). Metadata elements describing data sources belong, at the same time, to the data sources and to the KB, as part of the catalog descriptions. The application ontology is implemented in OWL DL, a decidable sublanguage of OWL that allows the use of DL reasoning services. Based on $\mathcal{SHOIN}(\mathcal{D})$, OWL DL supports transitive properties, role hierarchies, nominals, unqualified number restrictions, and data types (Baader et al., 2005).

19.4.3.2 Tools

The technical infrastructure is based on the Protégé OWL editor (Knublauch et al. 2004) and the Racer reasoning system (Haarslev and Möller, 2001). Protégé is an open-source development environment for ontology building in OWL DL and KB systems, supporting $\mathcal{SHOIN}(\mathcal{D})$. Its extensible architecture allows plug-ins. Protégé can be used with a DL reasoning system through a standardized XML common interface. Racer is a DL reasoning system that automatically classifies ontology and checks for inconsistencies. It implements highly optimized algorithms for the very expressive description logic $\mathcal{SHIQ}(\mathcal{D})$.

At the local level, XML schema are used to represent data source schemas and constraints, and XQuery is used to query it.

19.5 APPLICATION

19.5.1 A SIMPLE EXAMPLE

We consider a set of target applications related to health risks management. For each risk, we want to correlate, over a geographic territory (GT), the demands of services for dependent older people (DOP) and other vulnerabilities with the offer of services (hospital, beds ...). A lot of these social data (as data related to DOP) are collected on administrative territories (departments), while scientific data are in general related to geographic territories. For each target, an example of query is:

Q: For a GT concerned by a risk, and for each department in GT, how many are DOP? Q is formulated from a fragment of the global schema represented in Figure 19.4.

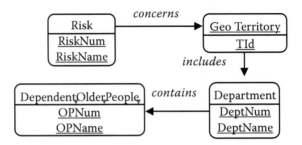

FIGURE 19.4   A global schema fragment.

19.5.1.1   Running Data

T_1, T_2, and T_3 represent, respectively, heat wave, cold wave, and inundations. With the notation of Section 19.4.1, we suppose $rids(T_1) \equiv \{S_1, S_2, S_3, S_4, S_8\}$, $rids(T_2) \equiv \{S_1, S_2, S_3, S_4, S_5, S_6, S_8\}$, and $rids(T_3) \equiv \{S_1, S_2, S_3, S_4, S_7, S_8\}$. These sources are linked to the risk management activities; for instance, S_1 relays risks and departments; S_2 represents hospitals; S_3 and S_4 represent DOP in two departments, respectively, AT_1 and AT_2; S_5 and S_6 represent homeless in the same departments; and S_7 represents camp sites in all departments. S_8 gives relations between geographic and administrative territories, for instance, GT contains AT_1 and AT_2.

We note that (i) all sources, except S_6, belong to the catalogs; and (ii) existing sources verify quality criteria, except S_7, which violates freshness criteria.

19.5.1.2   Results

We detect automatically that:

T_1 is well described (all sources exist and verify the quality constraints).
T_2 is not described because S_6 is not available.
T_3 is only described, because of the violation of quality criteria on S_7.

Queries on contents can be performed by T_1. T_2 and T_3 require actions to reduce impedance mismatches linked to requirements for S_6 and S_7. These actions could concern one or more aspects (theme, geography, time) of the required datasets.

19.5.2   STEP-BY-STEP QUERIES

The first iteration of the approach transforms Q into queries associated to each step:

Step1—Q1 (Existence): Which are the described targets on a geographic territory GT? For targets not described, reduce the mismatch impedance in the next iterations.
Step2—Q2 (Quality): Which are well-described targets on GT? For badly described targets, reduce the mismatch impedance in the next iterations.
Step3—Q3 (Contents): For well-described targets on GT, give contents about DOP and departments in GT.

19.5.2.1   Classification of Targets (Step 1 and Step 2)

We start by checking for correspondences in formula 19.1 (Section 19.4.1), which are identities. Hence, step 1 is limited to checking for existence, in the catalogs, of sources required by the targets.

Step1—Q1 requires a sequence of concepts:

- AvailableSource $\equiv$ Source $\sqcap$ $\exists$ belongs.Catalog
- MissingSource $\equiv$ Source $\sqcap$ $\neg$ AvailableSource
- TerritoryGT $\equiv$ Territory $\sqcap$ $\exists$ contains$^-$.{GT}
- SourceGT $\equiv$ Source $\sqcap$ $\exists$ covers.TerritoryGT
- AvailableSourceGT $\equiv$ AvailableSource $\sqcap$ SourceGT
- DescribedTargetGT $\equiv$ Target $\sqcap$ $\exists$ manages.(Risk $\sqcap$ $\exists$ concerns.{GT}) $\sqcap$ $\forall$ requires.(AvailableSourceGT $\sqcup$ $\neg$SourceGT)

DescribedTargetGT = $\{T_1, T_3\}$

Step2—Q2 requires a sequence of concepts:

- QualifiedSource $\equiv$ AvailableSource $\sqcap$ $\forall$ satisfies.RespectedCriteriaBound
- NotQualifiedSource $\equiv$ AvailableSource $\sqcap$ $\neg$ QualifiedSource
- QualifiedSourceGT $\equiv$ QualifiedSource $\sqcap$ SourceGT
- WellDescribedTargetGT $\equiv$ DescribedTargetGT $\sqcap$ $\forall$ requires.(Qualified-SourceGT $\sqcup$ $\neg$ SourceGT)

WellDescribedTargetGT$_1$ = $\{T_1\}$

19.5.2.2   Queries about Contents (Step 3)

The following elements illustrate the integration approach for querying about contents.

19.5.2.2.1   Metadata and Data Source Structures
Metadata elements are the same for all sources. They are described using XML schema. Figure 19.5 represents these metadata and the schema of a local source (DOP).

19.5.2.2.2   Mappings and Algorithms
The correspondences between the local data source schemas and the global schema are expressed by a set of mapping rules, using XPath, for example, rules below map paths in the source S_3 (DOP), augmented with metadata, to paths in the global schema:

$R$1: http://www.pa01.fr/S3.xml/Source3 as u1 $\rightarrow$ Department
$R$2: u1/MetaData/Coverage as u2 $\rightarrow$ DeptNum
$R$3: u1/DOP   as u3 $\rightarrow$ Contains
$R$4: u3/PNum as u4 $\rightarrow$ OPNum
$R$5: u3/PName as u5 $\rightarrow$ OPName

Global queries requiring many data sources are broken into a set of local queries, executed by the local systems (Xquery). Then, the partial results are merged to compose a global result.

```
<xs:element name="metaData">
  <xs:complexType>
    <xs:all>
          <xs:element name="SId" type="xs:anyURI"/>
          <xs:element name="Object" type="xs:string"/>
          <xs:element name="PubDate" type="xs:date"/>
          <xs:element name="Extension" type="sx:string"/>
          <xs:element name="Coverage" type="xs:string"/>
    </xs:all>
  </xs:complexType>
</xs:element>

...

<xs:element name="DOP" maxOccurs="unbounded">
  <xs:complexType>
    <xs:sequence>
      <xs:element name="PNum" type="xs:integer"/>
      <xs:element name="PName" type="xs:string"/>
    </xs:sequence>
  </xs:complexType>
</xs:element>
```

FIGURE 19.5 XML Schema code of Metadata for all sources, and of a particular source (DOP).

19.6   CONCLUSION

This chapter addresses impedance mismatch problems, occurring when two systems are plugged for business activities requiring datasets from heterogeneous systems, within autonomous organizations. We introduced an impedance mismatch metaphor in geographic information fusion and we proposed a metadata-based interoperability approach. The impedance mismatch has been broken down into three subclasses, related to existence, quality, and contents aspects. We classify target systems with respect to the requirements about these aspects, in order to decide if and how to reduce impedance mismatch. We aim to investigate how such awareness could drive the design of an interoperable architecture.

ACKNOWLEDGMENT

Work supported by Regions PACA and Midi-Pyrénées and by Marne-la-Vallée University is gratefully acknowledged.

REFERENCES

Amann, B., Beeri, C., Fundulaki, I., and Scholl, M. 2002. Ontology-based integration of xml web resources. In Horrocks, I., Hendler, J. (Eds.). *The Semantic Web–ISWC 2002. First International Semantic Web Conference, Sardinia, Italy, June 9–12, 2002, Proceedings.* Springer Berlin/Heidelberg, *LNCS,* 2342, pp. 117–131.

Ambler, S. W. 2001. Agile modeling: A brief overview. In Evans, A., France, R. B., Moreira, A. M. D., Rumpe, B. (Eds.). Practical UML-Based Rigorous Development Methods, *Workshop of the pUML-Group held together with the UML2001 Conference, October 1st, 2001, Toronto, Canada.* Proceedings. GI, Bonn, Germany, LNI series, vol. 7, pp. 7–11.

Baader, F., Horrocks, I., and Sattler, U. 2005. Description Logics as Ontology Languages for the Semantic Web. *In Hutter, D., Stephan, W. (Eds.).* Mechanizing Mathematical Reasoning. *Essays in Honor of Jörg H. Siekmann on the Occasion of His 60th Birthday.* Springer Berlin/Heidelberg, *LNAI,* 2605, pp. 228–248.

Calvanese D., McGuinness, D., Nardi, D., and Patel-Schneider, P. 2004. *The Description Logic Handbook: Theory, Implementation and Applications.* Cambridge Univ. Press, Cambridge, UK.

Guemeida, A., Jeansoulin, R., and Salzano, G. 2007. Quality-aware, Metadata-based Interoperability for Environmental Health. 5*th International Symposium Spatial Data Quality,* Enschede, The Netherlands, 13–15 June 2007.

Haarslev, V., and Möller, R. 2001. RACER System Description. In Gore, R., Leitsch, A., Nipkow, T. (Eds.), Automated Reasoning. *First International Joint Conference, IJCAR 2001 Siena, Italy, June 18–23, 2001 Proceedings.* Springer Berlin/Heidelberg, *LNCS,* 2083, pp. 701–705.

Knublauch, H., Fergerson, R. W., Noy, N. F., and Musen, M. A. 2004. The Protégé OWL Plugin: An Open Development Environment for Semantic Web Applications. In McIlraith, S. A., Plexousakis, D., Harmelen, F. V. (Eds.). The Semantic Web–ISWC 2004, *Third International Semantic Web Conference, Hiroshima, Japan, November 7–11, 2004 Proceedings.* Springer Berlin/Heidelberg, *LNCS,* 3298, pp. 229–243.

Lenzerini, M. 2002. Data integration: A theoretical perspective. In Proceedings of the 21st ACM SIGMOD-SIGACT-SIGART Symposium on Principles of Database Systems PODS'2002. Madison, Wisconsin, USA. June 3–5, 2002, pp. 233–246.

Parent, C., and Spaccapietra, S. 2000. Database Integration: The Key to Data Interoperability. In *Advances in Object-Oriented Data Modeling*, Papazoglou, Spaccapietra, & Tari, Eds., MIT Press, Cambridge, MA, 2000, pp. 221–254.

Van Heijst, G., Schreiber, A. Th., and Wielinga, B. J. 1997. Using explicit ontologies in KBS development, *International Journal of Human-Computer Studies*, 46(2–3), Feb. 1997, pp. 183–292.

Visser U. 2004. Intelligent Information Integration for the Semantic Web. In Intelligent Information Integration for the Semantic Web. Springer, Berlin/Heidelberg, LNAI, 3159, pp. 13–34.

20 Analyzing and Aggregating Visitor Tracks in a Protected Area

Eduardo S. Dias, Alistair J. Edwardes, and Ross S. Purves

CONTENTS

20.1 INTRODUCTION

20.1.1 MOTIVATION AND CONTEXT

In recent decades, recreational use of natural areas has grown rapidly from low-intensity and relatively passive use to a situation where tourism is the dominant force driving change in many rural areas and their associated communities (Butler et al., 1998). However, excessive use of natural areas can have significant direct and indirect negative impacts. These include both environmental degradation (Farrell and Marion, 2001) and diminishing quality of the visitors' recreational experience (Lynn and Brown, 2003). Mobile information services have been suggested as one means of supplying park managers with the possibility to monitor and manage visitor distribution within parks and, concurrently, help visitors achieve a fuller awareness of the richness of the natural and cultural resources they visit. In this chapter we analyze data collected using the prototype of such an information tool and assess its usefulness in monitoring and influencing the whereabouts of the visitors.

Location-based services (LBS) allow access to information for which the content is filtered and tailored based on the user's location. We tend to spend the majority of our time in known or familiar environments, where we either do not require information or know where to obtain it. LBS may therefore be particularly useful in tourism and leisure where visitors are both eager for information and unfamiliar with a locale (Dias et al., 2004). LBS can provide a wide variety of useful information, for example, answering questions such as (Edwardes et al., 2003):

- What birds of prey can be found here? (presence)
- Where can sea holly be found? (distribution)
- Can orchids be found in these dunes? (confirmation)
- Are these elderberries? (identification)
- Are these lichens always found on southerly dune slopes? (association)

In the context of this work, previous research from three different domains is relevant: that exploring how users behave and impact upon natural spaces; techniques to analyze GPS tracks from individual users; and methods to visualize, explore, and analyze large volumes of so-called moving point objects.

Previous research addressing issues of visitors' spatial distribution and behavior within natural areas has been carried out from the context of crowding, visitor density, and visitor simulation modeling (Elands and van Marwijk, 2005; Manning 2003). Such research is typically centered within the field of recreation management and aims, for example, to model the carrying capacities of natural areas. As technologies allowing tracking of individual paths have developed, researchers have started to apply conceptual research concerned with the analysis of space and time (e.g., the space-time aquarium suggested by Hägerstrand [1970]). However, as real, high-volume data describing geospatial lifelines (Mark, 1998) have become available, the inadequacies of techniques such as the space-time aquarium as more than a simple visualization tool for a limited number of paths have also become apparent (Kwan, 2000). These limitations have in turn led to the emergence of so-called geographic knowledge discovery techniques (for a full review, see Laube et al., 2006), which seek to allow both the qualitative and quantitative exploration of motion tracks. Laube et al. (2005) introduced a set of methods for analyzing relative motion in groups of objects, while Mountain and MacFarlane (2007) discuss methods for predicting an object's likely position based on previous fixes and describe example uses such as the filtering of queries to a geographic information retrieval system. One of the key limitations identified by Laube et al. (2006) is the lack of availability of real data with multiple geospatial lifelines for analysis. For this work, we collected data specifically to allow exploration of the behavior of visitors to a natural area, thus overcoming this problem. In contrast to previous work, park users were constrained to the same path, with few chances to leave the network, thus vastly simplifying the role of space in our work and allowing us to focus on users' behavior along this constrained track. We developed a set of techniques aimed at investigating how the spatial behavior of visitors to a protected area changes in response to information being supplied to them in differing forms. This problem is framed within the following research questions:

- How can tracks of multiple visitors to a park be used to explore visitor behavior?
- Is the geographic behavior of visitors altered by the provision of information?
- Do different forms of information media alter the geographic behavior of visitors?

20.2 METHODOLOGY

20.2.1 EXPERIMENTAL DESIGN

A controlled experiment was designed to measure the influence that location-based information had on the behavior of visitors to natural areas. In the experiment all subjects were issued global positioning systems (GPSs), which recorded their positions regularly, and divided into control and test groups. The test groups were each issued different forms of information, ranging from location-based services to traditional paper-based information. The control group was provided with no additional information. The tests were carried out between August 22 and September 9, 2005.

20.2.1.1 Study Area

The National Park "Dunes of Texel" located on an island in the north of The Netherlands served as the testing ground for this work. Part of the dune park is only accessible via the EcoMare museum and visitor center, which is visited by a large number of tourists during the summer period. EcoMare, together with Geodan b.v. and Camineo Systems, developed a location-based service to serve the visitors to the dune park. This system has two main components:

1. A cross indicating the exact location of the visitor while walking in the dune park on a map.
2. Information content is pushed to the visitor when they are at specific locations. A soft cuckoo-song-sound is emitted by the device at these locations and the relevant content page is automatically shown.

Random visitors to the EcoMare museum were approached and asked if they would be interested in participating in this research. In order to test four different information media, the test subjects were divided into four groups: no information, paper booklet, digital information, and LBS. All three groups, other than "no information," had access to the same information, but delivered using different media. In the case of the "LBS" group this information was enhanced with the location sensitivity explained above. The compositions of the groups were controlled to ensure their profiles were as similar as possible. In addition, all subjects set out to follow the

FIGURE 20.1   Map of the trail given to visitors.*

same route, in similar weather conditions. A GPS receiver was given to every participant irrespective of the group they were in. GPS tracks were recorded at a rate of one position fix every five seconds in order to analyze the subjects' spatial behavior.

20.2.1.2   Information Content

The information provided to the test group subjects was comprised of a map of the route with the locations of a number of points-of-interest (POIs) displayed (see Figure 20.1). Detailed information about each of these was supplied in the subsequent information. This content consisted of a prominent title, a photo of the feature, and a text description. The POIs were classified into four categories: "directions" (indicating the path the subject should follow); "plants" (information about a particular plant visible from the path); "animals" (information about animals relevant at a particular point on the path); and "landscape" (information about landscape features visible from a certain location).

20.2.2   ANALYSIS TECHNIQUES

The passage of each visitor traversing the dune park was recorded by a unique GPS track. While analysis of these tracks independently could yield valuable information about individual movements, the purpose of the analysis here was to investigate

---

* Color version of figure can be found at http://staff.feweb.vu.ne/edias/gasdm/

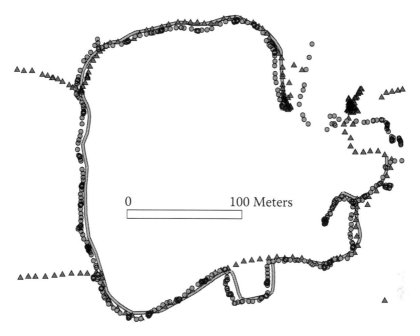

FIGURE 20.2 Example of GPS tracks for two visitors superimposed on the digitized path.*

whether significantly different behaviors occurred across groups as a result of the introduction of information in different forms. As such, our first task was to develop a method to aggregate the data. As shown in Figure 20.2, GPS tracks vary both as a function of the precision of the device and differences in subject behavior. The main types of variability include:

- Uncertainty introduced by imprecision in the GPS coordinates recorded
- The visitor leaving the prescribed path
- Missing GPS data for periods of traversal
- Individual differences in walking pace
- Differences in the period of time spent stopping at particular locations

In order to allow data analysis, two main methods were employed: linear referencing and aggregation. The purpose of linear referencing was to associate all individual GPS fixes with a single common baseline. In our case, the path provided the obvious reference to perform this function. It was therefore extracted as a linear geometry using a 1:10,000 topographic base map (the TOP10 vector dataset of the Dutch National Mapping Agency). GPS fixes were referenced by projecting them onto their closest path position. Aggregation involved the definition of a sampling frame segmenting the path, into which the referenced positions could be aggregated. To achieve this, the path was indexed at 5-meter intervals and the number of fixes occurring in each interval recorded. The size of the interval was chosen because it reflected the approximate precisions of the GPS receivers. A number of issues were

* Ibid.

encountered in performing these tasks. During aggregation, situations were found where the GPS fixes were not representative of the visitor's movement along the path with, for example, fixes occurring a considerable distance from the path. To handle these situations a filter was employed to reject fixes that were projected over a distance of more than 10 meters. This value represented twice the theoretical GPS precision and was validated by visual inspection of the tracks.

A second problem was that at one point the path forked, taking visitors up to a viewpoint, indicated by the POI labeled 34 in Figure 20.1. This presented a difficulty in defining a single linear reference. To handle this, the stretch of path leading to the viewpoint was duplicated within the linear reference, once for each direction. The closest fix to the viewpoint, measured along the path, was then used to discriminate which of the duplicated path segments should be referenced. Fixes within the segment that occurred before the closest position were assigned to the first segment and those thereafter to the second.

Two additional aggregations were also performed to consider sources of errors that might influence the data quality. To investigate the errors arising from the two different GPS receivers used, the dispersion of fixes allocated to each interval was recorded. This involved computing the centroid of the fixes assigned to a particular interval and the mean distance of the points to this centroid. To consider errors in the digitization of the path, the average projection distance to an interval for every segment was also calculated. This value was signed according to the side of the path that the fixes fell on.

After indexing each valid fix to its corresponding path interval, fix frequencies were calculated for each interval. Using these results, the tracks were graphically visualized and statistically analyzed. One issue emerged from this analysis: For a particular track, an interval could have zero recorded fixes. This situation could be indicative of one of two possibilities: either the visitor had moved rapidly through the 5-meter interval and there were truly no fixes, or there were no data available for the segment due to receiver issues. Since it was relatively unlikely that a visitor could move fast enough that there were no fixes over more than two segments (since the frequency of fixes was 5 seconds, this would represent a speed of more than 7 kilometers/hour), consecutive intervals with no fixes were selected and their values set to null. The average number of fixes on each interval for each visitor was calculated and used as a measure of time spent at an interval. Aggregated values for each information medium were also calculated and used for intergroup comparisons.

20.3 TRACK ANALYSIS

20.3.1 GENERAL OBSERVATIONS

The main goal of this research was to uncover differences in the spatial behavior caused by the provision of different information media to visitors of protected areas. The characterization of behavior was simplified into the variables time and place, represented by segments and the time spent in them. When the visitors spent 15 seconds or more in a segment, then it was considered that they had stopped or significantly slowed down.

TABLE 20.1

Time Statistics regarding the Time the User Spends per Segment

	Mean (sec.)	SD (sec.)	Min. (sec.)	Max. (min.)	N (# segs)
No info	7.3	27.5	0	23	4999
Paper	8.7	22.2	0	23	6684
Digital	11.9	24.7	0	12	6896
LBS	11.3	21.6	0	20.8	12,228

TABLE 20.2

Average Number of Stops (15 Seconds or More in a Certain Place) per Visitor per Group

	Mean	SD	Min.	Max.	N
No info	16.6	10.5	0	42	38
Paper	26.6	17.7	3	82	49
Digital	39.2	15.0	15	69	46
LBS	48.6	14.6	16	85	75

Table 20.1 and Table 20.2 summarize the overall influence that the different information media have on the behavior. Table 20.1 shows the average time each group spends per interval. This value is indicative of the overall time spent in the park; therefore, we can conclude that the technology has some effect since it is visible that visitors who had access to information via the PDA (the digital and the LBS groups) spent on average more time (around 45%) than the other groups (the no info and paper groups). The maximum amount of time that a visitor spends on a certain segment is also displayed in the same table—for all groups, visitors can be found that have spent long amounts of time in a segment (more then 10 minutes for a visitor with the digital info and more then 20 minutes for visitors in all the other groups). These values are indicative of activities such as picnicking or reading.

Table 20.2 indicates the number of stops ($t \geq 15$ seconds) each visitor made during their visit, averaged over the group. Visitors without information stopped on average in 16.6 places. For visitors with paper information, the average number of stops increased to 26.6, with digital information increasing to 39.2 stops and for those visitors receiving location-based information 48.6 stops.

20.3.2 VISUAL ANALYSIS OF RESULTS

The previous results demonstrate the influence of information in the number of stops, but we also wanted to analyze where the stops occur and if these stops are correlated in space. Figure 20.3 shows the information on spatial behavior for all the segments and for all the visitors grouped by information medium. POIs are shown at the top of the figure, indicating places where visitors were provided with information. Information categories are shown at the bottom of the figure using the same pictograms

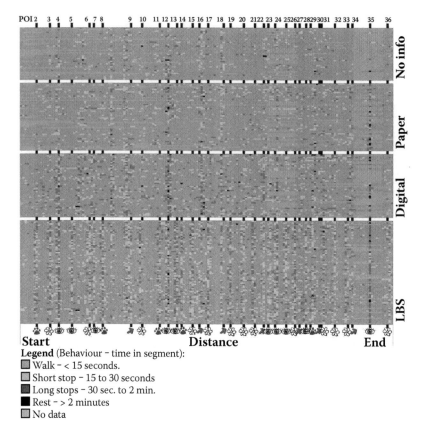

FIGURE 20.3 Visualization of the frequency of fixes per interval of path for every track grouped by information type.*

as in Figure 20.1. In order to simplify the visual analysis, segments were classified according to the time spent at the segment into four classes: rest locations (more than 2 minutes at a location; long stops (between 30 seconds and 2 minutes at a location); short stops (15–30 seconds at a location), and walking. The segments for which there was no data collected (due to either extreme inaccuracy of the GPS receiver or to the visitor taking a shortcut) were given a "no data" value. This method of presenting the data drew on the technique for identifying relative motion patterns suggested by Laube et al. (2005). The visualization reveals the stops that are spatially auto-correlated among the visitors; these are indicated by the darker vertical bars. The smeared areas (where the darker cells are not aligned along vertical structures) are indicative of low autocorrelations. This figure is also helpful in revealing shortcuts where the visitors did not take the correct path. Two areas of common shortcuts are clearly visible in the second half of the path, indicated by continuous missing data for about 13 segments. Scattered missing values that are not correlated in space (not vertically aligned) are due to GPS inaccuracy if they occur singly, or if temporally autocorrelated (i.e., horizontal bands of null values) indicate individual users leaving

---

* Ibid.

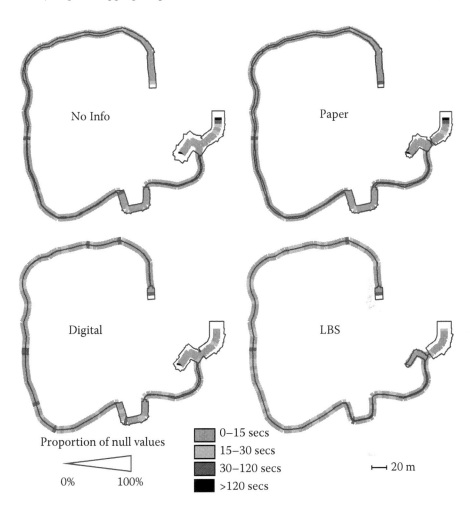

FIGURE 20.4   Average number of fixes per interval shown along the path for each information medium: (a) no info; (b) paper booklet; (c) digital info; and (d) LBS.*

the path. Figure 20.3 also indicates "natural" stopping places where all groups stop irrespective of the information medium. An interesting observation is the fact that the group with location-sensitive digital information appears to display more correlated stopping places (clearly defined darker bars).

These data were then averaged according to information media and then plotted along the path in order to visualize the coordinated stops in space (Figure 20.4).

Figure 20.4a shows that for the visitors with no access to information, there are, nevertheless, places that were common stopping points. This is indicative that the control group does not move at a constant pace along the entire route. It is also noticeable that most of the stops defined by the control group are also to be found in the other groups. A visual analysis of the aggregated tracks shows little difference between the control group (Figure 20.4a) and the paper booklet group (Figure 20.4b).

* Ibid.

Although the digital info and the LBS groups show some similarities, the LBS group in particular has more stopping points and these stopping points are more uniformly scattered along the path.

20.3.3 ANALYSIS OF ERRORS

As introduced in the methodology, the collected data (GPS fixes for moving visitors) had different possible sources of errors and uncertainty, primarily related to GPS positional error through canyoning effects and multipath reception, and the representation of the base path (onto which the fixes were being projected). In order to visualize these errors and identify biases or systematic errors in the data, Figure 20.5 was produced. It presents, for all the visitors' tracks (grouped by information medium) and for all segments, the average distance of the fixes to the base path. This distance was classified as positive for the fixes measured on the left side of the path and as negative for the fixes measured on the right side of the path. Systematic error or GPS biases can be identified in the figure as the spatially autocorrelated bands of color (the same color vertically aligned), meaning that on those specific segments,

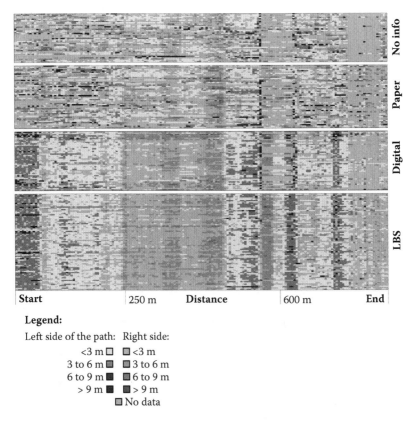

Legend:

Left side of the path: Right side:
<3 m □   □ <3 m
3 to 6 m ▨   ▨ 3 to 6 m
6 to 9 m ■   ■ 6 to 9 m
>9 m ■   ■ >9 m
▨ No data

FIGURE 20.5 Visualization of the average distance to the path for all fixes within a single interval for each track, grouped by information type.*

---

* Ibid.

all points for all tracks were being measured either on one side of the path or on the other.

Figure 20.5 also enables the identification of differences in the degree of uncertainty between the two types of GPS receivers used. The positional information for the non-tech groups (no info and paper booklet groups) was collected using a handheld Garmin12 GPS unit and for the Tech groups (the digital info and the LBS groups) positional measurements were made using a Bluetooth Globalsat receiver. The visitors from the non-tech groups show less autocorrelation than the tech groups, suggesting that the uncertainty related to the Garmin12 receiver is greater than for the Globalsat receivers. The spatial autocorrelation, for the information collected with the Globalsat receiver, is also much more apparent (vertical alignment of the same color patches). Figure 20.6 displays the distance data averaged and aggregated to path segments for each receiver. The average variance of the location data, represented by the delimiting lines on both sides of the path, is also shown. The variance was calculated as the mean radius of fixes per segment interval. To compute this, the mean position (centroid) of all fixes falling in a given interval was first calculated. The resulting point was therefore independent of the geometry of the interval itself. The variance was then given by the mean of the distances between each fix and this centroid.

It can be observed in Figure 20.6 that this variance is generally consistent in width along all the segments of the path for each receiver taken independently. The exceptions (segments where the variance is much greater) can all be explained by shortcuts (places where the visitors took a different way and therefore distanced themselves from the path, increasing the variance level). It can also be observed that the variance is higher overall for the Garmin GPS 12 receiver, compared to the Globalsat BT receiver. This is a reflection of differences in the positional error between the devices. Overall, Figure 20.6b shows a source of errors that is accountable to digitization (the path is shifted) rather than uncertainty in the GPS fixes. This is indicated by the fact that the distance values, which also consider the side the path fixes fall on, contain autocorrelation. However, since the variance of the GPS error is constant along the path, we can conclude that this autocorrelation must be due to a mismatch between the path on the ground and the digitized path. This divergence is less apparent for the Garmin receivers because the positional error of the fixes there is in a similar range to that of the positional error of the path digitization (Figure 20.6a). The uncertainty analysis (variance and distance to the path) also allows validation of the method used in projecting points to segments. The average distance from the path was normally distributed with a mean of 0.05 meters and a standard deviation of 3.02 meters.

Such results give confidence in the choice of both buffer size (10 meters) and segment length (5 meters) and indicate that the potential positional and digitizing errors did not significantly affect the location counts and the resulting classifications.

20.3.4 STATISTICAL ANALYSIS

In this section we set out to quantify the influence that information and its delivery mode has on the movement behavior of visitors. In an attempt to create "artificial" stopping places, information was provided to the three test groups (paper booklet,

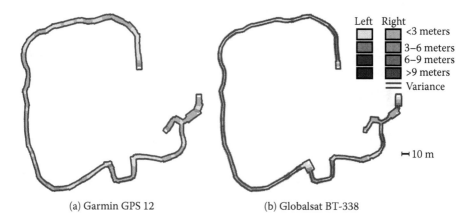

(a) Garmin GPS 12   (b) Globalsat BT-338

FIGURE 20.6 Average distances of fixes to the path with outline showing mean variance among fixes allocated to each interval. Results are aggregated by GPS receiver: (a) Garmin GPS 12 and (b) Globalsat BT-338.*

digital info, and LBS) that was relevant to the locations along the path indicated in Figure 20.1.

Figure 20.7 illustrates the average number of stops per segment for each information type, classified according to whether locations were POIs or not. Both the no info and the paper booklet groups spent roughly the same amount of time at all segments on the path. This finding was expected for the no info group because these visitors do not have knowledge of the information at certain segments, but is more surprising for the paper booklet group, where it was expected that the visitors would spend more time at the POIs exploring these places and the information. By contrast, the group issued with digital info show a significant difference in their behavior at POIs, even though the only difference between them and the paper booklet group was in the method of information provision. Finally, the LBS group displayed similar behavior to the digital info group, once again spending significantly more time at POIs. These results suggested that the method of providing information had an influence on visitors' behavior. In a second step, we wished to examine whether the type of information also influenced behavior. As explained in Section 20.2.1.2, the information available could be classified into four categories (POIs related to navigation, animals, plants, and landscape).

Table 20.3 presents the results of four binary logistic regressions between stops (defined as more than 15 seconds in a segment) and four information types that originated four different spatial behaviors. In the first column, below the information type, are the overall model statistics. χ^2 and M.Sig are the chi-square statistic and its significance. They result from the Omnibus Tests of Model Coefficients and measure how well the model performs. Only the model for the LBS group has a high performance, meaning that the stops and the information provision places are correlated for this group. For the other groups, a correlation could not be found. N is the number of valid segments included in the regression and the Nagelkerke R^2 is an approximation of the proportion of the variation in the response that is explained by the model

* Ibid.

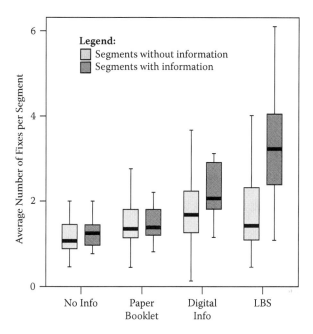

FIGURE 20.7 Box plot of average number of fixes per path segment grouped by information medium and whether the interval was related to a POI location or not.*

(comparable to the R^2 in linear regressions). As expected, the LBS information provision explains a bigger proportion of the stops than any of the other groups. Also presented in Table 20.3 are the specific results for the variables performance within the models. Exp(B) is the predicted change in odds for a unit increase in the predictor. The Wald and Variable Sig. columns provide the Wald chi-square value and two-tailed p-value used in testing the null hypothesis. Coefficients that have V. Sig. (p-values) less than alpha = 0.01 are statistically significant at the 1% level.

For the control group, which was given no information, there is nonetheless a significant correlation with the landscape POIs—this suggests that these POIs are in locations where park users might naturally stop. For both groups that were provided with information passively, no significant correlations were found. Finally, the group that was pushed information showed significant correlations with all POIs except for the navigation information. It is suggested that this is because, when pushed information, users stop to read it. However, at navigation points given the simplicity of the route the users were on, it was not necessary to travel significantly slower.

20.4 DISCUSSION

In order to obtain knowledge of the spatial behavior of visitors, it is necessary to capture fine-grained spatio-temporal data, but the collection of these high-resolution data leads to a problem in itself: Individual tracks contain too much variation (in terms of data quality and actual movement) to allow direct comparisons between

---

* Ibid.

TABLE 20.3

Logistic Regression Results for the Influence of POI Push Positions in the Spatial Behavior, Represented by Stops (Longer than 15 Seconds, Freq ≥ 3)

Spatial Behavior	POI Category	Exp(B)	Wald	V.Sig.
No info	Navigation	0	0	0.999
$\chi^2 = 9.029$; M.Sig = 0.060	Animals	0	0	0.999
Nagelkerke $R^2 = 0.154$	Plants	0	0	0.999
$N = 166$	Landscape[a]	8.929	7.364	0.007
Paper booklet	Navigation	0	0	0.999
$\chi^2 = 5.328$; M. Sig = 0.255	Animals	0	0	0.999
Nagelkerke $R^2 = 0.086$	Plants	0	0	0.999
$N = 169$	Landscape	3.938	2.478	0.115
Digital info	Navigation	0.897	0.01	0.922
$\chi^2 = 5.026$; M.Sig = 0.285	Animals	0	0	0.999
Nagelkerke $R^2 = 0.049$	Plants	0.978	0.001	0.978
$N = 169$	Landscape	3.587	3.449	0.063
LBS[a]	Navigation	0	0	0.999
$\chi^2 = 33.688$; M.Sig = 0.000	Animals[a]	19.304	6.728	0.009
Nagelkerke $R^2 = 0.268$	Plants[a]	5.63	8.25	0.004
$N = 169$	Landscape[a]	19.304	12.935	0

[a] Significant at the 1% level.

them. To deal with this issue, several techniques were applied to extract useful information and identify trends. The first step was to define when to accept or reject data points. A distance-based filter was developed, such that only the points close enough (within 10 meters) to the path were considered. The choice of tolerance was validated by analysis of the data. The second step aggregated data to a common baseline, by warping the highly variable individual GPS tracks onto the path. In addition, because often the datasets were not complete (due to inaccuracies of the receivers or to visitors' shortcuts), the analysis was not performed over the full tracks (which would require complete datasets), but rather by averaging datasets over single path intervals, which allowed null values to be ignored. It was still necessary to characterize such errors through a variety of visualization methods in order to contextualize their effects on the results and analysis (Figure 20.5 and Figure 20.6).

Providing visitors with information was expected to have an influence on their spatial behavior. Comparing only the no information and paper information groups, there is some evidence to support this hypothesis, though it is far from compelling. The average number of stops >15 seconds, shown in Table 20.2, is significantly higher (T-test $p > 0.001$). However, the visual difference in the patterns shown in Figure 20.3 and Figure 20.4 is negligible. More importantly, the interpretation of the box plot (Figure 20.7) indicates little difference in behavior, both between groups and between segments with and without information for the paper group. Likewise, the logistic regression shown in Table 20.3 was unable to find evidence that the positions of POIs influenced the stopping behavior for this group.

An unexpected difference in behavior between the digital info and the paper groups, for whom the information content was identical, was found. The visitors with the digital info not only stopped more (see Table 20.2) overall, but the places they stopped at were correlated along points of the path not investigated by the paper group. This can be seen in Figure 20.4. However, interpretation of the box plot in Figure 20.7 would suggest this difference should not be stressed too strongly. Indeed the logistic regression shown in Table 20.3 was unable to correlate the places that visitors stopped at with the POI information for the digital information group. Two reasons can be hypothesized to explain these finding: (1) The visitors from this group needed to interact more when handling the device, causing them to stop more, and (2) the technology had a "novelty effect," i.e., the visitors were more motivated to explore the information because it was presented in a media that was unfamiliar to them.

It is important to consider the potential impact of granularity—for example, the sensitivity of the results to the chosen length of stopping time (15 seconds)—and further work is required to explore this issue. Equally, the chosen segmentation length (5 meters) and GPS sample rate (5 seconds), although to some extent validated by the experiments on GPS uncertainty, is another example of variable granularity whose influence on the results should be explored. Previous work from Laube and Purves (2006) has shown that seemingly significant results can be artifacts produced as a function of granularity.

In terms of the overall results, it was possible to observe a clear difference between the non-tech (the no info and the paper booklet) groups and the tech groups (the groups that accessed the information via a PDA). One can assume that this difference indicates that the technologies have an intrusive effect on the behavior of visitors. Although both tech groups spent more or less the same amount of time on the route (see Table 20.1), two main differences were observable. The visitors with LBS information stopped more (see Table 20.2). Visual inspection of the data presented in Figure 20.3 and Figure 20.4 clearly shows more frequent autocorrelated stops for the LBS group when compared with the other groups. In addition, Figure 20.7 indicates that there is a significant difference in the behavior around path segments where the POIs were positioned and those without information, and the logistic regression of Table 20.3 is able to detect that this behavior is significantly influenced by the animal, plant, and landscape POIs. These findings indicate that location-sensitive information provisions can alter the spatial behavior of visitors. It appears that information about plants and animals introduced stopping points at locations where visitors with no info did not stop, by contrast to lanscape information, where all visitors appeared to stop. Thus, information about plants at the right place, for example, can lead people to direct experiences of nature, stopping to see plants about which they are receiving information.

The collection of anonymous-aggregated movement data allowed two additional qualitative behavior analyses: 1) Do visitors leave the trail and trample the protected dunes? and 2) Do visitors accept the park management advice to visit

particular places? Regarding the latter, the information provided to the three information groups was intended to help visitors fully explore and become more aware of
the park's natural richness (e.g., it recommended the visitors walk through a south
loop [POIs 23–26] and to see a breathtaking park (over)view by climbing to a dune
top [POI 35]).

The spatial data show that for the paper booklet group, 43% did not walk through
the loop, 39% did not see the viewpoint, and 31% went off-path in one or more
places, with similar values for the digital group. By contrast, within the LBS group,
only 4% took the shortcut, 20% did not visit the viewpoint, and only 7% were found
off-path. These results suggest that delivering location-based information is a potentially efficient channel for the park managers to communicate and influence visitors'
behavior towards eco-friendliness.

20.5 CONCLUSIONS

The research described in this chapter has three key outcomes that we believe should
influence future research.

Firstly, it has shown the value of spatio-temporal data collected according to a
rigorous experimental protocol in exploring behaviors that are unlikely to become
apparent through more common approaches to evaluating such technologies that are
grounded in psychology and usability. The importance of geography on influencing
behavior when dealing with location-based services cannot be understated.

Secondly, the research described has developed a set of techniques for aggregating
high-resolution track data and, in so doing, for dealing with uncertainty. We have illustrated how a combination of visualization and statistical methods is necessary to fully
explore such data and emphasized the importance of such a combined approach.

Finally, we have presented a case study where we have shown how behavior
was influenced by the provision of information, but not always as expected. Tourists provided with a paper booklet differed little in their actions from those with no
information, while those provided with digital information of any form spent longer
in the park. Furthermore, those to whom information was pushed were less likely to
stray from the official route and stopped more often at features that were not directly
related to features in the landscape.

It will be important in future work to control for the effect of novelty and determine whether it is undesirable, transient, or useful in terms of encouraging visitors
to explore natural environments.

While aggregation was useful to smooth out local variations among the singular
tracks and so explore the more general trends of the data, it also caused much potential interesting information about individual behavior to be lost. Future work will
thus also aim to explore disaggregated data.

ACKNOWLEDGMENTS

The authors gratefully acknowledge the study's participants; EcoMare, Camineo,
and Geodan for allowing the use of the LBS system in this research; and Patrick
Laube for his insightful comments. The first author would like to acknowledge the

support from the Portuguese National Science Foundation (FCT/MCT) under the PhD grant SFRH/BD/12758/2003. Additionally, the authors gratefully acknowledge the European Commission through their funding of the IST FP6 project TRIPOD (045335), which supported parts of this work.

REFERENCES

Butler, R., Hall, M., and Jenkins, J., 1998. *Tourism and Recreation in Rural Areas*. Wiley, Europe, pp. 1–274.

Dias, E., Beinat, E., Rhin, C., and Scholten, H., 2004. Location aware ICT in addressing protected areas' goals. In: *Research on Computing Science*, vol. 11 (Special Edition on e-Environment), Prastacos, P. and Murillo, M. (eds.). Centre for Computing Research at IPN, Mexico City, pp. 273–289.

Edwardes A., Burghardt D., and Weibel R., 2003. Webpark location based services for species search in recreation area. In: *Proc. 21st Intl. Cartographic Conference (ACI/ICA)*, Durban, South Africa, pp. 1012–1021.

Elands, B. and R. van Marwijk, 2005. Expressing recreation quality through simulation models: useful management tool or wishful thinking? In: *Proc. 11th International Symposium on Society and Natural Resource Management*, June 16–19, 2005. Östersund , Sweden.

Farrell, T. A. and Marion, J. L., 2001. Identifying and assessing ecotourism visitor impacts at eight protected areas in Costa Rica and Belize. *Environmental Conservation*, 28(3), pp. 215–225.

Hägerstrand, T., 1970. What about people in regional science. *Papers of the Regional Science Association*, 24, pp. 7–21.

Kwan, M.-P., 2000. Interactive geovisualization of activity-travel patterns using three dimensional geographical information systems: a methodological exploration with a large data set. *Transportation Research Part C*, 8(1–6), pp. 185–203.

Laube, P., Imfeld, S., and Weibel, R., 2005. Discovering relative motion patterns in groups of moving point objects. *International Journal of Geographical Information Science*, 19(6), pp. 639–668.

Laube, P. and Purves, R. S., 2006. An approach to evaluating motion pattern detection techniques in spatio-temporal data. *Computers, Environment and Urban Systems*, 30(3), pp. 347–374.

Laube, P., Purves, R. S., Imfeld, S., and Weibel, R., 2006. Analysing point motion with geographic knowledge discovery techniques. In: *Dynamic and Mobile GIS: Investigating Change in Space and Time*, Drummond, J., Billen, R., Forrest, D., and João, E., Eds., Taylor & Francis, London, pp. 263–286.

Lynn, N. A. and Brown, R. D., 2003. Effects of recreational use impacts on hiking experiences in natural areas. *Landscape and Urban Planning*, 64(1), pp. 77–87.

Manning, R. E., 2002. How much is too much? Carrying capacity of national parks and protected reas In: *Proceedings of Monitoring and Management of Visitor Flows in Recreational and Protected Areas,* Bodenkultur University Vienna, Austria, January 30–February 2, 2002, pp. 306–313

Mark, D. M., 1998. Geospatial lifelines, Integrating Spatial and Temporal Databases, Dagstuhl Seminars, no. 98471.

*Mountain, D. M. and MacFarlane, A., 2007. Geographic information retrieval in a mobile environment: evaluating the needs of mobile individuals. *Journal of Information Science*, 33(5), pp. 515–530.

Section V

Communication

INTRODUCTION

This section addresses communication aspects of the spatial data quality, which involves presenting and reporting quality about the spatial datasets that one is interested in. The contents in the metadata should be designed according to users' needs. Furthermore, the formal languages need to be developed for reporting the quality of metadata. The methods for communication information about spatial data quality vary from one application area to another.

It is important for a spatial data provider to provide information about spatial data quality that is indeed interesting to and needed by the spatial data consumers. The research by Boin and Hunter, presented in Chapter 21, investigates the experiences of the consumer of spatial data in determining whether a dataset is fit for use. They found that consumers are more interested in finding out what the data contained and how it matches with other information than statistical metrics of internal quality. Therefore, there is a need to redesign the metadata about spatial data quality according to users' needs.

In Chapter 22, Huth, Mitchell, and Schaab present a research development on judging and visualizing the quality of spatio-temporal data and an application on the multiple data source for the Kakamega-Nandi forest area in West Kenya. For the six data quality parameters (lineage, positional accuracy, attribute accuracy, logical consistency, completeness, and temporal information), a five-rank data quality system is defined for ranking the six quality levels qualitatively, and the method is designed for data quality visualization purposes. The traffic light system of visualization is selected as the best option for five-quality parameters while a slider is chosen to present the completeness parameter.

In Chapter 23, Lechner, Jones, and Bekessy present an investigation of the relationship between the scale-dependent factors and change in landscape pattern as measured by total area and landscape metrics. The scale-dependent factors may include, for example, pixel size, study extent, and the application of smoothing filters. It is found that changes in scale-dependent factors affected the level of patchiness, however, the total area remained constant.

In Chapter 24, Watson introduces a study on formal languages for expressing spatial data constraints and implications for reporting of quality metadata. It has been demonstrated that distributed geospatial data validation and data quality reporting are feasible within an open Web Services environment. When the logical rules is uniquely identified, this technology opens up the possibility of establishing standardized spatially semantic models within specific application domains.

21 What Communicates Quality to the Spatial Data Consumer?

Anna T. Boin and Gary J. Hunter

CONTENTS

21.1 INTRODUCTION

While the quality of spatial data is important within our industry, such preoccupations are not always reflected in research fields related to end-user applications (Goodchild, 2006). Consequently, there is a missing link between the spatial data quality our industry aims to communicate to the consumer, and the information consumers use to overcome the consequences of imperfect data.

In the following section, we review the debates in communicating spatial data quality, and we conclude there has been little empirical research conducted into how consumers determine fitness for use in a practical sense. Therefore, a broad, exploratory research technique was required. The technique is predominantly inductive because it starts by asking consumers about their experiences and then uses the findings to induce theories. The method section then explains that consumer opinions from two sources were used. These are interviews with data users from a range of backgrounds and existing feedback emails. The section on consumer experiences shows the results of the study and defines certain themes to describe the consumers' experiences to the reader.

Components of these experiences are summarized in the conceptual model section, which explains that consumers tended to determine a perception of product reliability through their own experiences rather than relying on quality information metrics from the data provider. Finally, the chapter closes with the suggestion that data providers could more effectively communicate the quality of their products if quality information were in the form of descriptive data content and opinions from other consumers.

21.2   BACKGROUND

Spatial data are inherently uncertain (Couclelis, 2003). It is the nature of our society that everything is interpreted and that the same reality is inherently likely to be modeled differently by different people (Bédard, 1987). It therefore follows that data consumers will always be exposed to data uncertainty in some form. The issue is how they determine a given set of data is fit to be used.

21.2.1   SPATIAL DATA QUALITY FOR THE CONSUMER

In an aim to express uncertainty, standards such as ISO 19113, 19114, and 19115 (ISO 2002, 2003a, 2003b) summarize quality into the well-known elements of lineage; positional accuracy; semantic, thematic, or attribute accuracy; temporal accuracy; logical consistency; and completeness. Little research has been conducted, however, into how these match with data consumers' concepts of quality or their understanding of the terminology itself. Devillers and Jeansoulin (2006) depicted these elements of spatial data quality as descriptions of internal quality—that is, they relate to the integrity of the spatial database. In contrast, external quality is concerned with the needs of the consumer and is hence related to fitness for use (Chrisman, 1984). There is, however, little empirical evidence of data consumers making practical use of these metrics.

Indeed, we believe there is a shortage of empirical research relating to how people perceive and use spatial data quality information for individual datasets in a real-world environment. Accordingly, the first challenge in this research is to determine what information conveys fitness for use to a data consumer who has not necessarily been educated in spatial information theory. The second challenge is in offering improved communication strategies.

Devillers and Jeansoulin (2006) argue that fitness for use relates specifically to each individual use case, and other anecdotal debates in the spatial industry consequently suggest that providing generic quality information is usually unhelpful to consumers. This research, however, contends there are ways that the details of spatial datasets can be made more understandable, even if there is no single overarching solution to the problem.

Frank (1998) suggests the burden of interpretation is high when expressing lineage, but this chapter will reveal that other aspects of information quality are similarly hard to understand for the studied consumers. Indeed, the terminology that describes the spatial information itself can be confusing. Furthermore, we believe that quantities expressing the accuracy of data often fail to contribute to consumer understanding.

21.2.2 CHOICE OF APPROPRIATE RESEARCH TECHNIQUES

Research into quality visualization (McGranaghan, 1993; Kardos et al., 2005) has included methods for clearly displaying extra dimensions of measurements. These assume that quality is quantifiable, which is conceivably true for positional accuracy but quickly loses relevance for the other semantic aspects of the data and data model. Indeed, these aspects cannot be narrowed down to an elegant list of independent variables on which to base a statistical assessment. Consequently, this research contrasts itself with more traditional deductive, experimental designs by employing a qualitative research strategy. Similar approaches have previously been used to explore map making and map use (Suchan and Brewer, 2000; Wealands et al. 2007) as well as selecting datasets in a controlled environment (Ahonen-Rainio and Kraak, 2005). Our study, however, investigates the subjective phenomena of consumers' actions and perceptions within their workplace, thus capitalizing on collecting data in an uncontrolled environment.

While the questions asked by the interviewer are semistructured, the interview has a conversational atmosphere and may include themes from previously collected data to enrich understanding (Bryman, 2004). In this way, the interviewing process aims to accumulate themes such that each interview is dependant on those previously completed. In this way, a high occurrence of a theme is not a measure of prevalence. Moreover, sampling is theoretical (rather than random), where the aim is to interview participants who are likely to contribute new themes. Approaches like this one can thus form robust foundations for identifying potential, valid statistical variables (Creswell, 2003; Suchan and Brewer, 2000).

Creswell (2003) offers primary strategies for validating qualitative data, and the following tasks have been incorporated into this research accordingly:

- Triangulating data from different sources by analyzing each source independently of the others to verify the cohesion of overall findings
- Member-checking by returning written interpretations of the interview to the participant and asking if they feel they were accurately represented
- Using rich, thick descriptions so that consumers' experiences are imaginable to the reader, to the extent that the reader can make judgments on generalizing the findings

- Including negative or contrary information because not all participants agree
- Clarifying bias of the interviewer. The qualities and inherent mannerisms and expectations of even the most experienced interviewer introduce biases into data.

Qualitative data can quickly show signs of theoretical saturation as each new interview yields less new themes. Determining an appropriate number of interviews depends on the homogeneity of the responses, which could be as few as 6 or more than 12 interviews (Guest et al., 2006; Nielsen and Landauer, 1993).

Various interview techniques were found that would be partially suitable. Semistructured interviews (Bryman, 2004) were flexible enough to include the basic think-aloud protocol (Hackos and Redish, 1998) or critical incident technique (CIT) (Flanagan, 1954; Chell, 1998). Most importantly, however, a qualitative researcher should adopt the role of an apprentice learning from each participant who is the expert of their own perceptions and opinions (Beyer and Holtzblatt, 1997).

21.3 METHOD

21.3.1 FEEDBACK EMAILS

In November 2005, more than 500 emails were inspected that dated back to 2002 and were received from customers by the major state-based mapping agency, the Department of Sustainability and Environment (DSE), Victoria, Australia. These were customers (a) replying to the email containing their ordered dataset, and (b) using the feedback link on the data producer's website. About 100 emails were found relating specifically to what we considered to be quality issues, half of which concerned the systematic absence of attribute information, leaving the remaining 50 emails for qualitative analysis.

21.3.2 SAMPLING OF INTERVIEW PARTICIPANTS

In addition to the emails examined, we conducted interviews with spatial data consumers recruited from (a) the distribution list provided by the DSE and (b) a call for interview participants placed on the Land Channel website of the DSE.

The aim was to make contact with data consumers who did not have expertise in spatial information. Six of the resulting participants (detailed in Table 21.1) fit into this category; yet, the four with more spatial data awareness had comparable practical attitudes to those without it. Indeed, the responses were quite homogeneous given the rate of new themes had declined significantly after ten interviews, thus indicating we were reaching theoretical saturation.

21.3.3 INTERVIEW PROCEDURE

The participants began the semistructured interview process with the knowledge that the interview would be about spatial data quality. Typically, interviews were audio recorded but field notes only were collected in two cases. Our first interview

TABLE 21.1

The Backgrounds of Data Consumers Interviewed:

An architect who has been working in the field for five years and habitually uses a few data sources to create plans.

A social researcher from an environmental science background who now needs data to study people and their geographic location in relation to retail outlets.

An acoustic analyst who has used the Internet and university libraries to understand noise emitted by machinery and sometimes requires large-scale map information to determine the shape of the land and possible noise sources.

A municipal council employee in charge of land contamination data in the United Kingdom. He has ongoing access to historical data for at least the last 100 years and is also a data producer.

A real estate agent in a fast-growing suburb who needs data about properties to estimate sale prices and is subject to disciplinary action if his estimates are incorrect.

A cartographer who grew up in the United States and used to publish maps to illustrate government policies there. He is now producing a statewide paper map in Australia to be used for a specific recreation while being attractive enough to frame.

An ecological researcher working in regional Victoria.

An archaeologist whose most resent interest was matching shipping routes with evidence of human presence. His experience with datasets has evolved over time and various projects to the extent he now has comprehensive practical knowledge of coordinate systems and GPS.

A technician in a university who transforms and disseminates data to students and is trained in nautical navigation.

A land owner planning to build a house who is required to submit plans to the council. He is competent with technical drawing software and is therefore using electronic data to create the plans.

schedule used the terms "fitness for use" and "quality" in the interview questions, however, we quickly discovered that even these terms, which were seemingly simple to us, can be mysterious to a data consumer.

Participants were asked what datasets they have been using without restricting them to discussing data from any one particular agency. With some datasets in mind, they were then asked how they determine whether a dataset is suitable or meets their needs or is good enough for their purposes. Care was taken here that the interviewer's initial use of vocabulary was restricted, allowing terminology to be first introduced by the participant.

While these discussions started with conversations about general interactions, the interviewer would also prompt using CIT for past incidences, switch to the think aloud protocol for current queries, or introduce theories or opinions to encourage more articulated comments.

21.4 EXPERIENCES OF THE DATA CONSUMERS

Findings indicate that the consumers' goals predominantly relate to finding out about the data content, then using the data. While perceptions and issues relating to quality play a role in this activity, they tend to be more of a consequence rather than being the primary aim of the user. These results are intentionally organized to reflect the approaches and vocabulary recorded in the interview data. All quoted data comes either direct from email [E], verbal interview data [Q], or interview notes verified by the interviewed consumer [V].

21.4.1 What the Consumers Looked for

Comparing between the dataset and other forms of data was a prevalent theme. When asked how she determined the data were good enough, the architect responded that quality had not been a conscious concern or problem: "Multiple people are involved in each project so crosschecking should uncover data problems, and the information will be merged with other sources so anything problematic will show up naturally" [V]. Similarly the cartographer stated: "[I] know where to find secondary sources to correct [a] problem" [Q], and even though the acoustic analyst did not have ongoing access, he would "buy it, download it and then work out" [Q] whether it was suitable. In fact, all consumers reported anecdotally cross-referencing or ground truthing data with other sources so as to "visualize where I actually am" [Q] or "so you can see where all the pathways are" [Q].

Interviewed consumers also looked for **data content** information on feature and attribute definitions with varying success. Customers sending feedback emails made requests for data in their own words, typically summarizing their requirements and the quality required in one sentence such as: "I am searching for a comprehensive gazetteer of Victorian place names that includes up-to-date gazetted localities as well as superseded place names" [E]. Similarly, one email from an engineering company began by asking: "Do you have a sample of what a map [from a particular product line] looks like?" [E].

In fact, **expectation** became an overarching theme, and this last quote summarizes not only a desire to know the extent of the data coverage, but also what to expect—a theme that 10 of the 50 emails fall under. Indeed, both comparing and expectation are summarized here: "[The missing walking tracks] are clearly marked on the [other agency's] documents ... there is NO WAY I can tell [whether] the walking tracks ... are marked on your maps or not until after I have purchased them" [E].

21.4.2 Conclusions Consumers Have Drawn

For those who had chosen to obtain a dataset, their conclusive opinions, again, predominantly come from **comparisons**. Both the municipal council employee and real estate agent could compare data of the same area over time, and both reported noticing inconsistencies. Moreover, issues related to "merging" [E] and datasets not "matching up" [E] appeared in feedback emails. Similarly, in response to being asked how good the data are, the council employee turned to his computer and "indicate[d] that the data is therefore 'poor' because the 'angles' of the road are 'different', 'don't line up'" [V]. He then considered a second set of aerial photography and "conclude[d] this [was] high quality aerial photography because they 'match' and because the [vector data] is a 'close fit'" [V].

The council employee, however, cautioned that "maps can fit together well because they are from similar [original] sources. Need to know sources well" [V]. Indeed, this was the first endorsement from a data consumer for some form of lineage information, though from a consumer perspective it is better termed **source of the data**. Indeed, perceptions of quality appeared to become tangible when an explanation of how the data were created could be used to describe the resulting characteristics: "There was not a high incidence of correlation until they found out the ... sites had been jotted in pencil on a map with 40m accuracy" [V]. Similarly, the ecological researcher attributed his understanding of the limitations of tree configurations to the following metadata, which he had found by clicking on the map layer title within a freely available interactive map: "Scattered tree cover boundaries will not necessarily be physically obvious at ground level ... it allows for minimum gaps in tree cover of 0.1 hectares."

Indeed, questions and discussions of **reliability** in relation to a purpose were also evident. For instance: "I have a concern that not all survey information shows up on the system. ... The system is not reliable for searching survey information if that is the case" [E]. The archaeologist used reliability to define the term "fitness for use": "Suitability to the task ... and of course the reliability is going to depend on whatever number of factors" [Q]. While the technician did not mention reliability until the interviewer raised it as a term, it triggered a new discussion about reliability charts on navigation maps in which he used the term repeatedly: "It's about the reliability of the data ... it says the data in this particular area was taken in 1853 and therefore fits into the not-too-reliable category" [Q]. The real estate agent, however, had a surprising attitude toward the data he uses on a daily basis: "[I] rely on it to the fact that it should be right. But in fact it isn't and I can't rely on it" [Q]. Indeed, this paradox is one of the key reasons why this chapter discusses conclusions and deciding factors individually.

21.4.3 Some Deciding Factors for Consumers

To decide to use the data, consumers ultimately need to discover them and then be able to obtain them, so this leaves interaction and cost as dominating themes. We use **interaction** to refer to the practical ability to make use of a computer interface to learn about and access the data, and success at this ultimately is a prerequisite to any other aspects of suitability. After all, for those people who search through the Internet, the website is the window to the data and without an adequate window, the data remain almost invisible.

Even the acoustic analyst, who regularly visited university engineering libraries to search for complementary information, found himself in a predicament with spatial data terminology when attempting to determine **data content**. After all, "if it starts talking jargon ... it's lost me because I can't translate jargon for [my client] ... if I don't understand it myself" [Q]. Similarly, the layout of an interface can undermine the ability for a consumer to obtain data within a reasonable timeline. Indeed, after using more reliable information a few times, the real estate agent stated: "[I would] like to have the time and energy to use more of it ... If I had unlimited hours in a day, I'd be right" [Q] because he would have to donate days rather than hours to obtaining the data he needed.

In cases, however, where the web interface had little impact, **reputation** played a significant role in the choice: The cartographer declared, "I'll use their data because they are the authority ... even though there are errors and I have reported some errors [to them]" [Q]. Similarly, the technician combined authority with his own perceptions of reputation, given he "judges the quality of the data by how 'authoritative' the provider is. When asked if there was any information on the web he used to work out whether he could trust the data, he thought for a while, and then said:" [V] "'No, it's trusted by the name of where it comes from. [The provider] is in charge of blocks of land, [I've] read enough about surveying to know it's ... precise ... so I just believe it's right. All I have to do is check my own work.'" [Q] Moreover, the successful GPS coordinate check he performed on his own property supported his reasoning. Regardless of whether this perception is correct, this data consumer showed no intention of looking further for information about the quality.

Finally, the presence of **cost** played an inconsistent role. The architect had a financial threshold to spend on data without further paperwork while the real estate agent was willing to pay for the data he "can rely on" [Q] because it was convenient. The cartographer, however, asserted "data that has a cost can undermine the financial feasibility of publishing a map or map series" [V].

21.4.4 REACTIONS TO METADATA ON THE INTERNET

Toward the end of each interview, existing metadata were brought to the attention of the participant if they had not already raised it as a topic. The authors are surprised that no consumer mentioned lineage information, attribute accuracy, or logical consistency. Furthermore, there was no express frustration regarding the metadata themselves, but rather a tendency to automatically ignore them as just another confusing web page. When the architect was encouraged to look at it, "she said under normal circumstances, she would have left the page after a few seconds because it made no sense to her" [V]. Similarly, the social researcher noted the attribute accuracy statistic and responded "I'm probably not up with it enough to know what is and what isn't high quality" [Q].

These findings gain significance given that they contradicted our expectations. Gillham (2005) suggests the researchers describe their biases by revealing what they expect to, hope to, and would prefer not to find. Accordingly, when we started the data collection process, we expected to find at the very least that (1) quality was a data concept that the consumer was aware of, and (2) a frustration that the current quality statements in metadata were hard to understand. The forecast challenge was to suggest more understandable language and include graphical representation. Although interviews were conducted with an impartial approach, we would have preferred to find that quality was important to the consumer and would have hoped not to find that people have already found other ways to decide whether data are fit for their use and are satisfactory to them. These findings have therefore evolved in spite of our biases.

21.5 A CONCEPTUAL MODEL FOR REAL-WORLD DETERMINATION OF SPATIAL DATA QUALITY

Surprisingly, standardized metadata have not played a significant role in the consumers' perception of the data; yet, they still have established opinions of the data quality. In an effort to make some sense of consumer reasoning, we have constructed a conceptual model to reflect the perceptions, actions, and goals that arose from studying these consumers (Figure 21.1). Indeed, it is quite evident that the process of determining whether a dataset is fit for use is a process influenced by a number of subjective perceptions that can change with experience.

The model is probably best described by three discrete paths a consumer might take:

- Interaction-as-a-barrier: Consumer needs data, uses the Internet and interacts with a spatial data website. He or she finds the terminology or the site architecture confusing or time consuming. He or she gives up and decides the data are not suitable.
- Content-and-cost: Consumer needs data, uses the Internet, determines data content, and decides whether the data are suitable. If suitable, data are used, compared, or added to other information. This influenced perceived data reliability.
- New-consumer: The consumer consults friends and colleagues or other queues from their environment to choose data by their reputation. Meanwhile, previous consumers can influence this reputation.

Indeed, as consumers follow these paths, there are two major goals, namely, to determine the data content and to make use of it. In the process of achieving these two goals, consumers will gain perceptions of both fitness for use and internal quality, however, they will do so using information and reasoning that may or may not be supplied by the data provider. In this way, fitness for use is a perception that is influenced by aspects beyond the control of the provider.

The contrast between the first and second paths, however, emphasizes the fundamental importance of terminology and website architecture. This means the provider can have some influence if consumers can determine what the dataset contains from web pages they can find that deliver timely search results in language they

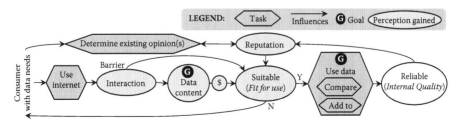

FIGURE 21.1   A conceptual model of how consumers determine spatial data quality.

understand. It is fundamental, however, to cater to their goals, and their goal is determining data content. So expressing quality as data content is the prime opportunity for quality to be communicated.

21.5.1 EXPECTATIONS

Indeed, communicating spatial data quality in terms of consumer goals is managing expectations. To give consumers realistic expectations, they need the sorts of information shown in Figure 21.2a: a concrete illustration of the coverage they are purchasing, the volume of data to expect, a thumbnail image of the data, bounding coordinates expressed in the coordinate system being purchased, and the number and volume of files or tables.

21.5.2 OPINIONS OF OTHER PURCHASERS

Data use is the second goal and occurs after the Internet has been utilized; but use influences impressions of reliability, which in turn influence reputation. For providers to be involved, data use therefore needs to have a presence earlier on in the conceptual model. Websites could thus include consumer opinions, similar to the book reviews on Amazon.com as suggested by Gould (2005), and Figure 21.2a accordingly has a link to "opinions from other purchasers."

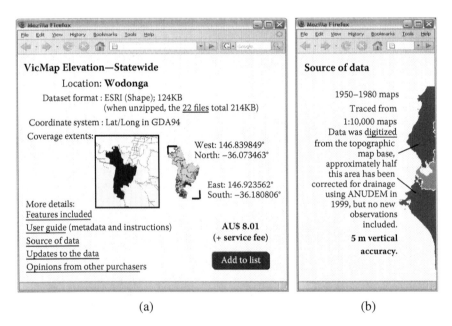

(a)   (b)

FIGURE 21.2 A prototype for managing expectations: On the left (a) is the confirmation page for obtaining data that links to more information such as, on the right (b), a concise statement about the source of the data.

21.5.3 Source of the Data and Updates

Source of the data, how complete, and how up to date the data might be are further details in expressing the data content, and a prototype for the former is shown in Figure 21.2b. This differs from lineage because it does not itemize each stage of creation. Rather, it offers anecdotal, detailed examples of how such origins manifest themselves in the composite end product.

21.6 CONCLUSIONS

This work has used exploratory research approaches to investigate the innately subjective experiences of how the everyday consumer of spatial data determines whether a dataset is fit for use. The research led to conversations with consumers from both spatial and nonspatial backgrounds and found similar attitudes between both groups. Overall, participants were more interested in finding out what the data contained and how it matched with other information than the statistical metrics of internal quality. We therefore argue that "fitness for use" information should aim to manage the expectations of the consumer as they undergo the data purchasing process. Consequently, two approaches need to be taken for the data quality information from providers to be influential. First, quality information should form enhanced descriptions of the data content. Second, insights into others' experiences of using the data need to be made available by including previous consumers' opinions.

ACKNOWLEDGMENT

This work has been supported by the Cooperative Research Centre for Spatial Information, whose activities are funded by the Australian Commonwealth's Cooperative Research Centres Programme.

REFERENCES

Ahonen-Rainio, P., and M.-J. Kraak. 2005. Deciding on fitness for use: evaluating the utility of sample maps as an element of geospatial metadata. *Cartography and Geographic Information Science* 32 (2):101–112.

Bédard, Y. 1987. Uncertainties in land information systems databases. Paper presented at the Eighth International Symposium on Computer-Assisted Cartography, 29 March–3 April 1987, Baltimore, MD, 175–184.

Beyer, H., and K. Holtzblatt. 1997. *Contextual Design: A Customer-Centered Approach to Systems Designs.* 1st ed. London, New York: Morgan Kaufmann.

Bryman, A. 2004. *Social Research Methods.* 2nd ed. New York: Oxford University Press.

Chell, E. 1998. Critical incident technique. In *Qualitative Methods and Analysis in Organizational Research.* London: Sage.

Chrisman, N. R. 1984. The role of quality information in the long-term functioning of a geographic information system. *Cartographica* 21 (2&3):79–87.

Couclelis, H. 2003. The certainty of uncertainty: GIS and the limits of geographic knowledge. *Transactions in GIS* 7 (2):165–175.

Creswell, J. W. 2003. *Research Design: Qualitative, Quantitative, and Mixed Methods Approaches.* 2nd ed. Thousand Oaks, CA: Sage.

Devillers, R., and R. Jeansoulin. 2006. Spatial data quality: Concepts. In *Fundamentals of Spatial Data Quality*, R. Devillers and R. Jeansoulin, Eds., 31–42. London: ISTE.

Flanagan, J. C. 1954. The critical incident technique. *Psychological Bulletin* 51 (4):327–358.

Frank, A. 1998. Metamodels for data quality description. In *Data Quality in Geographic Information: From Error to Uncertainty*, M. F. Goodchild and R. Jeansoulin. Eds., Paris: Hermes, 15–30.

Gillham, B. 2005. *Research Interviewing: The Range of Techniques.* New York: Open University Press.

Goodchild, M. F. 2006. Forward. In *Fundamentals of Spatial Data Quality*, R. Devillers and R. Jeansoulin, Eds., 13–16. London: ISTE.

Gould, M. 2005. Geospatial Metadata Part 2. GEO:connexion. http://www.geoconnexion .com/magazine/article.asp?ID=2253 (accessed 12 October, 2005).

Guest, G., A. Bunce, and L. Johnson. 2006. How many interviews are enough? An experiment with data saturation and variability. *Field Methods* 18 (1):59–82.

Hackos, J. T., and J. C. Redish. 1998. *User and Task Analysis for Interface Design.* New York: John Wiley & Sons, Inc.

ISO 19113. 2002. *ISO 19113:2002 Geographic Information—Quality Principles.* Geneva, Switzerland: International Organization for Standardization.

ISO 19114. 2003a, ISO 19114:2003, *Geographic Information—Quality Evaluation Procedures.* Geneva, Switzerland: International Organization for Standardization.

ISO 19115. 2003b. *ISO 19115:2003 Geographic Information-Metadata.* Geneva, Switzerland: International Organization for Standardization.

Kardos, J., G. Benwell, and A. Moore. 2005. The visualisation of uncertainty for spatially referenced census data using hierarchical tessellations. *Transactions in GIS* 9 (1):19–34.

McGranaghan, M. 1993. A cartographic view of spatial data quality. *Cartographica* 30 (2&3):8–19.

Nielsen, J., and T. K. Landauer. 1993. A mathematical model of the finding of usability problems. *Proceedings of the SIGCHI Conference on Human Factors in Computing Systems,* 206–13.

Suchan, T. A., and A. M. Brewer. 2000. Qualitative methods for research on mapmaking and map use. *Professional Geographer* 52 (1):145–154.

Wealands, K., S. Miller, P. Benda, and W. E. Cartwright. 2007. User assessment as input for useful geospatial representation within mobile location-based services. *Transactions in GIS* 11 (2):283–309.

22 Judging and Visualizing the Quality of Spatio-Temporal Data on the Kakamega-Nandi Forest Area in West Kenya

Kerstin Huth, Nick Mitchell, and Gertrud Schaab

CONTENTS

22.1   INTRODUCTION

Modeled predictions for the year 2100 reveal that the largest impact on biodiversity is expected to be due to land use/cover change (LUCC), this being especially true for the tropics (Sala and Chapin 2000; Chapin III et al., 2000). Within the BIOTA East Africa research framework, funded since 2001 by the German Federal Ministry of Education and Research (BMBF) (Köhler, 2004; http://www.biota-africa.org), the influence of fragmentation and anthropogenic use on the biodiversity of East African rainforests is investigated. With 15 subprojects at present, BIOTA East is following an integrated and interdisciplinary approach. Research is related to the vegetation structure, ecological interactions, certain animal groups (emphasizing invertebrates), and, since 2005, also to socio-economic issues in order to work toward a sustainable use of biodiversity (Schaab et al., 2005). The focal site is Kakamega Forest in west Kenya, with Mabira and Budongo Forests in Uganda also selected for comparative purposes with research mainly based on 1 km² biodiversity observatories (BDOs).

The BIOTA East Africa subproject E02 at Karlsruhe University of Applied Sciences supports this biodiversity research with geographic information system (GIS) and remote sensing activities aiming at an extrapolation of the field-based findings in space and time (Schaab et al., 2004). Here, E02 considers the analysis of longer-term forest cover changes in the three East African rainforests as one of its major research tasks. Data sources range from satellite imagery and historical aerial photography via old topographic maps, official governmental records, and forestry maps to oral testimonies by the local population, with place names giving evidence for much earlier forest extents (Mitchell and Schaab, 2006, in print). The analysis of such information will lead to a detailed picture of the forest use history for the different forests. The Nandi Forests are also included here as the development of the land use/cover time series brought them to light as having once been connected to Kakamega Forest (Mitchell et al., 2006). The data gathered reflect approximately the last 100 years, coinciding with the start of commercial-scale exploitation of forests in East Africa.

The geodata processing so far is most advanced for the Kakamega-Nandi forest complex. A total of 132 data layers are directly visually compared as well as jointly analyzed via their spatial reference by means of a GIS. The reliability of the spatio-temporal information must be accounted for and differences in geodata quality must be assessed by the scientist in order to draw correct conclusions. The method presented here represents a ranking of dataset quality levels for visualization purposes, rather than a quantitative method for determining data quality precisely. As such, it allows the quality of the time-stage-dependent spatial data to be documented and preferably visually cognizable for simply describing the underlying data layers and for the presentation of conclusions. This has lead to the concept of a visualization tool including a feature that, since these data would be available, enables a consideration and visualization of geodata quality.

This chapter (Figure 22.1) starts with a description of the data sources and methodology applied for analyzing the longer-term forest cover change. It reviews and discusses strategies for the visualization of geodata quality. And, finally, based on a crisp literature review, we will conclude which data quality parameters are considered of importance for the purpose of our work. Next, our system of judging these parameters for every geodataset is introduced and a statistical summary is given on the judgments of all the Kakamega-Nandi datasets. Subsequently evolved designs for diagrams exposing the quality aspects are presented and discussed. An assessment of the alternatives leads to the agreed version for illustrating the group of geodata quality parameters. The outlook will stretch to the implementation in the visualization tool.

22.2   ANALYSIS OF LONGER-TERM FOREST COVER CHANGE

In order to analyze the long-term change in vegetation cover datasets were sought to cover a 100-year time period. This necessitated the acquisition of spatial data in many

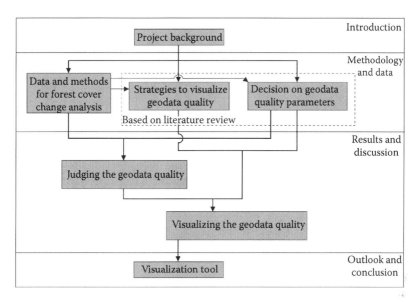

FIGURE 22.1   Structure of the chapter.

TABLE 22.1

Geodatasets Available for the Analysis of Longer-Term Forest Cover Change in the Kakamega-Nandi Area

Data Source	Combined Date Range[a]	Scale Range	Image/Raster	Vector
Satellite imagery	1972–2003 (31 years)	30 to 60 m	14/8	–
Aerial photography	1948–1991 (43 years)	1:25 k to 1:55 k	3/–	2
Topographic maps	1900–1970 (70 years)	1:50 k to 1:1 mill	7/–	10
Topographic drafts	1896–1970 (74 years)	1:62.5 k to 1.5 mill	8/–	9
Forestry maps	1937–1995 (58 years)	1:10 k to 1:50 k	8/–	10
Forestry sketch maps	1933–1977 (44 years)	1:10 k to 1.62.5 k	9/–	18
Sketch maps	1924–1949 (25 years)	1:62.5 k to 1:300 k	3/–	9
Thematic maps	1899–2000 (101 years)	1:25 k to 1:2.5 mill	3/–	10
Fieldwork	2002–2006 (4 years)	1:50 k	–	1
Totals	**1896–2006 (110 years)**	**1:10 k to 1:2.5 mill**	**63**	**69**

[a]   Date range refers to the "date on map/geodataset."

different formats and consequently their integration within a GIS (see Table 22.1). Thus, the most recent period is covered by Landsat satellite imagery (MSS, TM, and ETM+), which was purchased to allow analysis of the forest cover from the present day back to 1972 in eight approximately 5-year time-steps. A supervised multispectral classification was performed for each of the time-steps of the Landsat satellite imagery. This process distinguished 12 land cover classes, 6 of which are forest formations and 2 of which are grassland (Lung, 2004; Lung and Schaab, 2004).

Historical aerial photography was also acquired to extend the time series back to 1965/67 and 1948/52, although the latter represents only 65% coverage of the forests. The photographs were scanned, orthorectified, and mosaiced, and from this land cover classes were distinguished by visual interpretation and were digitized on-screen. Vegetation classes were assigned in keeping with those derived from the satellite imagery (Mitchell et al., 2006).

Extending the series beyond remote sensing has required topographic maps from archives in Kenya and libraries in Britain. The search produced 15 topographic maps or map drafts that pertain to the Kakamega-Nandi forest area across the period 1896 to 1970 (ranging in scale from 1:50,000 to 1:1.5 million). In some cases, the original map was acquired, but for the most part, they exist as photocopies or scans and occasionally as amateur digital photography. These were georeferenced for inclusion in the GIS and their relevant features, such as forest cover, were also digitized on-screen.

Forestry maps and logging records were painstakingly located in the forest offices of the Kenyan Forest Department. These forestry maps relate to the period 1933 to 1995 and can show both areas of logging and planting of trees. Some of these are printed maps while others are here termed forestry sketches. Logging concessions, for instance, were often sketched onto other maps by a forester using a colored crayon, while in other cases they exist as tracings, hand-drawn sketches, or even as written descriptions of the boundary with reference to local landmarks. Logging records have been extracted from forestry archives and are incomplete but are linked to the concession maps. All the maps have been scanned, georeferenced, and digitized to include relevant depictions or textual annotations of vegetation cover.

Other maps labeled here as sketch maps are present as hand drawings by, e.g., anthropologists and date from between 1924 and 1949. There are 13 thematic maps that range from 1899 to 2000 and depict various themes from land-use cover and population density to tribal locations. Fieldwork represents some of the most recent datasets and includes oral histories that were obtained by means of 69 semistructured interviews with old people living adjacent to the forests. The interviews' locations have been established while a summary table of the main issues investigated is linked to the point layer. Other datasets derived from fieldwork include place-name evidence and ground truth information.

To date, a total of 132 datasets are stored in the GIS covering the past 110 years (Table 22.1), but it should be emphasized that many of the geodatasets have incomplete coverage of the Kakamega-Nandi forests. Attention should also be drawn to the fact that several vector datasets can be derived from the digitizing of the features of a single scanned map. In the case of satellite imagery, scenes from different seasons have in some cases been combined to create a single timestep in order to enhance the classification of land cover types.

22.3   REVIEWING AND DISCUSSING STRATEGIES TO VISUALIZE GEODATA QUALITY

With such diverse geodatasets at hand, a literature review was performed in order to gain ideas for the presentation of their quality. Different sources on the topic of geodata quality served to find out which aspects of geodata quality can be treated and how others have visualized geodata quality in their projects.

Many examples in different contexts were found (for the complete compilation see Huth, 2007). In classifying the existing visualizations, we first consider what kind of information is presented. There are those that show only one criterion, for example, positional accuracy, while others just refer to the overall data quality. For a summary, see van der Wel et al. (1994). In particular, interactive information systems can present information on several parameters, sometimes even with different levels of detail. With such features, the software Quality Information Management Model (QIMM; Devillers et al., 2005) is already approximating a GIS. Others allow the user to choose between different, sophisticated visualization alternatives (QIMM or RVIS; see below). Such interactive tools are not only rather complex in development but also in their correct use.

For simple visualization of a single quality feature or the overall geodata quality, methods known from traditional cartography are quite common. These are, e.g., the reliability diagram or the indicatrix by Tissot (van der Wel et al., 1994). Adapted to the electronic presentation of geodata are the methods using sound or blinking but also the well-known dot animation by Fisher (1994). An example of a more extensive tool is the software Reliability Visualization System (RVIS) developed by MacEachren et al. (1996). Here the user can choose between different visualization alternatives for the same dataset focusing on spatial, temporal, and attribute quality aspects. Even more complex is the aforementioned QIMM by Devillers et al. (2005). Within this tool it is possible to show quality information for different levels of detail and six distinct quality parameters. Their display can be realized in the main map or as a quality dashboard next to the map.

Many of the examined methods are only suitable for particular types of geodata. RVIS, for example, is only designed for one special dataset and is thus restricted in its application. It is therefore difficult to transfer these particular methods to other geodatasets.

Geodata quality information can be either visualized within the map or map display or it can be placed independently from the map face. For the first option, it is necessary to have differentiated quality information available for different areas in the map, e.g., applying transparency for depicting uncertain areas by MacEachren's variable "clarity" (MacEachren, 1995). This spatially differentiating information is also necessary for the display of quality in a separate map, as this is, e.g., the case with reliability diagrams. The two maps can be arranged either next to each other or can be presented alternating, as in the case of an electronic display. If visualizing quality information for several datasets in comparison, one should only make use of geodata files of the same data type.

The characteristic of the project described in this chapter is the processing and handling of both numerous and varied geodata types, e.g., scanned topographic maps, satellite imagery, vector layers, or GPS readings. For this reason visualizing geodata quality differences within the mapped extent per dataset would be far too ambitious. This is especially so considering the collection of data depicting former time stages as the exact circumstances of map creation are often simply not known.

Therefore, a visualization in the form of a diagram next to the map is the only feasible option. However, to give a single overall quality statement per geodataset would not only be rather disappointing but would also not suffice for the user. Splitting the illustration of quality into several parameters provides a more detailed overview of the quality of a geodataset.

22.4   DECIDING ON QUALITY PARAMETERS TO DESCRIBE THE GEODATA USED

In this chapter, the term data quality does not refer to error as the opposite of accuracy. We do not use the term error, because it can have different meanings (see Zhu, 2005). "Fitness for use" is not an issue either, although we are aware that unsuitable data can lead to wrong analysis results. Uncertainty can be seen as an overall term for data that are not an absolute exact image of the objects in reality. Thus, a very high data quality requires a low uncertainty.

The term data quality can be split into different aspects, all contributing to the quality of a geodataset. One has to be familiar with these aspects before their visualization can be tackled. In the literature, five parameters are often mentioned: positional accuracy, attribute accuracy, logical consistency, completeness, and lineage (e.g., van der Wel et al., 1994; Slocum, 1999). Comber et al. (2006) name them the "big 5."

Lineage describes the development of the dataset to the current state (Slocum, 1999). Although there may be several steps to the actual state, we only pay attention to the state or data source type before its integration in our GIS, because quite often more of the dataset's history is not known to us. Positional accuracy deals with the difference of the geodata object to its true geographic position. This can also include a third dimension, like the accurate height of a mountain (Slocum, 1999). By attribute accuracy one can express whether the thematic variables were classified in a correct way (Buttenfield and Beard, 1994). A logical, consistent dataset must not have geometric, topologic, or thematic contradictions (Navratil, 2006). That includes the relation of the objects in the map to each other having to be correct. In a complete dataset, no object must be missing (Slocum, 1999). Due to aspects of generalization, one has to be aware of minimum sizes for mapped objects in order to judge completeness correctly.

Besides these five we consider the temporal information aspect to be of importance, too. For a correct joint analysis and interpretation of the geodatasets, it is important to know whether or not, e.g., the date mentioned on the map corresponds with the content of the map. If the analysis is based on the wrongly perceived date, this can lead to incorrect conclusions on the forest cover. This temporal aspect must not be confused with the often-mentioned quality parameter "currency," which refers to how up-to-date a dataset is (see Navratil, 2006). Instead, our study is to be based on a broad range of geodatasets covering more than the last 100 years in order to investigate the change in forest extent and state due to forest use practices.

The six selected geodata quality parameters are listed and explained in Table 22.2. Besides a judgment on each of the parameters, we put additional information beside each one, e.g., scale or resolution in case of positional accuracy (see Table 22.2). An

TABLE 22.2

The Six Selected Geodata Quality Parameters with Their Additional Information

Geodata Quality Parameter	What Is Judged?	Additional Information
Temporal information (TI)	Difference between year on map/ geodataset and year of content, plus knowledge of these dates	Year on map/geodataset
Lineage (Li)	Reliability of original (parent) dataset	Data type of original (parent) dataset
Positional accuracy (PA)	The georeference of the objects	Scale or resolution
Attribute accuracy (AA)	Quality of attributes	Number of attribute classes
Completeness (Co)	Not a judgment, but percentage of completeness	Referring to forest boundaries or to study area
Logical consistency (LC)	Mainly geometric contradictions in dataset	(none)

exception is the parameter "logical consistency," which has to make do solely with a judgment. "Completeness" can be better described than simply with a judgment, i.e., it can be specified as a percentage. Here, as additional information, the choice between making reference to the forest boundaries or to the complete study area is given, because for quite a number of datasets information mapped within the forest extents is clearly sufficient. The temporal information has another peculiarity as it consists of three different types of information: The judgment is based on the difference between the year with which the map is labeled and the year of the content. The year on the map is given as additional information. In addition, the judgment includes an assessment on how well the actual date are known. In case of a satellite scene, both dates coincide and therefore the reliability judgment is very high, but for old maps it is necessary to address the temporal aspect in a detailed manner.

The parameters can be regarded independently from each other, but they are at the same time mutually conditional. When two datasets of different sources are graphically overlayed by means of a GIS, it is certain that imprecision in positional accuracy will appear and will often result from their creation by different organizations for different purposes (see also Longley et al., 2001).

22.5 JUDGING THE GEODATA QUALITY

Each geodataset was assessed for data quality as a whole against the six different categories as listed in Table 22.2. For consistency of interpretation, all datasets were judged by the same person who had the greatest working knowledge of the datasets and the area.

The lineage scale is related to the purpose of the product used to derive the described dataset for inclusion in the GIS, and thus it is an impression of its process of emergence or heritage. The products used can be related to nine categories (see Table 22.1 for geodatasets available). The grading is an ordinal scale of 1 to 5, with

5 being the best. In general, satellite imagery as the source is ranked high, while forestry maps gain a higher grading than forestry sketches. Considering the 132 datasets in total, the five gradings show a fairly normal distribution but skewed to its higher end (see Figure 22.2a).

The positional accuracy was also ranked by "factors" on a scale of 1 to 5. Here, the scale or resolution of the graphic enabled a rough ranking as a starting point. For example, datasets of a scale larger than or equal to 1:10,000 received a score of 5, while those of 1:1 million or less scored a value of 1. This grading has been further refined by also taking into account knowledge of the georeferencing process or the fitting of features in a visual overlay by means of GIS. In the case of forest logging geodatasets, the ratings for positional accuracy are typically adjusted downward by a value of 1 in order to compensate for the inaccuracy of the actual logging, which is known to often stray beyond the boundaries of marked logging concessions. Typical scales or resolutions of the different geodata types used can be found in Table 22.1. The statistical graph (Figure 22.2b) reveals the distribution of the scores. Overall more high scores have been given, which reflects the aim of the study, namely, to investigate differences in local forest use. A general but worthwhile pattern to mention is that the older the stage represented by the geodataset, the lower it has been judged for its positional accuracy.

Attribute accuracy was assessed again on the basis of a purely ordinal scale of 1 (inaccurate) to 5 (accurate). This judgment is independent of the number of attributes and is solely related to a judgment on the accuracy with which the attributes were assigned. As additional information, the number of attributes or datafields are shown and this excludes the default datafields. In the case of imagery, scans of maps or vector datasets showing a single class, for example, forest cover only, no attributes, or datafields, are present. Scanned maps and such vector datasets can nevertheless be judged regarding their attribute accuracy. Figure 22.2c shows the generally high ranking for our data pool.

Logical consistency normally considers geometric, topological, and thematic aspects. In our case, every dataset has been carefully checked and corrected as required (in particular for topology), and this quality parameter is predominantly judged on the basis of the correct positioning of the landscape objects in relation to each other. Here a scale of 1 to 3 is used, representing low, medium, and high consistency levels. In our case, of all the parameters treated separately, logical consistency is the most difficult to handle by a differentiating judgment. This is because it requires the most detailed knowledge of a dataset, which is often not available in the case of geodatasets representing much earlier stages. The judgment has generated a very limited range (Figure 22.2d) with most datasets scoring the highest class. Only five of the datasets appear to be inconsistent in terms of positioning of objects in the landscape in relation to each other, and four of these represent official boundaries of forests and administrative units.

The judgment of completeness is derived from a percentage coverage of either the official forest boundaries (for purely internal forest datasets) or by the percentage represented of the whole 60 × 65 km Kakamega-Nandi study area (for more general datasets). It is the only quality measure that is derived directly from factual numbers without an element of judgment, although at present in most cases these are only

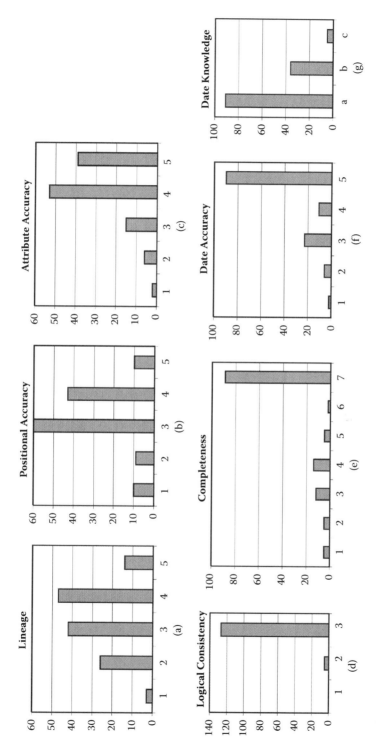

FIGURE 22.2 Summary of the geodata quality judgments performed for the 132 datasets on the Kakamega-Nandi area. For a description, see Chapter 5.

estimated visually. There are 89 datasets that have a top ranking of class 7 and which represent between 95% and 100% complete coverage of either of the two extents mentioned above (Figure 22.2e). In those cases in which completeness is based on the forest boundaries, most of the geodatasets do not have full coverage, such as a map solely dedicated to the South Nandi Forest. While the forestry sketch maps tend to be complete as they relate to isolated forest areas, the datasets resulting from formal forestry maps are the most fragmentary of all since they often have covered the whole forest in several adjoining map sheets. Their partial coverage here reflects the fragmentary nature of the forestry archives from which most of the forest related datasets were acquired. In the case of the datasets that relate to the extent of the study area, the coverage is more frequently complete (52 of 65 as compared to 37 of 67).

Date reliability is measured in relation to the number of years between the date as specified on a map and the date of its actual content. Five classes are created wherein class 5 represents no difference between the two dates, and similarly the awarding of a class 3 score, for instance, reflects a difference of 6 to 20 years. Since some of the maps hold historical information, the scale has been set to include those cases of large time spans and thus class 1 represents a discrepancy of at least 100 years (two datasets). However, for most geodatasets, the "date of the map/geodataset" is consistent with the date of its content (90 of the 132 cases; see Figure 22.2f). In particular, the satellite imagery is, as would be expected, very high scoring. While forestry sketches also score highly since they represent snap-shots in time, the forestry maps are poorly rated here since they attempt to locate multiple data of differing phases of forestry on the same map. A further scale of "a" to "c" is used to reflect the state of our knowledge of these dates. Thus, "a" is awarded to a dataset for which the relevant dates are known (being the case 91 times) and "c" to one for which they are known only vaguely (5 times). The high numbers of datasets for which we have high scores (see Figure 22.2f and Figure 22.2g for rankings "5" and "a") is not surprising since the dates for all the imagery and derived products are well known. It is the historical anthropological data that have the greatest discrepancy but also vagueness.

22.6   VISUALIZING THE GEODATA QUALITY

As discussed in Chapter 3, our geodata's quality is to be visualized in the form of a diagram next to the map. Here six distinct geodata quality parameters have to find space with their specific ways of being judged (see Chapter 5). Five of them will be complemented by additional information (see Chapter 4). A major aim of the visualization is that the feature be memorable. But, at the same time, the details of the information given should be easily and fully grasped by the user.

22.6.1   GENERAL CARTOGRAPHIC METHODS FOR VISUALIZING GEODATA QUALITY IN A DIAGRAM

Table 22.3 helps to evaluate different general methods widely used in cartography regarding their appropriateness for visualizing the geodata quality in diagrammatic form. Here, we dicsuss only the most convenient of these.

TABLE 22.3

Evaluation of General Methods Used in Cartography for Visualizing Geodata Quality in Diagram Form

Method	Suitable for All Parameters	Memorable/ Overview	Adding Text Possible	Understandable	Reference Needed	Division Too Exact
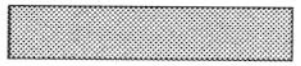 **Shaded rectangle**	OK	OK	Good	OK	Yes	No
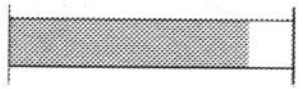 **Bar**	OK	Good	OK (horizontally)	OK	Maybe	Yes
 **Slider**	OK	OK	Bad (e.g., to the right)	Good	No	Yes
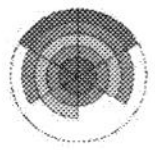 **Segment of a circle**	Bad	Good	OK (perhaps a little too tight)	OK	No	No
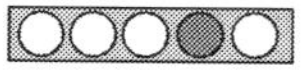 **Traffic light**	Good, with adaptations	Good	OK (e.g., to the right)	Good	No	No
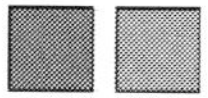 **Graphic variable**	Good	OK	OK (perhaps a little too tight)	Moderate	Yes	No
 **Line**	Bad	OK	OK	Moderate	Maybe	No
 **Arrow**	Bad	Good, with length	OK	Moderate	Maybe	No
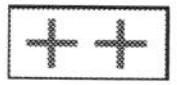 **Plus/minus symbol**	OK	Good	OK (to the right)	OK	No	No
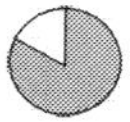 **Circle**	Bad	OK	OK (perhaps a little too tight)	OK	No	Yes

positive assessment

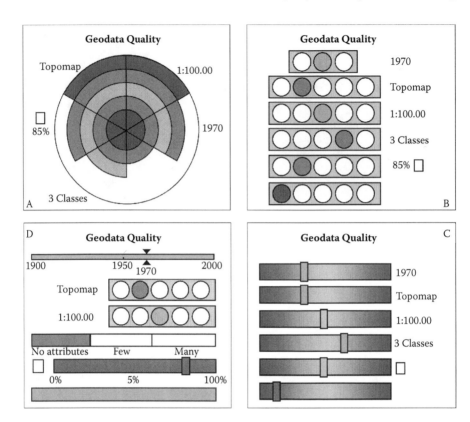

FIGURE 22.3 Four alternatives for visualizing the six geodata quality parameters in a complex diagram.

A visualization making use of graphic variables suits all quality parameters. Here, for example, color saturation could be applied, with a variable number of saturation steps or even a continuous range, depending on the differing judgment scales. For a more complex geodata quality judgment, as in the case of temporal information, saturation could be combined with the graphic variable of color hue. The implementation via segments of a circle would not fit consistently unless the number of evaluation classes is equal (see Figure 22.3a). A slider would work well only with more or less continuous judgment scales. Here, the introduction of interval markings could enhance its suitability for visualizing the distinct parameters. Traffic lights can be easily adjusted to differing numbers of ranking classes. However, the number of colored lights should preferably be uneven.

The best option for keeping the quality judgment in mind is judged to be the segments of a circle diagram where assessments for all parameters are arranged in a closed shape. Also, the options displaying a position along a scale range are easily memorable (see bar and traffic light). Assistance can also be given by a color ramp scheme (see slider). Whether the graphics can be accompanied by text depends mainly on the space available, either inside or close to the diagram. In order to demonstrate the connection between the judgment and additional, mostly textual information, the latter should be placed in the immediate vicinity of the diagram. Within a colored rectangle there is plenty of space to add textual information, even in the case of longer words as used for the lineage information. However, diagrams like the traffic lights require the placement of additional text or figures next to the diagram.

Both the comprehensibility of the diagram and in particular the necessity of showing the minimum and the maximum as reference points contribute to how easily the diagram can be understood. If the ends of a scale are not obvious, as is the case with a colored rectangle, the user might misinterpret the valuation shown. This is the reason that the slider and traffic lights are so easily understood: Here the actual score is presented in relation to the highest and lowest ranks. In addition, the traffic lights' interpretation is intuitive, with a red light giving a warning, etc. However, the appearance of the slider encourages the user to presume the rating schemes behind it are continuous. For the same reason, the circle should also only be used with caution.

22.6.2 COMBINING THE QUALITY PARAMETER INFORMATION IN A COMPLEX DIAGRAM

Having discussed the suitability of general cartographic possibilities, we can now discuss the adoption of an effective combined presentation of the six quality parameters in one complex diagram.

The order of the single parameters displayed in the diagram is influenced by a rating of importance for the particular project aim. The quality judgment's level of exactness affects its position in the final diagram too. Temporal information is considered to be the most important parameter because geodatasets, including data of the past 100 years and more, are used for analyzing forest cover change. Considering the long-term nature of the forest research, the older documents gain an added status even though their positional accuracy might not be as good. Therefore, temporal information is placed at the top, while the judgment on logical consistency, the least objective and least detailed criterion, is moved to the end.

Four options for combining the different parameters are presented in Figure 22.3 and are shown simply as graphic concepts, i.e., they are not linked to real geodatasets. In general, there are two alternatives for the visualization. On the one hand, a complex representation can be realized, with each parameter visualized being customized to its information content. This complex form places emphasis on the most exact communication of information but is less concise (Figure 22.3d). On the other hand, the data quality information can be presented in a simpler way where the parameters are visualized similarly (see Figure 22.3a to Figure 22.3c). These diagrams have the advantage of providing a faster-to-grasp overview that the user can easily keep in mind. From the cartographic point of view, it would be best to treat every parameter differently, finding the optimal representation for its specific characteristics. But this would require the user to regularly consult a detailed description and would necessitate a lengthy learning period. As the later users will not necessarily be cartographers, the simplest visualization strategy is chosen here.

A major difference between using segments of a circle and the presentation by traffic lights, aside from the arrangement, is the space available to add textual information (see Figure 22.3a and Figure 22.3b). The segments of a circle make it difficult to position text of differing length. Traffic lights can be presented in a very small size, allowing for plenty of space for even longer text lines. Here the text should be positioned next to the diagram and not above or below the traffic light sign, because this would disturb the overview of the constellation of the colored lights. When making use of sliders (Figure 22.3c) one has to be aware that the color representing the judgment has to be visually highlighted in contrast to the colors of the complete color ramp. When using traffic lights, this requirement is easily achieved by simply varying the color of the particular light.

22.6.3 THE FINAL DIAGRAM ADJUSTED TO THE QUALITY JUDGMENT AT HAND

It has become clear that keeping the visualization simple is advantageous. Visualization via traffic lights is seen to be adequate for five of the parameters as their judgment considers five or three ranks. As illustrated by the gray shading in Table 22.3, this method gained the best overall assessment among all the alternatives demonstrated. It is only for the completeness parameter, which does not provide a judgment but a factual measure, that the decision was made to adopt a slider.

In order to link the temporal information parameter with three kinds of information, the variable crispness as introduced by MacEachren (1995) is planned in order to reflect the state of our knowledge on the dates on which the judgment is based. While the color and position of the light indicate the concept for visualizing geodata quality for six distinct judgments on date reliability, three variations in crispness applied to the light reveal the degree of certainty. The redundant expression of data quality information by color and position prevents interpretation problems due to possible color deficiencies.

This final concept for visualizing geodata quality for six distinct parameters separately is shown in Figure 22.4. The additional, mostly textual, information is always placed next to the diagram on the right-hand side while an abbreviation for the parameter's name is placed on its left. This arrangement contributes to a consistent

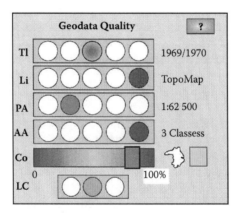

FIGURE 22.4 The agreed concept for visualizing the six selected geodata quality parameters. For explanation of abbreviations see Table 22.2.

and clear overall picture. Using just two letters occupies the minimum space while enabling even a new user to quickly link the correct diagram with each of the six geodata quality parameters. For assistance, a help button placed in the upper right corner will lead the user to a comprehensive description of the visualization concept.

22.7   OUTLOOK AND CONCLUSION

The concept of visualizing quality for varied geodatasets as introduced, described, and discussed in this chapter is currently implemented in a visualization tool for displaying and working with spatio-temporal data of the Kakamega-Nandi forest area in west Kenya. The tool will consist of two windows for changing between a scientific report on forest use history in this area (see Mitchell, 2004) and a display of the geodata available. Hyperlinks in the text will open the map window loading relevant geodatasets or centering the map field to a specific location. Further navigation within the map view is enabled by buttons arranged in a toolbar. Several vector datasets can be displayed at the same time, and raster datasets can be viewed one at a time. Here a table-of-contents list will provide the required versatility to toggle between the datasets. The display of the geodata quality diagram per dataset is also controlled from here and can be viewed in succession. The programming (see Huth et al., 2007) is based on XHTML for the text window and SVG for the map window. The database behind is MySQL with access enabled by PHP. Interaction is realized by JavaScript. The tool will not only offer the opportunity to the scientists already familiar with the geodata to gain new insights, but can be of use to a wider audience for simple documentation and presentation of results and provides them the opportunity of working with the gathered data and information. Including a presentation of the quality aspect helps to enhance the understanding of the characteristics and usefulness of the geodatasets and thus allows a judgment of the descriptive text in relation to the geospatial information.

To conclude, geodata quality has been visualized before. Our research, however, provides the opportunity to visualize data quality for a substantial collection of geodatasets of very different origin, data type, and quality. The geodata quality judgment carried out in combination with the actual geodata gives a thorough example of real use visualization in which the geodata quality is not spatially visualized but instead six distinct quality parameters are differentiated. The concept gives special consideration to the temporal aspect of the geodata, which covers a period of 110 years. As such, it is particularly useful for describing and visualizing the quality of geodata collections that include a historical dimension and is readily transferable to such data pools.

REFERENCES

Buttenfield, B. and M. K. Beard 1994. Graphical and geographical components of data quality. In: Hearnshaw, H. M. and D. J. Unwin, Eds., *Visualization in Geographical Information Systems*, John Wiley & Sons, Chichester, pp. 150–157.

Chapin III, F. S., E. S. Zavaleta, V. T. Eviner, R. L. Naylor, P. M. Vitousek, H. L. Reynolds, D. U. Hooper, S. Lavorel, O. E. Sala, S. E. Hobbie, M. C. Mack, and S. Diaz 2000. Consequences of changing biodiversity. *Nature*, 405, pp. 234–242.

Comber, A. J., P. F. Fisher, F. Harvey, M. Gahegan, and R. Wadsworth 2006. Using metadata to link uncertainty and data quality assessment. In: Riedl, A., W. Kainz, and G. Elmes, Eds., *Progress in Spatial Data Handling, 12th International Symposium on Spatial Data Handling*, Springer-Verlag, Berlin and Heidelberg, pp. 279–292.

Devillers, R., Y. Bédard, and R. Jeansoulin 2005. Multidimensional management of geo-spatial data quality information for its dynamic use within geographical information systems. *Photogrammetric Engineering and Remote Sensing*, 71(2), pp. 205–215.

Fisher, P. 1994. Animation and sound for the visualization of uncertain spatial information. In: Hearnshaw, H. M. and D. J. Unwin, Eds., *Visualization in Geographical Information Systems*, John Wiley & Sons, Chichester, pp. 181–185.

Huth, K. 2007. *Entwicklung eines prototypischen, SVG-basierten Tools zur Visualizierung von Geodaten für das Waldgebiet Kakamega-Nandi in Westkenia unter besonderer Berücksichtigung ihrer Qualität*. Unpublished Diploma thesis, Studiengang Kartographie und Geomatik, Hochschule Karlsruhe–Technik und Wirtschaft.

Huth, K., O. Schnabel, and G. Schaab 2007. SVG-based visualization of geodata quality. Taking the Kakamega-Nandi forest area as an example. In: *Proceedings of the XXIII International Cartographic Conference* 2007, 4–10 August 2007, Moskow, Russia.

Köhler, J. (2004) Was hat Biodiversitätsforschung mit, nachhaltiger Nutzung' zu tun? *Tier und Museum*, 8(3), pp. 82–91.

Longley, P. A., M. F. Goodchild, D. J. Maguire, and D. W. Rhind 2001. *Geographic Information Systems and Science*. John Wiley & Sons, Chichester.

Lung, T. 2004. Landbedeckungsänderungen im Gebiet "Kakamega Forest und assoziierte Waldgebiete" (Westkenia)—Multispektrale Klassifikation von Landsat-Satelliten-bilddaten und Auswertung mittels Methoden im Raster-GIS. *Karlsruher Geowissen-schaftliche Schriften*, A 15, G. Schaab, Ed.

Lung, T. and G. Schaab 2004. Change-detection in western Kenya: The documentation of fragmentation and disturbance for Kakamega Forest and associated forest areas by means of remotely-sensed imagery. In: *ISPRS Archives Vol. XXXV Part B* (DVD), Proceedings of the ISPRS XXth Congress, 12–23 July 2004, Istanbul, Turkey.

MacEachren, A. M. 1995. *How Maps Work. Representation, Visualization, and Design*, Guilford Press, New York.

MacEachren, A. M., D. Howard, D. Askov, T. Taormino, and M. von Wyss 1996. Reliability visualization system (RVIS). http://www.geovista.psu.edu/publications/RVIS (accessed Dec. 10, 2006).

Mitchell, N. 2004. The exploitation and disturbance history of Kakamega Forest, Western Kenya. *Bielefelder Ökologische Beiträge,* 20, BIOTA Report 1, B. Bleher, and H. Dalitz, Eds.

Mitchell, N., T. Lung, and G. Schaab 2006. Tracing significant losses and limited gains in for-est cover for the Kakamega-Nandi complex in western Kenya across 90 years by use of satellite imagery, aerial photography and maps. In: Kerle, N. and A. K. Skidmore, Eds., *Proceedings of ISPRS (TC7) Mid-Term Symposium "Remote Sensing: From Pixels to Processes,"* 8–11 May 2006, ITC Enschede, The Netherlands.

Mitchell, N. and G. Schaab in print. Developing a disturbance index for five East African forests using GIS to analyse historical forest use as an important driver of current land use/cover. In: *African Journal of Ecology*.

Mitchell, N. and G. Schaab 2006. Assessing long-term forest cover change in East Africa by means of a geographic information system. In: *Hochschule Karlsruhe–Technik und Wirtschaft, Forschung aktuell* 2006, pp. 48–52.

Navratil, G. 2006. Data quality for spatial planning—An ontological view. In: Schrenk, M., Ed., *Sustainable Solutions for the Information Society, Proceedings of the 11th Inter-national Conference on Urban Planning and Spatial Development in the Information Society (CORP)*, 13–16 February 2006, Vienna, Austria, pp. 99–105.

Sala, O. E. and T. Chapin 2000. Scenarios of global biodiversity for year 2100. *GCTE News. Newletter of the Global Change and Terrestrial Ecosystems Core Project of IGBP*, 16, pp. 1–3.

Schaab, G., T. Kraus, and G. Strunz 2004. GIS and remote sensing activities as an integrating link within the BIOTA-East Africa project. In: *Sustainable Use and Conservation of Biological Diversity—A Challenge for Society. Proceedings of the International Symposium Berlin*, 1–4 December 2003, Berlin, Germany, pp. 161–168.

Schaab, G., T. Lung, and N. Mitchell 2005. Land use/cover change analyses based on remotely-sensed imagery and old maps as means to document fragmentation and disturbance for East-African rainforests over the last ca. 100 years. In: *Proceedings of the International Cartographic Conference* 2005, 9–16 July 2005, A Coruña, Spain.

Slocum, T. A. 1999. *Thematic Cartography and Visualization*. Prentice-Hall, Upper Saddle River, NJ.

Van der Wel, F. J. M., R. M. Hootsmans, and F. Ormeling 1994. Visualization of data quality. In: MacEachren, A. M., and D. R. F. Taylor, Eds., *Visualization in Modern Cartography*, Serie Modern Cartography, Vol. 2, Elsevier Science Ltd., Oxford, pp. 313–331.

Zhu, A.-X. 2005. Research issues on uncertainty in geographic data and GIS-based analysis. In: McMaster, R. B., and E. L. Usery, Eds., *A Research Agenda for Geographic Information Science*, CRC Press, Boca Raton and London, pp. 197–223.

23 A Study on the Impact of Scale-Dependent Factors on the Classification of Landcover Maps

Alex M. Lechner, Simon D. Jones, and Sarah A. Bekessy

CONTENTS

23.1 INTRODUCTION

23.1.1 SCALE-DEPENDENT FACTORS

Scale-dependent factors such as pixel size, study extent, and the application of smoothing filters affect the classification of landcover. These factors are dependent on the remote sensing data, classification techniques, and class description used. Landcover maps will vary in their extent, patchiness, and accuracy of classified areas based on the relationships between these factors. Many studies have investigated these factors using empirical data and have come to conclusions on the basis of unique case studies investigating one factor in isolation (Hsieh et al., 2001). This study holistically investigates the impact different scale-dependent factors had on the classification of landcover maps to better understand their interactions and their relative importance.

In many studies, data are collected at the most appropriate scale; however, for studies using remote sensing data, users are often limited to specific scales available. The most appropriate scale for a study is a function of the environment (its spatial arrangement), the kind of information that is to be derived, and the classification technique used (Woodcock and Strahler, 1987). Numerous combinations of these factors are possible and their effects are usually interrelated and scale dependent.

At different spatial scales, landscape composition and configuration will change. Area and spatial pattern will change when spatial-dependent factors such as grain and/or extents are altered (Wiens, 1989). Unfortunately, knowledge of how these spatial patterns change is limited (Wu et al., 2002).

The primary aim of this project is to investigate the relationship between scale-dependent factors and landscape pattern, as measured by total area and landscape metrics in the context of vegetation extent mapping. The project is not aimed at solving the problem of uncertainty in spatial dependent factors, but rather attempts to quantify its nature. Although the development of an integrated model is not new to the field of remote sensing (e.g., Ju et al., 2005; Hsieh et al., 2001), many previous studies have investigated scale-dependent factors and reached conclusions on the basis of site-specific evidence, without considering the interactions between these various factors (Hsieh et al., 2001). This chapter aims to provide greater understanding of how these factors interact and to examine their relative importance.

Interactions between scale-dependent factors were investigated from the user's perspective through examining a number of landscape metrics. These metrics were chosen because they are simple and they summarize important patch characteristics. They have straightforward practical uses such as the measurement of total area and mean distance between patches rather than purely characterizing fragmentation such as the fractal dimension index.

This study is novel in that it uses real landscapes with a large study area and sample size. The majority of previous studies have either used simulated landscapes (e.g., Li et al., 2005) or real landscapes with small study areas and sample size (e.g., De Clercq et al., 2006; Wu et al., 2002).

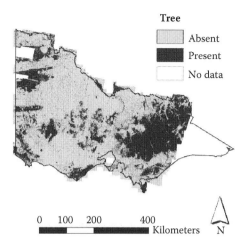

FIGURE 23.1 Map of the study area and Tree25; tree presence/absence data set overlaid.

23.1.2 LANDCOVER MAPS

This study utilizes the Tree25 presence/absence tree cover data set produced for the Department of Sustainability and Environment's Corporate Geospatial Data library (DSE, 2006) (Figure 23.1). This data set is typical of woody/nonwoody vegetation data layers used worldwide in land-use planning and habitat mapping.

Although the uses of this data set are varied, its initial purpose was to provide a comprehensive and consistent data set for tree cover monitoring for the state of Victoria (Australia). Furthermore, it is expected to provide an excellent source of data for applications that require the identification of remnant tree cover, such as connectivity analysis and habitat modeling (DSE, 2006).

23.1.3 CHANGING SCALE-DEPENDENT FACTORS

Pixel size (or spatial resolution) and extent were manipulated, and a smoothing filter was used to examine the differences in classification. All variables were manipulated to simulate a range of conditions and determine how patchiness and patch area changed accordingly.

Pixel size is an important variable to investigate as using the default pixel size (i.e., sensor resolution) will result in a view of the world that relates to the sensor but may not necessarily be relevant to the question being asked (Fassnacht et al., 2006). Pixel size is one of the most important elements determining how other scaling factors will change. Pixel size controls the limit of the smallest feature that can be extracted from an image. For areas in which vegetation is highly fragmented, such as urban areas and where patches appear as small as median strips and backyards, Jensen and Cowen (1999) concluded that at least 0.5 to 10 m spatial resolution is required. Resolution was altered to simulate differing sensor resolutions by degrading the original classified image.

The second factor investigated was the use of a smoothing filter. Pixel-based landscape classification can result in a salt-and-pepper effect because spatial autocorrelation is not incorporated in the classification technique (Ivits and Koch, 2002). A common practice used in remote sensing is smoothing the image by aggregating pixels to reduce classification error caused by this effect. The use of a smoothing filter will often result in the removal of edge complexity as well as an increase of the minimum mappable unit (MMU). The MMU tends to be larger than the pixel size, so that spatial and/or content information may be lost (Fassnacht et al., 2006). Larger MMUs may result in patches of interest being falsely combined within adjacent patches (Fassnacht et al., 2006). For this study, the smoothing algorithm used was a majority filter. However, other filters can be used for the similar purposes, such as mean or low-pass filters.

The final variable investigated was extent, which is the total physical area covered by the data source. As the extent increases, so does the probability of sampling rare classes (Wiens, 1989). Furthermore, if grain size is fixed, fragmentation increases with increasing extent (Riitters et al., 2000). The effect of extent was investigated by comparing many landscape samples at different extents.

Landscape metrics were used to analyze the effects on landcover classification of varying pixel sizes, applying smoothing filters, and changing extents. These metrics were chosen because they describe simple patch characteristics that users of the Tree25 data layer in Victoria often utilize. Users of landcover maps need a practical understanding of how scale-dependent factors affect classification. For example, in the region of Victoria it is important to measure correctly the area of native vegetation, as a permit is required to remove, destroy, or modify native vegetation from a landholding greater than 0.4 hectares (Cripps et al., 1999). Understanding the landscape metric "mean patch area" is therefore critical when assessing the suitability of a particular landcover map for this purpose. Another example is to understand how the mean distance between patches changes as a result of altering scale-dependent factors. An understanding of distance between patches is useful for population modelers to calculate the probability of dispersal between populations based on this distance (e.g., RAMAS [Akcakaya, 2002]).

23.1.4 DATA

The study area encompasses most of the state of Victoria, which is approximately 227,416 km². The study area is dominated by broad acre cropping and crop pasture, vegetation and dryland pasture (Figure 23.2). There are a variety of abiotic and biotic processes occurring at multiple scales, resulting in a complex landscape composition and configuration.

Comparison of the effects of scale between landscapes as well as within landscapes is important as the relationship between spatial patterns and scale may not be linear. Each landscape will vary with respect to the different processes operating at various scales (Wu et al., 2002). For example, disturbance can operate at many different scales from housing development to large fires to tree falls. Simulating landscapes at different scales that concurrently reflect reality is likely to be very difficult.

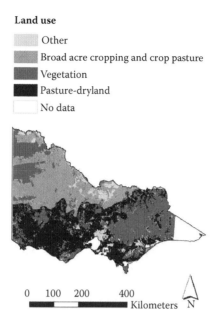

Land use

Other

Broad acre cropping and crop pasture

Vegetation

Pasture-dryland

No data

FIGURE 23.2   Map of land use in the study area.

Numerous studies have investigated scaling effects, but most of these studies have been confined to a few metrics or cover a narrow range of scales (Wu et al., 2002). Studies that have a large sample size tend to use simulated landscapes (e.g., Li et al., 2005). Real landscapes are used in this study, as opposed to computer-generated simulated landscapes, such as the those created by software programs such as Rule (Gardner, 1999) and SimMap (Saura and Martínez-Millán, 2000). Although simulated landscapes are a useful tool in terms of overcoming the impracticalities of replicating landscape scales, commentators such as Li et al. (2004) have suggested that simulated landscape models are insufficient in their ability to capture in detail the characteristics of real landscapes. This study is unusual in that the large study area allows for multiple replications at the landscape level of real landscapes.

23.2   METHOD

23.2.1   Data

The original classified data were derived from SPOT panchromatic imagery with a 10 m pixel size through a combination of digital classification and visual interpretation (DSE, 2006). No smoothing or filtering was applied at this layer-creation stage. Tree cover is defined by the producers of the original data set as woody vegetation over 2 m with crown cover greater than 10%.

23.2.2   Postprocessing

The original data were postprocessesed to test the effect of resolution, extents, and applying a smoothing filter on classification. All processing was performed using

ArcGIS 9.1. The original image was first degraded to different pixel sizes. A filter was applied to the degraded images to smooth the image. Finally, each combination of filtered and degraded images was clipped to different extents.

23.2.3 PIXEL SIZE

Pixel size was changed by degrading the original image through an interpolation technique based on a nearest-neighbor assignment using the center pixel of the original image. This technique is particularly suitable for postprocessing of discrete data as it will not change the values of the cells (ESRI, 2007).

The original image was degraded from 10 m to 100 m at 10 m increments. In this chapter, a decrease in resolution is analogous to an increase in pixel size and vice versa.

23.2.4 SMOOTHING FILTER

A majority filter was used to smooth the image. The majority filter replaces cells in a raster based on the majority of their contiguous neighboring cells. The majority filter process has two criteria to fulfill before a replacement occurs. The number of neighboring cells of a similar value must be in a majority, and these cells must be contiguous around the center of the filter kernel (ESRI, 2007). A 3 × 3 kernel was used for this process. A majority filter is useful for postprocessing as it works with discrete data.

23.2.5 EXTENTS

Subsets of this image were randomly clipped at 3,000, 10,000, and 20,000 m replicating landscapes of different extents (Figure 23.3). The extents represent the distance

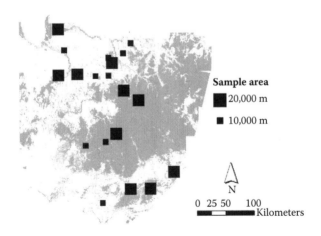

FIGURE 23.3 Clipped areas for western portion (50% of total area) of study area for extents 10,000 and 20,000 m.

of a single side of a square. The image was clipped so that no replicant overlapped. Twenty samples were taken for each combination of smoothed image, extents, and resolution with a total sample size of 600.

The lower bounds of the sampling size was set at 3 km as suggested by Forman and Godron (1986), although it is recognized that in principle landscape size is related to the scale at which an organism perceives their environment. The upper limit was based on the approximate area of a small catchment, at around 20 km. Furthermore, as the extents were increased beyond this amount, computer processing time increased markedly.

23.2.6 CALCULATING LANDSCAPE METRICS

Area was calculated based on pixels classified as either tree present or absent as identified by ArcGIS. Landscape metrics were then calculated using the Fragstats package (McGarigal et al., 2002). Five landscape metrics were used: patch number, mean patch area, mean patch density, mean nearest-neighbor distance, and mean perimeter-to-area ratio.

23.3 RESULTS

The total classified area remained relatively constant when the image spatial resolution changed. However, large differences in the patchiness of the image occurred as a result of altering the resolution and applying a smoothing filter. As image spatial resolution decreased (i.e., pixel size increased) or the smoothing filter was applied the subtle levels of patchiness declined. Small patches either aggregated into larger patches or completely disappeared (Figure 23.4). Although most measures of patchiness appeared to be nonrandom in relation to the spatial dependent factors, this was not uniformly the case. For most metrics used, it was impossible to test the effects of changing extent owing to the low sample size and high variability.

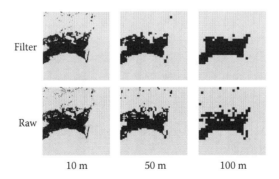

FIGURE 23.4 Example of processing. The original (raw) image at 10 m spatial resolution was degraded up to 100 m. For each degraded image, a majority filter was used to smooth the image.

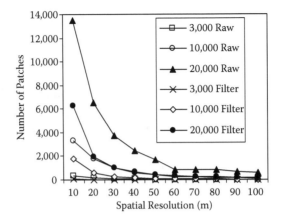

FIGURE 23.5 Comparison of the effect of changing extents, spatial resolution, and applying a smoothing filter on the number of patches.

23.3.1 MEAN NUMBER OF PATCHES

It was found that the greater the extent, the greater the mean number of patches, and the lower the spatial resolution, the lower the number of patches identified (Figure 23.5). Additionally, using the smoothing filter also resulted in a lower number of patches.

23.3.2 MEAN PATCH AREA

The relationship between mean patch area and the spatially dependent factors was the opposite to mean number of patches. Decreasing the spatial resolution and the application of the smoothing filter resulted in an increase in mean patch area (Figure 23.6a). The mean number of patches changed as a result of changing the

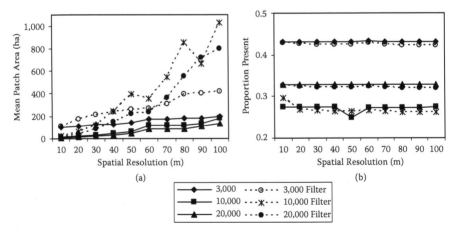

FIGURE 23.6 Mean patch area in hectares: (a) raw data and (b) data smoothed with a majority filter.

spatial resolution; however, the total area classified as tree or nontree remained constant (Figure 23.6b). Owing to the high standard error resulting from the small sample size ($n = 20$), a comparison between extents could not be conducted. The differences between the value of proportion classified as present or absent for different extents is the result of high variability in the landscape. However, the filtered data tended to have a significantly ($P < 0.05$) lower proportion of cells classified as present for both 3,000 and 20,000 m extents.

The relationship between patch area and spatial resolution was not perfectly linear. The overall trend was to increase the mean patch area with decreasing spatial resolution and the application of the smoothing filter (Figure 23.6a). However, applying the majority filter at lower spatial resolutions resulted in a greater increase in the mean patch area than at higher spatial resolutions. For 3,000 m extents, there was an increase in the mean patch area of 5% at 10 m spatial resolution compared to 115% at 100 m spatial resolution. For 20,000 m extents there was an increase in the mean patch area of 93% at 10 m spatial resolution compared to 505% at 100 m spatial resolution.

23.3.3 MEAN PATCH DENSITY

Patch density was calculated as the number of patches in the landscape divided by the total landscape area. As the spatial resolution decreased, the mean patch density decreased for all extents (Figure 23.7). This decrease was quite dramatic. At 10 m spatial resolution, there was a decrease in the mean patch density from 38.4 to 1.6 at 100 m resolution for 3,000 m extents and from 33.7 to 1.4 for 20,000 m extents. The results of applying a filter had a similar effect as decreasing resolution, that is, decreasing mean patch density. However, applying the filter resulted in a greater decrease at lower spatial resolutions. For 3,000 m extents at 10 m spatial resolution, there was a decrease in the mean patch density of 53% compared to 71% at 100 m spatial resolution. For 20,000 m extents at 10 m spatial resolution, there was a decrease in the mean patch density of 53% compared to 78% at 100 m spatial

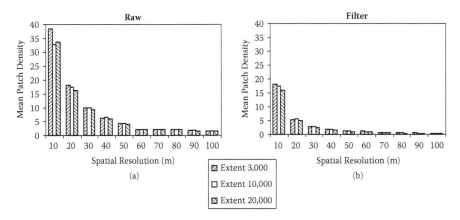

FIGURE 23.7 Mean patch density: number of patches in the landscape divided by total landscape area. (a) Raw data and (b) data smoothed with a majority filter.

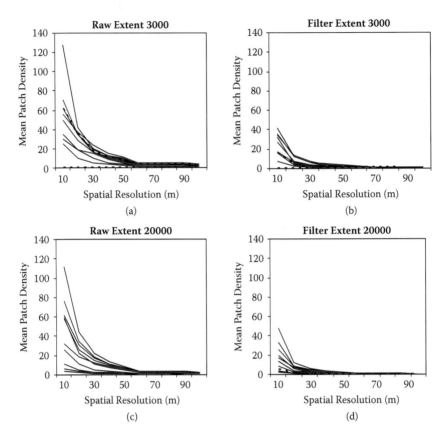

FIGURE 23.8 Mean patch density (number of patches in the landscape divided by total landscape) for 10 samples at extents 3,000 and 20,000 m for data before and after being smoothed with a majority filter.

resolution. Figure 23.8 shows the relationship between patch density and resolution for single samples compared to Figure 23.7, which shows the mean of all the samples. Figure 23.8 also shows that as spatial resolution decreases, patch density will predictably decrease. The relationship appears to fit an inverse exponential function.

23.3.4 ISOLATION AND PROXIMITY

Isolation and proximity were calculated using the nearest-neighborhood value based on the shortest edge-to-edge distance for a patch of the same type. As spatial resolution increased, the nearest-neighbor distance generally increased (Figure 23.9). However, of all the measures of patchiness, this appeared to be the least predictable. The variability appeared to be inconsistent and unrelated to spatial resolution. Furthermore, there appears to be no relationship between spatial resolution and the use of a smoothing filter (Figure 23.10).

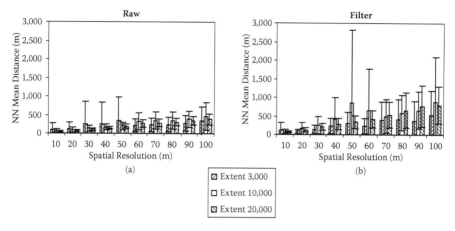

FIGURE 23.9 Mean Euclidian nearest-neighbor distance in meters. Error bars indicate standard deviation. (a) Raw data and (b) data smoothed with a majority filter.

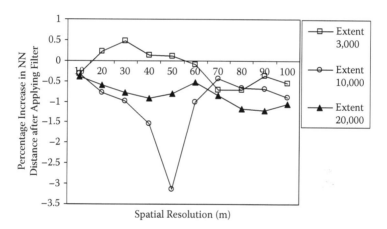

FIGURE 23.10 Percentage change in mean nearest-neighbor distance between patches after applying the majority filter to images from 10 to 100 m spatial resolutions.

23.3.5 Perimeter-to-Area Ratio

Perimeter-to-area ratio describes the relationship between shape and area. As spatial resolution increases, the ratio decreases (Figure 23.11). The mean perimeter-to-area ratio and spatial resolution is the inverse of patch area. By default, the mean perimeter-to-area ratio is strongly related to patch area. For example, if shape is held constant and patch size increased, there will be a decrease in the ratio. Applying the smoothing filter resulted in a predictable decrease in the mean perimeter-to-area ratio.

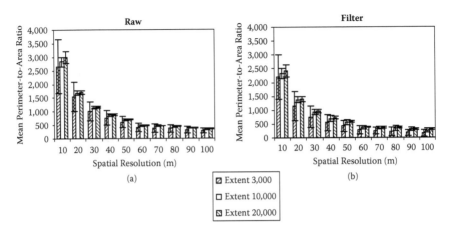

FIGURE 23.11 Mean perimeter-to-area ratio. Error bars indicate standard deviation for (a) raw data and (b) data smoothed with a majority filter.

23.4 DISCUSSION

This study clearly demonstrates that changes in scale-dependent factors affect the patchiness and total area of landcover maps classified. Although this study indicates that some relationships between factors were predictable, this was not always the case and not all metrics varied in the same way.

The effects of applying the smoothing filter are of particular interest. Applying the smoothing filter caused a greater increase in the mean patch area and greater decrease in the mean patch density at lower spatial resolutions. Furthermore, after applying the smoothing filter, significantly less area was classified as "tree present" at all extents and spatial resolutions compared to when the filter was not applied.

Owing to the small sample size and large variability, it was impractical to compare the effects of changing the study area extents. We would expect greater variability in smaller extents and that a larger extent will have a greater probability of containing all the variability within a landscape. Furthermore, if the sample size was increased, the mean of these samples should reflect the mean of the variability in the landscape. However, increasing the sample size or the sample area could be problematic as the area of real landscapes is finite.

23.5 CONCLUSION

The measurement of landscape pattern from landcover maps has become a common practice in various disciplines such as landscape ecology. However, many people are unaware of the scale dependency of this phenomena. This study demonstrates that the characterization of landscape patterns by landcover maps is the product of the interrelationship between a number of scale-dependent factors, such as spatial resolution, the application of smoothing filters, and the use of different study areas. Specifically, this study demonstrates that landcover maps will vary in terms of the extent and patchiness of classified areas on the basis of the interrelationship between these scale-dependent factors. For example, the effect of using a majority filter at low spatial resolutions will not be the same when used at high or low resolutions. Techniques that are used at one resolution are not necessarily transferable to different resolutions and may result in a very different classification. This has wide-ranging consequences for users transferring techniques used on medium-resolution imagery from sensors such as Landsat to high-resolution imagery from sensors such as IKONOS and Quickbird.

This study represents the first step in the development of a framework to quantify the magnitude of the effect of different spatial-dependent factors on landcover classification. This study demonstrated that there is considerable interaction between scale-dependent factors, indicating that investigations of spatial dependent factors need to be done simultaneously.

Future research is needed to assess the effect of these spatially dependent factors on accuracy as well as patchiness and area. Furthermore, as the landscape patterns found in the study area may be site specific, it is difficult to generalize to other areas. Thus, there is a need to perform the same spatial analysis for a wide range or spatial resolutions using different smoothing filters and extents in multiple real landscapes settings to create a significant volume of data. This will allow for wide-ranging generalizations to be made that will be the basis for the development of guidelines for map users.

ACKNOWLEDGMENTS

This work was supported by ARC Discovery Grant DP0450889 and by the Landscape Logic research hub. The authors would also like to acknowledge the help of Michael Conroy and John White at the Department of Sustainability and Environment and Bill Langford and Ascelin Gordon from Re-imagining the Australian Suburbs.

REFERENCES

Akcakaya, H. R. 2002. *RAMAS GIS: Linking Spatial Data with Population Viability Analysis*. New York: Applied Biomathematics.

Cripps, E., Binning, C., and Young, M. 1999. Opportunity denied: Review of the Legislative Ability of Local Government to Conserve Native Vegetation. Environment Australia, Canberra.

De Clercq, E. M., Vandemoortele, F., and De Wulf, R. R. 2006. A method for the selection of relevant pattern indices for monitoring of spatial forest cover pattern at a regional scale. *International Journal of Applied Earth Observation and Geoinformation* 8(2): 113–25.

DSE 2006. Corporate Geospatial Data Library, Melbourne, Australia. http://www.dse.vic .gov.au/dse/ (accessed August 7, 2006).

ESRI 2007. Arc GIS Desktop Help. http://webhelp.esri.com/arcgisdesktop/9.1/index.cfm? TopicName=welcome (accessed August 7, 2006).

Fassnacht, K. S., Cohen, W. B., and Spies, T. A. 2006. Key issues in making and using satellite-based maps in ecology: A primer. *Forest Ecology and Management* 222(1–3): 167–81.

Forman, R. T. T. and Godron, M. 1986. *Landscape Ecology*. New York: Wiley.

Gardner, R. H. 1999. RULE: Map generation and a spatial analysis program, in *Landscape Ecological Analysis: Issues and Applications*. New York: Springer.

Hsieh, P. F., Lee, L. C., and Chen, N. Y. 2001. Effect of spatial resolution on classification errors of pure and mixed pixels in remote sensing. *IEEE Transactions on Geoscience and Remote Sensing* 39(12): 2657–63.

Ivits, E. and Koch, B. 2002. Landscape connectivity studies on segmentation based classification and manual interpretation of remote sensing data. eCognition User Meeting. October 2002, München.

Jensen, J. R. and Cowen, D. C. 1999. Remote sensing of urban/suburb an infrastructure and socio-economic attributes. *Photogrammetric Engineering and Remote Sensing* 65(5): 611–22.

Ju, J., Gopal, S., and Kolaczyk, E. D. 2005. On the choice of spatial and categorical scale in remote sensing land cover classification. *Remote Sensing of Environment* 96(1): 62–77.

Li, X., He, H. S., Bu, R., Wen, Q., Chang, Y., Hu, Y., and Li, Y. 2005. The adequacy of different landscape metrics for various landscape patterns. *Pattern Recognition* 38(12): 2626–38.

Li, X., He, H. S., Wang, X., Bu, R., Hu, Y., and Chang, Y. 2004. Evaluating the effectiveness of neutral landscape models to represent a real landscape. *Landscape and Urban Planning* 69(1): 37–48.

McGarigal, K., Cushman, S. A., Neel, M. C., and Ene, E. 2002. FRAGSTATS: Spatial Pattern Analysis Program for categorical maps. Computer software program produced by the authors at the University of Massachusetts, Amherst. http://www.umass.edu/landeco/research/fragstats/fragstats.htm (accessed January 30, 2007).

Riitters, K., Wickham, J., O'Neill, R., Jones, B., and Smith, E. 2000. Global-scale patterns of forest fragmentation. *Conservation Ecology* 4(2). http://www.consecol.org/vol4/iss2/art3/. (accessed January 14, 2007).

Saura, S. and Martínez-Millán, J. 2000. Landscape patterns simulation with a modified random clusters method. *Landscape Ecology* 15(7): 661–78.

Wiens, J. A. 1989. Spatial scaling in ecology. *Functional ecology* 3(4): 385–97.

Woodcock, C. E. and Strahler, A. H. 1987. The factor of scale in remote-sensing. *Remote Sensing of Environment* 21(3): 311–32.

Wu, J. G., Shen, W. J., Sun, W. Z., and Tueller, P. T., 2002. Empirical patterns of the effects of changing scale on landscape metrics. *Landscape Ecology* 17(8): 761–82.

24 Formal Languages for Expressing Spatial Data Constraints and Implications for Reporting of Quality Metadata

Paul Watson

CONTENTS

24.1 INTRODUCTION

Data are typically collected for a specific use (Pira International Limited, 2008). Many aspects of the data are specific to the originating application, and these constraints act to limit the range of application of the data. The mere presence of particular data elements does not guarantee that they are fit for purpose in new applications. The interoperability of services that exchange and reuse spatial information is dependent on the interoperability of the underlying spatial information at both the syntactic and semantic levels. XML encoding languages such as Geographic Markup Language (GML) provide a good foundation for ensuring syntactic interoperability. However,

329

these syntactic constraints are not sufficient to ensure that the feature's meaning is correctly interpreted. It is necessary to describe the logical constraints within the domain in a formal way and test the features against these rules. Taken together, a set of constraints constitutes a logical model against which data consistency can be evaluated. Egenhofer (1997) thus defines consistency as a lack of any contradictions within a model of reality and points out that the model must therefore contain the necessary constraints among data elements in order to capture the intended semantics. Without these it is impossible to detect or assess inconsistencies. He identifies conveying consistency in relation to distributed data on the Internet as the fundamental need in the design of future spatial information systems.

Previous work has focused on performing an explicit linkage between disparate datasets (Haunert, 2005; Walter and Fritsch, 1999; Sester et al., 1998) that does not scale well to the very sparse connections between a large number of spatial datasets on the Internet. Conversely, work relating purely to the assessment of consistency (Sheeren et al., 2004; Egenhofer et al., 1994) has not previously considered the explicit communication of the results via metadata.

The ISO metadata standard for geographic information (ISO 19115:2003) develops a taxonomy of metadata elements relating to data quality. These elements address areas such as completeness (errors of commission or omission), logical consistency (e.g., topology rules), thematic accuracy (whether data elements are correctly classified), and positional and temporal accuracy. However, the standard does not address the precise contents of the metadata, specified simply as free text. This unstructured data model facilitates simple uses such as browsing. However, as the range of spatial information available online grows, it will become increasingly difficult to reliably locate relevant information and interpret its meaning correctly.

In this chapter, we define a set of requirements for such metadata content and describe the structure of a rules language that meets the requirements. The rules language is illustrated using examples and the integration of the quality rules with standard metadata descriptors is explained. Finally, an outline is given of a Web services implementation of this rules language and usage of the system is shown through a number of scenarios. Conclusions are drawn on the generality of the approach and the consequences of rigorous, semantically meaningful metadata for enriching data discovery capabilities and interoperability.

24.2 REQUIREMENTS

We begin by setting out the requirements of the rules language that will allow the logical constraints satisfied by a particular data source to be specified:

1. Unambiguous. Domain constraints must be expressed in a mathematically rigorous way. This allows the rules to be used as the basis of fair testing and means that the results are fully objective.
2. Logical and Portable. Keep a logical separation between the terms and definitions (ontology) of the feature application schema and the terms and definitions of the domain model to which the rules apply. Rules are genuinely abstract knowledge and should be decoupled from any particular

physical implementation of the instance data to allow reuse of the logical rules with any number of logically compliant feature sources.

3. Compact. The rules language should have a concise grammar. The number of concepts in the language should be minimized.

4. Intuitive. The language should be easy to learn. The transformation between constraints expressed using natural language and the formal rules language should be as simple as possible. In some cases, this may conflict with the requirement for simplest terms.

5. Quantitative. The language should support quantitative reasoning and formal metrics that summarize the level of compliance.

6. Web compatible. The language should be compatible with feature data that are scattered across multiple physical and organisational barriers. The principal implication of this requirement is that the rules language support namespaces.

7. Declarative and Refinable. The entire rules base is very rarely known completely at the start. New constraints are constantly being discovered. The rules language must make it simple to add and refine rules without disrupting the overall structure of the rules base.

24.3   CHOICE OF RULES LANGUAGE

Authoring and exploitation of abstract knowledge representations have received attention recently through the Semantic Web community (W3C, 2004b). A key objective of this initiative is to make the exchange of mathematically rigorous models of knowledge such as conceptual graphs possible. This has led to the development of the Web Ontology Language (OWL) (W3C, 2004a). This language has its foundations in Description Logic (Baader, 2002), and this has been used to formally classify the complexity of different sorts of logical expressions. OWL divides expressions into three kinds:

1. OWL Lite. This is a simple dialect suitable for expressing simple concepts and relationships.

2. OWL DL (Description Logic). This represents only concepts that are formally decidable (there exists a decision procedure whether a logic expression is true or false.)

3. OWL Full. This language permits a much richer range of expressions (e.g., concepts that may represent both instances and classes) that make the language formally undecideable (there exists no decision procedure).

Development to date has concentrated on OWL Lite and OWL DL because these are mathematically tractable and therefore general tools support is feasible. Some kinds of constraints, especially reasoning over relationships, are not supported using the concepts defined in OWL but can be expressed using a rules language layered on top of OWL. An early candidate draft of the Semantic Web Rules Language (SWRL) (W3C, 2004c) has been proposed for this purpose. However, SWRL contains built-in operators that address only basic XML schema data types and therefore have no support for geometric types. This makes SWRL unsuitable for the current purpose. Instead, a dedicated XML grammar based on first-order logic (predicate logic) was developed. It should be recognized that the approach here is conceptually similar to SWRL, with the addition of support for spatial operators.

The XML rules have a very simple vocabulary:

1. Predicate, an operator that returns either true or false
2. Constant
3. Variable—free or bound
4. Built-in function
5. Logical connective—NOT, AND, OR
6. Quantifier—universal, existential

24.3.1 PREDICATE TYPES

TABLE 24.1

Predicate Types

Relational Predicate
Exists Predicate
ForAll Predicate
Conditional Predicate
Referential Predicate
Range Predicate
And Predicate
Or Predicate
Not Predicate

See Table 24.1 for a list of predicate types. The simplest predicate type is the RelationalPredicate. It is used to check whether two Values (see below) have a defined relation. It consists of two Values, a LeftValue (Lvalue) a RightValue (Rvalue), and a comparison operator (Relation).

The ExistsPredicate is an existential quantifier. It contains a feature type, a numerical quantifier, a relation, and a child predicate. It allows expressions of the form, "There exist greater than 3 features of type B for which the following condition holds -> {child predicate}." This may be used to test for the existence or absence of features of a particular type, as in "For Lake features: There exist exactly zero forest features for which the forest geometry is contained within the lake geometry."

The ForAllPredicate is a universal quantifier. It contains a feature type and two child predicates. It allows expressions of the form, "For all features of type X which satisfy {first child condition}, verify that {second child condition} also holds true."

The ConditionalPredicate permits conditional evaluation of parts of a rule. It contains two child predicates. It allows expressions of the form, "If {first child condition} holds, then check that {second child condition} also holds."

The ReferentialPredicate tests whether a particular named association exists between two features. It contains two target feature types and an association name. It allows expressions of the form, "Check if there exists a relationship from {feature instance A} to {feature instance B} via the association {reference name}."

The RangePredicate tests whether a value lies in a range. It contains three Values and tests the first supplied Value to find whether it lies between the second and third supplied Values. It allows expression of the form, "Check whether {First Value} lies between {Second Value} and {Third Value}."

The logical predicates AndPredicate, OrPredicate, and NotPredicate allow for Boolean logic to be applied to any of the results returned by other predicate types. AndPredicate and OrPredicate take two child predicates and return the standard Boolean result. The NotPredicate logically inverts the sense of the child predicate result. Although, in a logic sense, these elements are connectives rather than predicates, they can be interpreted here as predicates because they are defined to be Boolean-valued operators that operate on the contained predicates, which are themselves Boolean-valued.

Note that the language is not minimal because it is possible to make alternative expressions (sometimes more complex) that are logically equivalent using, for example, existential and universal quantification and conditional expressions. These alternatives have been retained because of their more naturalistic mapping onto rules as expressed by subject matter experts, for whom the language is designed.

24.3.2   VALUE TYPES

See Table 24.2 for a list of value types. A StaticValue is a typed constant. Its value is assigned explicitly within the rule expression, and this value can then be used within other comparisons such as RelationalPredicates. An AssignableValue represents a variable in a rule expression that is one of two types—a DynamicValue is a typed attribute fetched from a feature instance, and a TemporaryValue is used to hold a derived result within a rule for comparison in a later clause.

A ConditionalValue is a value that may take one of two values, depending upon the truth of a child predicate. It contains two values and a predicate. If the predicate evaluates to true, the first value is returned else the second is returned.

An AggregateValue is used to return some Aggregated result (sum, average, concatenation, geometric union, etc.) from a number of features. It contains a feature type, a feature attribute name, an aggregation function, and a child predicate that holds true for the features to be aggregated. It allows expressions of the form, "For features of type {Type} which satisfy {Child Predicate}, compute and return the {Aggregation Function} from the attributes {Attribute Name}."

A BuiltinFnValue is used to derive one value from another using a specified algorithm. It contains a value and an algorithm name. A variety of algorithms are supported varying by the data type of the value supplied, including simple mathematical and string manipulation functions as well as geometric algorithms such as convex hull, buffer, or Douglas Peucker simplification. This functionality can be used, for example, to test whether a feature lies within a specified buffer of the geometry of another feature.

A ClassValue returns the class name of a feature.

The final set of Value types are simple arithmetic convenience types having the conventional meanings.

TABLE 24.2
Value Types

Static Value
Dynamic Value
Temporary Value
Conditional Value
Aggregate Value
Built-in Function Value
Class Value
Summed Value
Difference Value
Product Value
Quotient Value
Modulus Value
Negated Value

24.3.3   RELATION TYPES

See Table 24.3 for a list of relation types. Relation types are gathered into two groups, ScalarRelation and SpatialRelation. ScalarRelation specifies a relationship test between

TABLE 24.3
Relation Types

Scalar	Spatial
Equals Relation	Spatial Equals Relation
NotEquals Relation	Spatial Disjoint Relation
Less Relation	Spatial Intersects Relation
LessEquals Relation	Spatial Touches Relation
Greater Relation	Spatial Overlaps Relation
GreaterEquals Relation	Spatial Crosses Relation
Begins Relation	Spatial Within Relation
Ends Relation	Spatial Contains Relation
RegExp Relation	Spatial Within Distance Relation

two scalar values of an appropriate type. Numerical relationships have the conventional meanings. Character string relationships test whether a character string value begins or ends with the supplied fragment or tests whether a character string value matches a supplied fragment according to a PERL-compatible regular expression.

SpatialRelation types correspond to the ISO/OGC Simple Feature specification spatial interaction types (ISO 19125-2:2004) and take those meanings. In addition to the topological interaction types, SpatialWithinDistanceRelation can be used to test whether two geometries approach within a user-specified distance.

24.4 EXAMPLES

The first example represents the simplest spatial consistency test possible. It states the single constraint that, in most cases, the presence of forest within water areas is inconsistent. Therefore, forest features should be tested to ensure that their geometry does not intersect the geometry of any water body features. The illegal forest features can be depicted graphically, as shown in Figure 24.1.

The constraint might be expressed in prose as follows:

Check for **Coniferous Forest** objects that there are no **Water Area** objects for which **Coniferous Forest**.geometry overlaps **Water Area**.geometry.

The rule can be visualized using a predicate tree structure as shown in Figure 24.2.

FIGURE 24.1   Illegal coniferous forest-water area relationship.

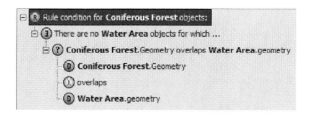

FIGURE 24.2 Predicate tree structure for coniferous forest–water area consistency rule.

This tree shows that the main rule structure is an ExistentialPredicate testing for the existence of Water Area features that meet a particular RelationalPredicate. The RelationalPredicate tests candidate Water Area features to check whether their geometries overlap the Coniferous Forest feature currently under test.

This predicate tree corresponds very closely with the XML serialization of this rule:

```
<?xml version="1.0"?>
<Rule>
 <RootPredicate classLabel="Coniferous Forest">
  <ExistsPredicate qualifier="exactly" n="0" classLabel="Water Area">
   <RelationalPredicate>
    <DynamicValueclassRef="Coniferous Forest" propName="geometry"/>
    <SpatialOverlapsRelation/>
    <DynamicValue classRef="Water Area" propName="geometry"/>
   </RelationalPredicate>
  </ExistsPredicate>
 </RootPredicate>
</Rule>
```

The target feature types appear as the classLabel and classRef attributes of the appropriate predicates and values, and the feature property names for each Dynamic-Value are given in the Value propName attribute.

In this second example, we show that some complex and powerful expressions may be constructed from the relatively simple building blocks of Predicates, Values, and Functions. This rule tests that the shoreline of Island features matches the corresponding limits of all of the Water Areas that border the Island.

We can portray the correct relationship between Island and Water Area as shown in Figure 24.3.

The Island at the center of the picture is surrounded by a number of Water Area features. The derived shoreline of the Island is highlighted as is the derived set of Water Area features that abut the island. The rule can be expressed as:

Check for **Island** objects that outer_ring(**Island**.geometry) equals intersection (**Island**.geometry,union(**Water Area**.geometry) over all **Water Area** objects for which (**Water Area**.geometry touches **Island**.geometry)).

The corresponding predicate tree looks like that shown in Figure 24.4.

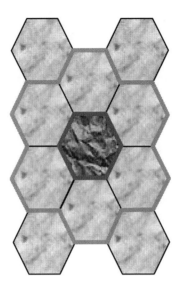

FIGURE 24.3   Island water area consistency rule.

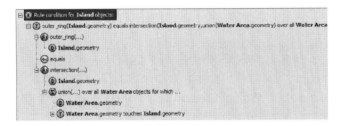

FIGURE 24.4   Predicate tree for island water area consistency rule.

This tree is a simple RelationalPredicate that compares a BuiltinFunctionValue (outer_ring) with another BuiltinFunctionValue (geometric intersection), which in turn nests an AggregateValue (geometric union over Water Areas touching the Island) and tests them for (geometric) equality. The resulting tree is very compact for such a sophisticated expression.

Once again, the XML encoding closely mirrors the predicate tree structure:

```
<?xml version="1.0"?>
<Rule>
 <RootPredicate classLabel="Island">
  <RelationalPredicate>
   <BuiltinFnValue fnName="outer_ring">
    <DynamicValueclassRef="Island" propName="geometry"/>
   </BuiltinFnValue>
   <SpatialEqualsRelation/>
   <BuiltinFnValue fnName="intersection">
    <DynamicValue classRef="Island" propName="geometry"/>
```

```
  <AggregateValue fnName="union" classLabel="Water Area">
   <DynamicValue classRef="Water Area" propName="geometry"/>
   <RelationalPredicate>
    <DynamicValue classRef="Water Area" propName="geometry"/>
    <SpatialTouchesRelation/>
    <DynamicValue classRef="Island" propName="geometry"/>
   </RelationalPredicate>
  </AggregateValue>
  </BuiltinFnValue>
 </RelationalPredicate>
 </RootPredicate>
</Rule>
```

An XSLT stylesheet allows for rendering this rule into pseudo-prose.

A further advantage of the strict hierarchical structure is that it is simple to parse the rule to determine its validity and feedback any syntactic inconsistencies (e.g., values out of scope) in the rule to the user.

24.5 METADATA PUBLICATION

The final results of the conformance tests are obtained in the form of metadata that are compliant to the conceptual model of ISO 19115 Metadata and encoded in the form recommended in ISO 19139:2007. The metadata may reflect either a single logical constraint or an extended set of related domain constraints collected together that are uniquely identified. In the latter case, the metadata give quantitative information as to the overall consistency of the data with respect to the logical model expressed in the complete set of constraints. The results are supplied within the DQ_ DataQuality metadata element as DQ_Element descriptors. The nameofMeasure and measureIdentification are taken from the corresponding rule or ruleset identifier. The dateTime is taken from the completion time of the conformance check, and the results (DQ_Result) are compiled from the appropriate summary statistics within the conformance checking session. The metadata can be published automatically to a compliant OGC Catalogue (OGC, 2005) for long-term archiving and to facilitate discovery of data with appropriate quality characteristics.

24.6 IMPLEMENTATION

We have made an online rules engine, Radius Studio, capable of evaluating logical rules expressed using the language on multiple sources of vector feature data from disparate locations.

The server has been implemented as a number of stateful Web services that have been exposed using a standardized SOAP binding (W3C, 2003).

A thin, Javascript, browser-based client was written to facilitate interaction with each of the service components.

A data store is an external repository for data that acts as an abstraction over feature services including OGC Web Feature Service (OGC, 2004). The user selects data by specifying the feature types and attributes of interest and a spatial extent. An optional schema mapping can be applied between the schema of a data store and the internal rules schema used by the server. This permits data from different stores to be compared using a common set of terminology for any given domain.

It is possible to define several data stores that access the same data through different schema mappings. It is possible to read data from several data stores into the service for processing against a set of rules that analyze relationships between datasets as well as within a dataset. It is also possible to input from one store and output to a different store. The service supports externally defined ontologies. This is achieved by interfacing with the open-source Jena ontology library (see http://jena. sourceforge.net/), allowing ontologies in various formats such as RDF and OWL to be read into the service and used for rules authoring and rules-based reasoning.

A rule is a logical expression that can be used to test the logical consistency of a feature. The rule is expressed using an XML encoding of first-order logic. A dedicated service allows rule expressions, along with suitable metadata, to be stored and their definitions retrieved and used within conformance checking and data reconciliation tasks. The rules are initially abstract representations not bound to any physical data model. There is a binding step which takes place at rule assertion time that makes an association between terms used within the rule and elements within the physical data model. This binding is part of the Datastore definition, not the rule, and this ensures that rules can be reused meaningfully in many data contexts. Rules can be collected together arbitrarily in user-defined Folders that can be used for rule assertion. Rules can appear in any number of logical Folders. Folders of Rules define sets of constraints that constitute an abstract logical model against which data consistency may be measured.

The SessionManager service allows the definition of an ordered sequence of tasks to process data. The service also manages the execution of these sequences against feature instance data and storage and retrieval of the resultant metadata. The task types are

- *Open Data*, which enables access to data from a defined data store. A session may choose to open data from a number of data sources and then check rules based on the relationships between the features stored in the different locations
- *Discover Rules*, which analyze data based on a defined discovery specification to identify candidate rules
- *Check Rules*, which checks a defined set of rules on the data and reports nonconformances
- *Apply Action*, which applies one or more actions to the data, which are encoded similarly to rules
- *Apply Action Map*, which checks a set of rules defined in an action map and applies the associated action to each nonconforming object
- *Commit Data*, which will incrementally commit any data changes back to the data store it came from. Typically, this occurs after a correcting action has been applied by an action or action map

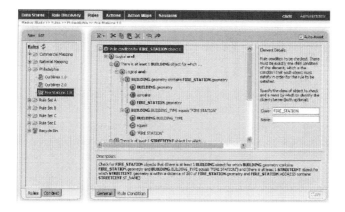

FIGURE 24.5 Rules browser form.

- *Copy to…*, which will copy data to a different data store
- *Pause*, which requests the service to suspend processing to allow results to be reviewed.

A session can be viewed as a specialized workflow template for data quality testing and reconciliation. The template can be stored, retrieved, and modified as necessary to work against different sources of feature data or to incorporate new rule definitions or actions.

The Rule Builder (Figure 24.5) allows the definition of complex rules with an easy-to-use, tree structured browser interface. The rule is built up using pulldown menus from the bar immediately above the graphical illustration of the rule.

The description at the bottom provides English text representing the currently selected clause. The element details are used to specify the parameters associated with the currently selected rule. An optional name label is used when the rule needs to distinguish between two different features of the same class. While editing a rule, it may temporarily be incomplete until a new clause is added or parameters are defined. These problems are highlighted clearly in red and a description of what is required displayed. Multilevel undo/redo is available to recover from mistakes while editing. Drag and drop can be used to re-order clauses of a rule. Cut and paste can be used to transfer all or part of a rule into another rule.

24.7 RESULTS

This section describes the use of the SOAP Web Services interface to validate features remotely. The Web Services will be used to first define a sequence of data processing tasks called a session. The session is then run and rules are asserted against the data. Progress is monitored and, finally, the results of the conformance test at both a summary and individual feature level are obtained.

The features are a pair of data layers (MOORLAND and LFA – Less Favoured Area) from different feature services with a geometric containment constraint between the feature instances. The rule is: There is exactly one LFA object for which MOORLAND.geometry is contained within LFA.geometry. The scenario will use just one of the service endpoints—the SessionManager. All other objects, data stores, and rules have been defined before the session commences and are created similarly.

The first task is to create a session in which the validation rules will be checked. This session consists of a number of tasks: a connection must be established to the relevant data source(s), and a set of rules will be indicated for checking. The SOAP message to create the session object within the system is as follows:

```
<soap:Envelope
xmlns:soap="http://schemas.xmlsoap.org/soap/envelope/"
xmlns:ns="http://laserscan.com/RadiusStudio">
 <soap:Body>
  <ns:create>
   <DTO_1>
    <id   xmlns:ns1="http://www.w3.org/2001/XMLSchema-instance"
ns1:nil="true"/>
    <metadataXml><![CDATA[<Metadata>
<Name value="Test Session"/>
<Description value="TQAS Web Service"/>
<Comments value="Deleted after use."/>
</Metadata>]]>
   </metadataXml>
   <name>Test Session</name>
   <parentId>0ac8dd7dc0a85abd01bd967519e60748</parentId>
   <referencesXml
xmlns:ns2="http://www.w3.org/2001/XMLSchema-instance" ns2:nil="true"/>
   <sequence>1</sequence>
    <xml><![CDATA[<Session>
<Sequence>
<Task label="1" type="Open Data">
<DataStoreRef ref_id="0ac3a94ac0a85abd0181f53b7fc2e033"/>
</Task>
<Task label="2" type="Open Data">
<DataStoreRef ref_id="0ac98ebdc0a85abd019732afc38ace6e"/>
</Task>
<Task label="3" type="Check Rules">
<RuleRef ref_id="1a714a8ec0a85abd01dafa55d327e7be"/>
</Task>
</Sequence>
</Session>]]>
     </xml>
    </DTO_1>
   </ns:create>
```

```
  </soap:Body>
</soap:Envelope>
```

The request has two sections: the metadata XML and definition of the sequence of tasks itself. This example states that the system should open and read data from two data sources (each known by a unique identifier) as indicated by the Tasks with type "Open Data" and that is followed by conformance checking a single rule (the "Check Rules" task, element) again known by its unique identifier. The Check Rules task can also refer to a folder (logical collection of rules) but the syntax is identical. The Datastore and Rule objects referenced are defined by their create() method, and may be queried and retrieved by invoking the get() and getByName() methods on the appropriate DatastoreManager or RuleManager service.

The response is in effect a copy of the input embellished with further metadata and internal references and crucially an identifier in the id element that can be used to refer back to this session definition. Once the session has been defined, it can be run using the run() method:

```
<soap:Envelope
xmlns:soap="http://schemas.xmlsoap.org/soap/envelope/"
xmlns:ns="http://laserscan.com/RadiusStudio">
  <soap:Body>
   <ns:run>
    <String_1>1a7331fbc0a85abd006c7fcd01bc0b08</String_1>
   </ns:run>
  </soap:Body>
</soap:Envelope>
```

Progress can be monitored at any time using the getSessionProgress() method:

```
<soap:Envelope
xmlns:soap="http://schemas.xmlsoap.org/soap/envelope/"
xmlns:ns="http://laserscan.com/RadiusStudio">
  <soap:Body>
   <ns:getProgress>
    <String_1>1a7154bfc0a85abd006c7fcd67afd715</String_1>
   </ns:getProgress>
  </soap:Body>
</soap:Envelope>
```

The session passes through a number of states while executing, and the sequence number of the current working task and % complete are reported while the session continues to execute. Once the session has run to completion, it is said to be in the "finished" state as given in the status attribute of the Progress element. Use the getSessionResults() method to obtain the session results of any task within a session:

```
<soap:Envelope
xmlns:soap="http://schemas.xmlsoap.org/soap/envelope/"
xmlns:ns="http://laserscan.com/RadiusStudio">
 <soap:Body>
  <ns:getResults>
   <String_1>1a7154bfc0a85abd006c7fcd67afd715</String_1>
   <String_2>3</String_2>
   <int_3>1</int_3>
   <int_4>10</int_4>
  </ns:getResults>
 </soap:Body>
</soap:Envelope>
```

This method takes four parameters: the identifier of the session, the task within the session for which the results are required, the index of the first result required starting at 1, and the index of the final result required (0 for all results). The response contains summary and detailed, per-feature metadata on the conformance levels achieved. In the example, the third task from the session definition (Check Rules) was queried and the report details the first ten per-feature rule non-conformances. The following is an extract of the getSessionResults() response:

```
<?xml version="1.0" encoding="UTF-8"?>
<env:Envelope
xmlns:env="http://schemas.xmlsoap.org/soap/envelope/"
 <env:Body>
  <ns0:getResultsResponse>
   <result>
    <Results count="10" finished="1164379794439" first="1" last="10"
started="1164379791002">
    <Summary count="228" error="0" label="3" total="1355" type="Check
Rules">
        <Object class="MOORLAND" count="228" error="0" total="1355"/>
        <RuleRef count="228" error="0" ref_id="1a714a8ec0a85abd01dafa55d
327e7be" total="1355"/>
      </Summary>
      <Object class="MOORLAND">
      <RuleRef ref_id="1a714a8ec0a85abd01dafa55d327e7be">&amp;lt;font
color=red&amp;gt;&amp;lt;u&amp;gt;there is exactly 1 LFA object for
which MOORLAND.geometry is contained within LFA.geometry&amp;lt;/
u&amp;gt;&amp;lt;/font&amp;gt;</RuleRef>
      <Attribute name="ID">62.0</Attribute>
      <Attribute name="geometry">
      <MBR x0="304193.000139739" x1="361489.001589227" y0=
"482708.995242653" y1="527026.999345939"/>
      </Attribute>
      </Object>
...... truncated ......
```

The Results element indicates the total number of individual feature results contained in the report and the start and end timestamps as milliseconds since UTC origin. It has two major subelements: the Summary gives overall conformance levels for all rules together and for each rule individually, and any number of Object elements that contain detailed per-feature metadata (i.e., which features failed which rule checks). The Rule identifier and text are included in each Object element with any unique key attributes and the bounding box (envelope) of the checked feature geometry attribute. This can be used in manual reconciliation processes to locate and correct data.

The Summary can also be used to make decisions about data quality in the context of workflow. This is useful in monitoring service-level agreements that govern the exchange of information between partners in a distributed data supply chain. These may stipulate that data must exceed 95% conformance to a logical model as expressed in a ruleset. In the Summary we can see that the service indicates that a total of 1355 Moorland features were tested using the containment rule (RuleRef element). Two hundred and eighty-eight Moorland features (16.8%) failed the containment check, giving an overall conformance level of 83.2%. Such data supplied under contract would fail the terms of a 95% conformance level in a service-level agreement and would therefore be rejected by the consumer. Automation of this type of process can be of significant operational benefit compared to manual verification.

24.8   CONCLUSIONS

We have shown that distributed geospatial data validation and reporting are feasible within an open Web Services environment. In addition, encoding quality rules in an XML-based form has been shown, and using a first-order logic formalism yields rules that are compact, generic, and implementation independent. By making logical groupings of such rules and asserting them together, consistency to a logical model, as expressed in the set of rules, can be measured. Implementing the conformance service using standard SOAP bindings makes it suitable for integration with enterprise workflow technologies like BPEL (OASIS, 2007).

The implications of collecting quantitative metadata relating to semantically rigorous constraints for spatial information are potentially far-reaching. As each of the logical rules or rulesets can be uniquely identified, it is possible to establish standardized spatial semantic models within specific application domains, such as link-node in transportation networks. Data providers may use them to certify their data and publish quantitative metrics within a metadata portal. As semantic metadata resources grow, conformance levels to a particular domain model can be searched, allowing data consumers to hone in on relevant data sources for reuse in novel applications.

Further work is needed to examine the relationship between GML application schemas and the underlying abstract conceptual models (ontologies) against which the rules are specified. Eliminating or automating model transformation steps will be key to improving the usability of an online quality assessment service and limiting the proliferation of terms. The number and value of standard rules domains (e.g., link-node transportation networks, cadastral land management), the mechanisms for industry standardization, and provision of a set of canonical validation rulesets for enhancing semantic interoperability between datasets (with distinct application schemas) should be explored.

A practitioner program is available for Radius Studio at http://www.1spatial .com/partners/practitioner.php.

REFERENCES

Baader, F., 2002 (ed.). *Description Logic Handbook*, Cambridge University Press, Cambridge.

Egenhofer, M. J., 1997. Consistency revisted, *GeoInformatica*, 1(4), pp. 323–325.

Egenhofer, M. J., Clementini, E., and Di Felice, P., 1994. Evaluating inconsistencies among multiple representations. In *Proceedings of the 6th International Symposium on Spatial Data Handling (SDS '94)*, pp. 901–920.

Haunert, J.-H., 2005. Link based conflation of geographic datasets. *8th ICA Workshop on Generalisation and Multiple Representation*, A Coruna, Spain.

ISO 19115:2003. Geographic Information—Metadata.

ISO 19125-2:2004. Geographic information—Simple feature access—Part 2: SQL option.

ISO 19139:2007. Geographic Information—Metadata—XML schema implementation.

OASIS, 2007. WS-BPEL 2.0—Web Services Business Process Execution Language. http:// www.oasis-open.org/committees/download.php/22036/wsbpel-specification-draft%20 candidate%20CD%20Jan%2025%2007.pdf.

OGC, 2004. Web Feature Service (WFS) Implementation Specification. http://portal .opengeospatial.org/files/?artifact_id=8339.

OGC, 2005. Catalogue Service. http://portal.opengeospatial.org/files/?artifact_id=5929& version=2.

Pira International Limited, 2000. Commercial Exploitation of Europe's Public Sector Information. http://ec.europa.eu/idabc/servlets/Doc?id=18390.

Sester, M., Anders, K.-A., and Walter, V., 1998. Linking objects of different spatial datasets by integration and aggregation, *GeoInformatica*, 2(4), pp. 335–358.

Sheeren D., Mustière S., and Zucker, J.-D., 2004. Consistency assessment between multiple representations of geographical databases: a specification-based approach, In P. Fisher, Ed., *Developments in Spatial Data Handling, Proceedings of the 11th International Symposium on Spatial Data Handling (SDH'2004)*, Springer-Verlag, pp. 617–628.

Walter, V., and Fritsch, D., 1999. Matching spatial data sets: a statistical approach. *International Journal of Geographical Information Science*, 13(5), pp. 445–473.

W3C, 2004a. Web Ontology Language (OWL). http://www.w3.org/TR/owl-features/.

W3C, 2004b. Semantic Web. http://www.w3.org/2001/sw/.

W3C, 2004c. SWRL: A Semantic Web Rule Language Combining OWL and RuleML. http:// www.w3.org/Submission/SWRL/.

W3C, 2003. SOAP – Simple Object Access Protocol. http://www.w3.org/TR/soap/.

Epilogue

Putting Research into Practice

Michael F. Goodchild

INTRODUCTION

The previous chapters of this book have described new advances in the constant effort to improve the quality of spatial data, and to make it possible for users of spatial data to understand the effects of data quality on their work. These results add to the already substantial literature on the topic, and it is clear that after two decades of work we now have a much better understanding of spatial data quality and its effects. While we are a long way from preventing future mishaps with spatial data, or lawsuits over their misuse, nevertheless we now have an excellent range of models and tools to employ.

But it is one thing to argue that specialists in spatial data quality have good theory and techniques at their disposal, and quite another thing to argue that these theories and techniques are accessible to all. Academics often make the mistake of assuming that once a discovery is published in the refereed journal literature, and appropriate entries have been made in CVs and annual reports, it is time to move on to new pastures—that implementation and use will inevitably follow. We know that this is hopelessly naïve, and particularly in dealing with spatial data quality, given the mathematical complexity of many of its models.

Essentially this problem is one of communication, and it concerns the ways in which producers of spatial data communicate what they know about quality to users—and the ways in which users share what they have learned on the same topic. Knowledge of spatial data quality is communicated through metadata, which must therefore be the key to any improvement in communication. It seems fitting, therefore, to devote at least part of this epilogue to the topic of metadata, and to examine whether our current arrangements, in the form of standard metadata formats, are adequate for the task. The next sections discuss these current formats, and examine them against the current state of knowledge about spatial data quality. The final section raises a number of other issues, and speculates on future directions for spatial data quality research.

GEOSPATIAL METADATA

As a term of technical English, *metadata* is still comparatively new, not yet appearing in the *Shorter Oxford English Dictionary* (Brown, 2002). Yet the concept of "data about data" is as old as language itself, and humans have struggled for millennia with the task it attempts to solve—the succinct description of the contents of a body

345

of information or *dataset*. Metadata must serve several somewhat independent purposes, all of them related to the ability of a potential user to assess the suitability of a body of information for a specific use. Metadata must provide sufficient information to allow the user to evaluate quality, the degree to which the contents match the requirements, the technical details of transporting or using the information, information about legal and ethical constraints on use, and sources of further information.

Unfortunately, the standards process is inherently conservative. It proceeds by consensus, and may therefore overlook the most recent research; and once standards are in place there is clearly an incentive not to change them, since the cost of doing so can be high, both in the effort required to write new standards and in the legacy of compliant descriptions that must be replaced. Ideally standards should be devised once research is complete and the domain is fully understood. But in areas such as this research is likely to continue almost indefinitely, whereas the need for effective means of data description and documentation is constant and immediate.

The need to document and describe geospatial data was well established by the nineteenth century, when national mapping agencies began to dominate the process of acquiring, compiling, and publishing geospatial data in analog form. Metadata began to appear in the form of marginalia (map legends, scales, publication history, and other information printed around the edge of the map) or sometimes on the back. Metadata about entire map series, including the details of the series' specifications and the series' index, typically appeared in separately published documents. In the computing world, data documentation began informally with such simple approaches as tape labels, which progressed to file names and file headers, and eventually became the structured, digital approaches we see in today's standards.

As our ability to share data through such media as CDs or the Internet has grown, so too, has the necessary complexity of metadata. Sharing of data with a colleague in the same department is comparatively easy, since both the custodian and the potential user probably share a common discipline, language, and set of expectations. But sharing data over the Internet with potential users in other countries, cultures, and disciplines is vastly more problematic, and in the extreme may even be comparable to the problems faced by Columbus in communicating with the native inhabitants of the Americas, or NASA in deciding what message to send with the Voyager spacecraft (http://voyager.jpl.nasa.gov/spacecraft/goldenrec.html).

One of the earliest attempts to devise a standard for geospatial metadata was made in the early 1990s by the U.S. Federal Geographic Data Committee (http://www.fgdc.gov), as part of a larger effort to establish the U.S. National Spatial Data Infrastructure. The first version of the standard, under the name Content Standard for Digital Geospatial Metadata (CSDGM), was adopted officially in 1994. Quality was recognized as a major part of the standard, and addressed in Section II using an earlier schema devised during the 1980s for the Spatial Data Transfer Standard (http://mcmcweb.er.usgs.gov/sdts/) and popularly known as the "5-fold way":

Attribute accuracy, or the accuracy of the attributes by which geographic features are characterized

Logical consistency, or the degree to which the contents of the dataset follow the rules by which they were specified

Completeness, or the degree to which the dataset reports relevant every feature present on the landscape

Positional accuracy, or the degree to which locations reported in the dataset match true locations in the real world

Lineage, or details of the processes by which the dataset was acquired and compiled

In Version 2 of the standard, adopted in 1998, a sixth optional component was added to allow cloud cover to be described for remotely sensed datasets.

The International Standards Organization adopted its standard for geospatial metadata, ISO 19115, in 2003, with the intention that standards in member states would eventually be brought into compliance (the U.S. compliance effort is being led by the Federal Geographic Data Committee; http://www.fgdc.gov). The ISO 19115 approach to geospatial data quality strongly resembles the earlier CSDGM, but following several commentaries (e.g., Guptill and Morrison, 1995) adds temporal accuracy, which had earlier been partially subsumed under completeness. Attribute accuracy was renamed thematic accuracy, but otherwise there is little effective difference between the two standards, and a simple cross-walk has been devised (http://www.fgdc.gov/metadata/documents/FGDC_Sections_v40.xls).

Substantial efforts have been made to accommodate the standards within GIS software and data formats, allowing metadata to be stored with the dataset itself, ingested along with it, presented in different formats, and made available to users. For example, ESRI's ArcCatalog supports several variants of both FGDC and ISO standards, allowing metadata to be imported in a variety of formats. Automated update of metadata during GIS manipulation is clearly desirable (Lanter, 1994)—it would be good if every new dataset created through such GIS operations as overlay or join could be automatically documented. In practice, however, this remains a largely elusive goal, particularly in the area of data quality, because of the difficulties associated with processing metadata that are largely text-based and because of gaps in our knowledge of error propagation (Heuvelink, 1998) and more generally data quality propagation.

The FGDC is composed of federal agencies with a long tradition of production and use of geographic data, and detailed knowledge of the processes used in production. On the other hand, the average user of geographic data may know comparatively little about the process of production, and may be more concerned with the effects of data quality on the user's particular analyses. As a producer-centric view of metadata, the FGDC standard emphasizes

Details of the production process, such as the measurement and compilation systems used

Tests of data quality conducted under carefully controlled conditions

Formal specifications of dataset contents

By contrast, a user-centric view would emphasize

Effects of uncertainties on specific uses of the data, ranging from simple queries to complex analyses

Simple descriptions of quality that are readily understood by nonexpert users

Tools to enable the user to determine the effects of quality on results

Goodchild, Shortridge, and Fohl (1999) have examined the alternative ways of describing data quality, and have argued that simulation provides a general and readily understood option. Comber, Fisher, and Wadsworth (2007) have also argued for a more user-centric approach to metadata, and one that places greater emphasis on spatial data quality.

ISSUES

The following seven sections discuss issues that I believe need to be addressed if geospatial metadata standards are to adopt a user-centric approach, reflecting both the state of research knowledge about spatial data quality and the practices of current technology.

DECOUPLING

As analog representations, paper maps are characterized by a scaling ratio, or *representative fraction* (RF), which is defined as the ratio of distances on the representation to their corresponding distances in the real world. The fact that no flat paper model of the Earth's curved surface could ever have a truly constant RF is a minor issue for maps representing areas of small extent, but significant for maps of substantial fractions of the Earth's surface. The representative fraction acts as a surrogate for the map's contents, as formalized in the map's specification, such that maps with a coarser RF portray only the larger features of the Earth's surface. It also acts as a surrogate for spatial resolution, since there is a lower limit to the sizes of symbols that can be drawn and read on a map, and thus to the real-world sizes of features that are likely to be portrayed. Finally, it also acts as a surrogate for positional accuracy, since national map accuracy standards commonly prescribe the positional accuracy of features on maps of a given RF.

In the 1980s comparatively few geographic datasets were available in digital form, and virtually all that were had been created by digitizing or scanning paper documents. Table digitizers and large-format scanners were regarded as an indispensable part of any GIS lab, and their use was included as a substantial part of any training program. The digitizing process introduces errors and uncertainties that add to those present in the paper document. For example, digitizing is found to introduce positional errors on the order of fractions of a millimeter in addition to those already present in the map, which are themselves of similar order. As a result, positional errors in datasets derived by digitizing paper maps are themselves directly related to RF, and the RF of the original map is an effective measure of positional accuracy for such datasets.

Unfortunately, this strategy fails for other datasets that have not been obtained by digitizing or scanning paper maps. In principle, RF is not defined for digital data, since there are no distances in digital media to compare to distances in the real world. Goodchild and Proctor (1997) argue that by the 1990s the coupling of content, spatial resolution, and positional accuracy under a single surrogate measure had broken down. Datasets created with newer technologies, such as digital orthophotos, were never in analog form and never had an RF to inherit. Spatial resolution is well defined for raster data, but its definition for vector data is often problematic, since for area-class maps it is related both to the minimum patch size (minimum mapping unit) and to the level of detail with which boundaries are drawn.

However, little of this is evident in the current metadata standards. The FGDC standard mentions RF once, as a parameter defining source documents in its section on lineage, but it is not mentioned in the ISO standard. Positional accuracy is mentioned as one of the five components of data quality in the FGDC standard, and the ISO standard allows for both absolute and relative positional accuracy. Spatial resolution is not mentioned in the FGDC standard; in the ISO standard it is mentioned once as an optional element of "Core Metadata," but no further detail is provided. In summary, the authors of the standards appear to be aware of the difficulties associated with RF as a surrogate for several aspects of data quality, but they have not fully adopted the decoupled approach that now seems needed in its place.

UNCERTAINTY

Early discussions of the quality of digital geographic datasets focused on the concepts of accuracy and error, perhaps reflecting the roots of GIS in area measurement (Foresman, 1998) and the earlier work of Maling (1989). Efforts were made to apply the theory of errors to the compilation of digital representations of features (Goodchild and Gopal, 1989), a practice that was already well established in surveying. By the 1990s, however, it had become clear that there was much more to data quality than error and accuracy, because for many types of geographic data it was unreasonable to assume that the observations reflected some real-world truth modified by the process of measurement. The principle that an observed measurement x^* could be modeled as a true value x plus an error δx clearly could not apply to the nominal data of soil, land use, or vegetation cover classifications, but neither could one define a probability $p_i(i^*)$ that the true class found at a point would be i if the observed class was i^*. Moreover, the definitions of the classes used in many mapping programs were clearly open to varying interpretation, and the maps compiled by two equally trained observers could not be expected to be equal.

Some progress was made using concepts of probability, but theoretical frameworks more compatible with vagueness, such as fuzzy and rough sets (Fisher and Unwin, 2005), proved very attractive for many forms of geographic data. The terms *error* and *accuracy* are now generally avoided in the research community, which tends instead to favor *uncertainty* as the umbrella term, along with *imprecision*, *vagueness*, and terms more closely related to the theories of evidence and non-Boolean sets. Yet neither standard shows any evidence of this significant change of thinking. The term *accuracy* occurs 7 times in the ISO standard and 85 times in the FGDC standard, while *uncertainty* occurs in neither.

SEPARABILITY

Both standards distinguish clearly between the accuracies of attributes and positions. It has long been known, however, that in the case of geographic variation conceptualized as a continuous field the two concepts are not readily separable. Consider, for example, a continuous field of topographic elevation, in other words, a mapping from location $\mathbf{x}$ to a single-valued function $z = f(\mathbf{x})$. Suppose that at some location $\mathbf{x}_0$ a value z_0 has been measured. Except under special circumstances it will be impossible to separate the errors in these two parameters—for example, to distinguish between one case in which the correct elevation has been measured at an incorrect location, and another in which an incorrect elevation has been measured at a correct location, or any combination of the two. Only when some independent means exists to specify location, such as the existence of a survey monument or a sharp peak, or if the measuring instruments have known error metrics, is it possible to separate the two sources. The first case implies a shift from a continuous-field to a discrete-object conceptualization, while the second implies that each measurement inherits levels of error that were known a priori. Similar arguments exist for area-class maps, where it is impossible to separate errors of boundary positioning from errors of class determination, except when boundaries follow well-defined features such as roads or rivers, again implying a shift of conceptualization. Given these arguments, it makes little sense to attempt to specify the positional accuracy of isolines, or to separate attribute and positional accuracies for area-class maps, as both the FGDC and ISO standards do.

GRANULARITY

In the earlier world dominated by paper maps, the body of information described by metadata was a single map, and an intimate association existed between a map's contents and its marginalia. In the digital world, however, the concept of a dataset is much more fluid. The digitized contents of maps can be separated into layers, based perhaps on print color or thematic divisions. They can be partitioned spatially as well into separate tiles, or integrated (*edgematched*) into larger, apparently seamless datasets. Agencies such as the Ordnance Survey of Great Britain have invested heavily in *on-demand* mapping, allowing the user to extract data for uniquely defined areas from seamless databases. To some extent, the metadata for such sub- or supersets can be inferred automatically. But it is clearly difficult to define a single positional accuracy for a combination of two layers of data.

In principle it is possible to define quality at a hierarchy of levels, from the single attribute or measurement of position through entire features, entire layers, and entire seamless databases. A topographic map is typically compiled from many sources using a complex process of analysis and inference, yet very little of this lineage is preserved when the finished product is made available for use. It is possible, for example, that the buildings have been obtained by photogrammetry, that the roads have been obtained by tracking vehicles, and that the rivers have been extracted from an existing digital source, and ultimately from topographic mapping at a different RF. Each of these sets of features has very different data quality characteristics that are difficult to capture in a single data quality statement prepared according to a metadata standard.

In the 1990s a significant shift occurred in the dominant paradigm of geographic data modeling. Today, projects are likely to be supported by integrated databases containing representations of many different types of features, linked by the full range of relationship types defined by object-oriented principles: specialization, association, aggregation, and composition. Existing arrangements for handling metadata attempt to describe the database at the level of the class or collection of features (the *feature dataset* in the terminology of ESRI's Geodatabase). But one could equally well argue that metadata are needed at the level of the entire database, and that the quality of the information about relationships between classes needs to be described.

COLLECTION-LEVEL METADATA

Metadata that describe the properties of individual datasets are sometimes termed *object-level metadata*, since they focus on a single information object within the larger framework of an entire collection. Many such collections exist in the form of geospatial data warehouses, digital libraries, or geolibraries (Goodchild, 1998), each containing potentially thousands of separate datasets. The U.S. Geospatial One-Stop (GOS; http://www.geodata.gov; Goodchild, Fu, and Rich, 2007) is an effort by the federal government to provide a single point of entry into this distributed resource, with a union catalog that describes each available dataset and points to its host server. GOS currently provides access to more than a thousand such collections, and its catalog includes tens of thousands of entries.

Consider a user faced with searching this distributed world for a dataset meeting specific requirements. While GOS attempts to be a single point of entry, inevitably the collections available through it are only a subset of the collections available in the entire universe of servers. Thus the user requires some form of guidance as to which collections to search: GOS or one of numerous alternatives, any of which may contain terabytes and even petabytes of information. Collection-level metadata (CLM; Goodchild and Zhou, 2003) is defined as data about the contents of an entire collection, describing such characteristics as geographic and temporal coverage, the set of themes that dominate the collection, and the general level of data quality. Efforts have been made to develop content standards for CLM (http://www.alexandria.ucsb. edu/~lhill/alex-imp/Metadiversity_narrative.html), but the task of describing collections is far more complex than the task of describing individual datasets.

AUTOCORRELATION

Tobler's First Law (Tobler, 1970; Sui, 2004) describes the tendency for "nearby things to be more similar than distant things." There is now abundant evidence that this principle applies to errors and uncertainties in geographic datasets. For example, we know that errors in elevation in a digital elevation model are strongly autocorrelated, such that nearby errors tend to be similar, and indeed if this were not so, our ability to estimate such properties as slope and curvature would be severely impaired (Hunter and Goodchild, 1997). Recently, such errors have been analyzed within the framework of geostatistics, which formalizes Tobler's First Law as the theory of regionalized variables.

There are many common causes of this pattern of autocorrelation. Any geographic dataset inherits errors and uncertainties from many parts of its compilation process. For example, misregistration of an image affects the positions of all of the features extracted from that image, and misclassification of an agricultural field from a rasterized aerial photograph affects the classes assigned to every pixel intersecting that field. Goodchild (2002) has discussed the implications of the common practice of storing every coordinate in a GIS in absolute form, and has proposed a radically different design that he terms a *measurement-based* GIS. By storing the uncertainties associated with each measurement and the process by which the database is obtained from the measurements, it would be possible to update automatically when improved measurements become available.

In summary, it is known that the correlations or covariances of errors of attributes and positions is as important, if not more important, than their variances— that the joint properties are at least as important as the marginal properties. Such covariances account for the widely observed tendency for *relative* errors in spatial databases to be less than *absolute* errors. For example, even though a road segment may be substantially out of place in absolute terms, its representation in a spatial database is likely to record its shape with a much higher degree of accuracy. While it is difficult to measure position on the Earth's surface in absolute terms to much better than a meter due to Earth tides, wobbling of the axis, poor approximation of the geoid, and tectonic movement, it is possible to measure relative position to millimeters over substantial distances.

Knowledge of the covariance of errors may be of only limited significance to the visualization of geographic data in maps, but it is critical to any analysis of the propagation of uncertainties during manipulation of spatial datasets. Virtually all interesting products of GIS analysis, from simple measures of slope or area to complex analyses, respond directly to the covariances of errors and uncertainties. Thus appropriately defined parameters should be an essential part of any attempt to describe data quality in metadata. Yet current standards focus entirely on marginal properties such as mean positional error.

CROSS-CORRELATION

This discussion of autocorrelation leads directly to the final issue, which is in many ways the most problematic. Although the ability to overlay disparate layers is often presented as a major advantage of a GIS approach, many users will have experienced the problems of misfit that almost always occur. If the positional uncertainties in two layers are other than perfectly correlated, and if both layers contain representations of the same features, then the result of overlay will be a large number of small slivers, formed by the two versions of each feature. On the other hand, if the two layers were both obtained from the same root, then uncertainties may be perfectly correlated and no misfits will occur.

While it is possible to describe the uncertainties of each dataset independently, the results of overlay cannot be obtained from this information—misfit is a *joint* property of a pair of datasets, rather than a *marginal* property of either of them. More broadly, one might define *binary* metadata as metadata describing the ability of two datasets to interoperate, and note that such metadata cannot be obtained from the separate metadata descriptions of each dataset (although perfect correlation of uncertainties might be inferred in some circumstances by comparing information about lineage).

Such information seems essential to the entire GIS enterprise, insofar as it is based on the ability to overlay, and to extract layers of data from widely disparate sources. Great effort has been expended over the past decade at making geographic technologies and datasets interoperable. Yet data quality has received very little attention in this drive to open, interoperable GIS (http://www.opengeospatial.org), and the approach to metadata reflected in the standards is uniformly unary.

THE WAY FORWARD

While the focus of this discussion has been on data quality, several other authors have commented on the need to reopen the metadata question. Schuurman (http://www.sfu.ca/gis/schuurman/research/onto.html), for example, has identified several ways in which changing practices, particularly the increasing sharing of data across widely disparate cultural and disciplinary divides, is prompting a demand for more comprehensive and thus more complex metadata. It has long been known that metadata have the potential to exceed data in sheer volume, and it is not unreasonable to expect that as much effort be spent documenting data as in compiling them.

That said, however, the willingness of data custodians to document and describe data is clearly an issue, and there is no doubt that the generation of metadata lags behind in many domains (National Research Council, 2001). Thus one can expect resistance to any effort to reexamine metadata standards if the result is likely to be greater complexity. Yet it is clearly naïve to expect that the costs of metadata creation should be borne entirely by the custodian, and models of geographic data dissemination that recover at least part of the costs of creation from users are attractive in this regard.

I believe that the responsibility for improving the description of data quality in metadata lies firmly with the research community, which must decide whether the results of research are sufficiently stable and conclusive to merit being embedded in standards. What is needed is a concerted effort on the part of this community to define a more enlightened and research-based approach; and if successful, to lobby the standards community for its adoption. This seems to me to be one of the most important things we can do to bring the results of our research into practical use, and to demonstrate its benefits.

AN ALTERNATIVE APPROACH: WEB 2.0

Recently a distinct sea-change has been evident in the online world of the Internet. Information on the early Web was essentially one-way, as users accessed sites to examine their contents. Today, however, users treat the Web as a far more interactive environment in which the accumulation of information *from* individuals is as important as the distribution of information *to* individuals. Sites such as Wikipedia are the result of a massive collaboration between volunteers, as are YouTube, MySpace, and sites devoted to geospatial data such as Wikimapia and OpenStreetMap. The term *Web* 2.0 has been suggested as a way of capturing this sea-change.

In the Web 2.0 world one thinks of metadata as something contributed by users, rather than something supplied by producers. Arguably the most useful item of geospatial metadata is the phone number (or email address) of the person who last tried to use the dataset. In this world comments from previous users would be readily accessible, and could be scanned and searched by potential new users. Instead of the highly structured, formalized, and producer-centric approach typified by the existing standards, a Web 2.0 approach to geospatial metadata would give access to unstructured, informal commentaries. Just as sites such as Wikimapia demonstrate the willingness of individuals to volunteer geospatial data, so too might a Web 2.0 approach to metadata address a longstanding issue, the reluctance to create metadata, by providing an environment in which users are more willing to contribute voluntarily.

THE FUTURE OF SPATIAL DATA QUALITY RESEARCH

As I noted at the outset, the past two decades have seen a very substantial accumulation of excellent research on spatial data quality. What then remains to be done, and what new discoveries will be made in the next decade? Any answers to such questions are inevitably personal and notoriously inaccurate in hindsight, but nevertheless it seems appropriate to attempt some kind of forecast as a way of closing this book. I would like to address three topics.

First, the concept of geospatial data as a *patchwork* or *mosaic* is increasingly important as we adopt more and more of the concepts of spatial data infrastructure (National Research Council, 1993) and as geospatial data volunteered by individuals become more significant. The quality of such data is characterized by substantial spatial heterogeneity, and models must therefore adopt a *piecewise* approach in which the quality of one part of the patchwork is largely independent of the quality of neighbors—and in which sharp changes occur along patch boundaries. Church et al. (1998) developed a piecewise model of positional accuracy to represent error in street centerline maps, but this research remains an outlier in a literature that has relied heavily on assumptions of spatial homogeneity.

Second, and related to the first, is the issue of modeling spatial data quality *across* layers. While the concept of overlay has been a cornerstone of GIS development for four decades, the ability to combine layers in services such as Google Earth has brought overlay to the masses, and has led to obvious problems of positional misfit. But while we know much about modeling error in single layers, the modeling of *joint* or *relative* errors in multiple layers is still a largely unexamined area, but one that will be increasingly important in the coming years.

Finally, the history of GIS has been dominated by the map, and by the use of maps to conceptualize GIS operations. But increasingly GIS is being used to process data that were never associated with maps, and often are impossible to visualize in map form. Geospatial data are now being obtained by intelligence agencies from email, telephone conversations, and news reports, through the use of sophisticated techniques for recognizing references to places. Data on financial transactions, events, and movements also fall into this category, as do the kinds of field observations obtained by explorers or by field scientists, and information on the Web that is georeferenced through the use of microformats. This new kind of geospatial data is comparatively unstructured, and more likely to exist in the form of XML or plain text than as organized tables or maps. It raises a host of interesting questions about quality and uncertainty that will be increasingly important as its supply increases, and as more and more interest is expressed in its analysis.

These are just three of what must be a multitude of interesting questions that will surely keep the research community active for many years, and form the basis for many more interesting International Symposia on Spatial Data Quality.

ACKNOWLEDGMENTS

Support from the National Science Foundation (Award 0417131), the Army Research Office, and the National Geospatial-Intelligence Agency is gratefully acknowledged.

REFERENCES

Brown, L., editor, 2002. *Shorter Oxford English Dictionary.* Fifth Edition. Oxford: Oxford University Press.

Church, R. L., K. M. Curtin, P. Fohl, C. Funk, M. F. Goodchild, V. T. Noronha, and P. Kyriakidis, 1998. Positional distortion in geographic datasets as a barrier to interoperation. *Technical Papers, ACSM Annual Conference.* Bethesda, MD: American Congress on Surveying and Mapping.

Comber, A. J., P. F. Fisher, and R. A. Wadsworth, 2007. User-focused metadata for spatial data, geographical information and data quality assessments. *Proceedings, Tenth AGILE International Conference on Geographic Information Science, Aalborg, Denmark.* http://www.plan.aau.dk/~enc/AGILE2007/PDF/71_PDF.pdf.

Fisher, P. F. and D. J. Unwin, editors, 2005. *Re-Presenting GIS.* New York: Wiley.

Foresman, T. W., editor, 1998. *The History of Geographic Information Systems: Perspectives from the Pioneers.* Upper Saddle River, NJ: Prentice-Hall PTR.

Goodchild, M. F., 1998. The geolibrary. In S. Carver, editor, *Innovations in GIS* 5. London: Taylor & Francis, pp. 59–68.

Goodchild, M. F., 2002. Measurement-based GIS. In W. Shi, P. F. Fisher, and M. F. Goodchild, editors, *Spatial Data Quality.* New York: Taylor & Francis, pp. 5–17.

Goodchild, M. F., P. Fu, and P. Rich, 2007. Sharing geographic information: an assessment of the Geospatial One-Stop. *Annals of the Association of American Geographers* 97(2): 249–265.

Goodchild, M. F. and S. Gopal, editors, 1989. *Accuracy of Spatial Databases.* Basingstoke: Taylor & Francis.

Goodchild, M. F. and J. Proctor, 1997. Scale in a digital geographic world. *Geographical and Environmental Modelling* 1(1): 5–23.

Goodchild, M. F., A. M. Shortridge, and P. Fohl, 1999. Encapsulating simulation models with geospatial datasets. In K. Lowell and A. Jaton, editors, *Spatial Accuracy Assessment: Land Information Uncertainty in Natural Resources.* Chelsea, MI: Ann Arbor Press, pp. 123–130.

Goodchild, M. F. and J. Zhou, 2003. Finding geographic information: collection-level metadata. *GeoInformatica* 7(2): 95–112.

Guptill, S. C. and J. L. Morrison, editors, 1995. *Elements of Spatial Data Quality.* Oxford: Elsevier.

Heuvelink, G. B. M., 1998. *Error Propagation in Environmental Modelling with GIS.* Bristol, PA: Taylor & Francis.

Hunter, G. J. and M. F. Goodchild, 1997. Modeling the uncertainty in slope and aspect estimates derived from spatial databases. *Geographical Analysis* 29(1): 35–49.

Lanter, D. P., 1994. A lineage metadata approach to removing redundancy and propagating updates in a GIS database. *Cartography and Geographic Information Systems* 21(2): 91–98.

Maling, D. H., 1989. *Measurement from Maps: Principles and Methods of Cartometry.* Oxford: Pergamon.

National Research Council, 1993. *Toward a Coordinated Spatial Data Infrastructure for the Nation.* Washington, DC: National Academies Press.

National Research Council, 2001. *National Spatial Data Infrastructure Partnership Programs: Rethinking the Focus.* Washington, DC: National Academies Press.

Sui, D. Z., 2004. Tobler's First Law of Geography: A big idea for a small world? *Annals of the Association of American Geographers* 94(2): 269–277.

Tobler, W. R., 1970. A computer movie simulating urban growth in the Detroit region. *Economic Geography* 46(2): 234–240.

Index

T

U

Cyberspace and Cybersecurity
Second Edition